SOCIOLOGICAL IDEAS

CONCEPTS AND APPLICATIONS

Fourth Edition

SOCIOLOGICAL IDEAS

CONCEPTS AND APPLICATIONS

Fourth Edition

William C. Levin
Bridgewater State College

Wadsworth Publishing Company
Belmont, California
A Division of Wadsworth, Inc.

Sociology Editor: Serina Beauparlant
Senior Editorial Assistant: Marla Nowick
Production: Sara Hunsaker/*Ex Libris*
Print Buyer: Barbara Britton
Permissions Editor: Jeanne Bosschart
Signing Representative: John Moroney
Text and Cover Designer: Christy Butterfield
Copy Editor: Elliot Simon
Compositor: G&S Typesetters, Inc.
Cover Photograph: © 1993, Spencer Jones

This book is printed on acid-free recycled paper.

International Thomson Publishing
The trademark ITP is used under license

Printed in the United States of America

3 4 5 6 7 8 9 10—98 97

Library of Congress Cataloging-in-Publication Data

Levin, William C.
Sociological ideas : concepts and applications / William C. Levin. — 4th ed.
p. cm.
Includes bibliographical references and indexes.
ISBN 0-534-20856-8
1. Sociology. I. Title
HM51.L3595 1993
301—dc20 93-3256
CIP

For Alice, Dan, Doug, and Joann,
my brothers and sisters; and for Flea and
Jack, who are not, but might just as well
have been.

CONTENTS

PREFACE

Since publication of the third edition of *Sociological Ideas,* I have spoken to or corresponded with many other teachers of introductory sociology. As with previous editions, their suggestions have led to changes for this current edition of the text. My colleagues have always been interested in how best to teach. However, in the last few years questions about the quality of education in America have made the issue one of general concern. This has led to extensive discussions with my students about what has worked best in their classes, both before and during their college years. These discussions have also influenced my work on this edition. Thank you to all the people whose ideas have shaped this book, and thank you also for confirming an idea I was taught early in my professional training—that we are richer as a field for our differences. Though the overall approach of the text seems to work well for students at a wide range of schools, it appears that no two instructors use this book the same way or even present the chapters in the same order. I hope that the changes in this edition improve on the previous editions without changing the book's character or usefulness.

The Concepts Approach

Sociology is an abstract discipline that uses the basic concepts of the field as building blocks. Once understood, these concepts have wide application to the analysis of social behavior. With one exception, each chapter (called a concept) in *Sociological Ideas* focuses on one of these basic sociological concepts and on its related ideas. This approach is suited to the conceptual nature of the discipline and has the advantage of giving the instructor the freedom to use his or her special interests and expertise to enliven and clarify the material. In addition, the concepts approach allows more time and money to be spent on additional materials in an introductory course. Inexpensive paperbacks, reprints, special readings, and other sources of information can provide the current, topical, and regionally pertinent materials normally made prohibitive by large, expensive introductory texts.

The last chapter in the book is the only one not focused on a sociological concept. This discussion of sex and gender in society is intended to serve as a sort of laboratory to demonstrate the usefulness of the concepts learned in the rest of the text. Any of a dozen other substantive issues—such as medicine, aging, education, and religion—would have served, but issues of sex and gender are currently of such wide and deep concern that reviewers independently suggested its inclusion.

After learning the basic concepts of the sociological approach, the student should be in a position to use them to analyze any of a range of topics, in much the same way that a skilled carpenter uses the tools of that trade to build, no matter what the project. I have used this approach in my own classes as the basis of a final paper in which students are asked to pick a substantive topic and demonstrate the use of the concepts they have learned. As it turns out, the most commonly chosen topic has been sex and gender.

Organization

Each concept chapter consists of three parts: (1) Definition, (2) Illustration, and (3) Application. This organization is intended to teach the concepts not only by explaining them but also by showing how they have been used by sociologists and how they can be used by students. Concepts that might otherwise be mere abstractions to be memorized for exams become useful tools for the analysis of human behavior.

Definitions

The first part of each concept chapter is a definition of the concept and its related ideas. For example, the discussion of role includes treatment of role sets, role conflict, and role strain. Definition sections emphasize the work of the classic theorists and researchers associated with specific concepts. Throughout, the writing is as direct and free of unnecessary jargon as possible. In the Definition sections, the key concepts printed in **boldface** are also briefly defined in a Glossary at the end of the text.

Illustrations

Following the Definition section in each concept chapter is an Illustration of the concept. Illustrations are intended to provide examples of how specific concepts have been used in social research. Because original studies are often difficult for the beginning student to understand or are too long to fit into the demanding schedule of an introductory course, the original research studies have been paraphrased and written in the style and at the level of the Definition sections.

Applications

The last part of each concept is an Application section, which gives students firsthand experience in using the concepts in research. Employing a variety of research techniques, such as participant observation and survey and content analysis, the Applications provide step-by-step instructions for the collection and analysis of data by the student. In this way, each student is given the tools to conduct small-scale studies independently, in which various concepts are central. The data analysis sheets from these studies are perforated so they may be turned in to the instructor as assignments.

Applications are designed to be completed by students with minimal assistance from instructors. As a result, the concepts become more meaningful and concrete to students because conducting empirical research on their own shows the concepts "at work." In practice, not all of the twenty-one Applications (one for each concept) can reasonably be assigned in one semester. In choosing which Applications to assign, instructors can tailor selections to their own areas of interest and expertise, special interests of students, and events of current and regional concern. In addition, some Applications, such as the observation of the workings of formal organizations (Concept 15), may be extended to cover several weeks or even an entire semester.

What's New in the Fourth Edition

Those data collection exercises (Applications) that presented some problems for students or instructors have been replaced. I also have broadened the range of sociological methods discussed that sociologists actually use and have taken into account the varying characteristics of students who use this book, such as differences by age, region, and campus residency. A number of the summarized readings (Illustrations) have been replaced by studies with more recent data on the same topic or by studies of more timely topics. The replacement of Illustrations was also influenced by a teaching issue that has come up repeatedly over the years: what sorts of readings, and at what level, should be assigned in introductory courses.

In working on each edition of this book, I have had many conversations about the importance of using original readings in an introductory course. (This subject also starts lively discussions among my friends in other disciplines. We sociologists are hardly alone.) Clearly, there is great value in having students read original sociological work, and no summary can capture the richness and complexity of the original work on which it is based. However, given the different characteristics of introductory students across the country, how much of the original work can the readers grasp, especially when they have, by definition, no training in the methods and vocabulary of the field? In addition, unless we are willing to include an unedited collection of original readings within the text, we authors face an impossible sampling prolem. What portion of published studies should be included to maintain the meaning of the original?

My approach is to rely on summaries of original studies in the text and also to assign some carefully chosen original material that students have already read in summarized form. Thus, new summarized readings for the fourth edition of *Sociological Ideas* have been taken from sources that are likely to be available in any college library in the country. Although a great deal of fine work is published in smaller, more specialized journals, the research summarized here is typically from journals such as the *American Sociological Review* and *Social Science Quarterly.* After reading a given summary of a published article, students can go on to read the original, which is referenced at the beginning of each Illustration section.

Also new to the fourth edition is a discussion of sex and gender in society. What might have been an organizational problem (how to fit one substantive chapter into a concepts text) disappeared quickly when I realized that in my classes this was one of the topics to which we turned repeatedly when attempting to illustrate how the concepts under discussion operated. I even admit to going back to student papers for ideas about how to write it. The discussion is not designed to be anything like a comprehensive treatment of the topic. Rather, it is meant to demonstrate the uses to which the concepts from the text can be put.

Finally, in preparing this edition of *Sociological Ideas,* I have included discussion of a range of methodological issues not already covered in the chapters on quantitative and qualitative methods in sociology. For example, the data collection Applications include brief treatments of issues such as content and secondary analysis, causality, longitudinal studies, use of census data, and operationalization of variables. Discussions of such topics are not comprehensive; they are meant to introduce each issue and explain it adequately to allow the reader to understand the research in which it is used. To help instructors find where these methodological topics are discussed, a separate methods index has been included.

Acknowledgments

I would like to give credit to those people who have contributed so importantly to the completion of this book. Careful reviews of the manuscript were provided for the first edition by Jeanne Ballantine, Wright State

University; Walter Clark, St. Louis Community College at Florissant Valley; James Dedic, Fullerton College; Vaughan Grisham, University of Mississippi; Michael Leming, St. Olaf College; and Craig Little, State University of New York at Cortland; for the second edition by Robert J. Dunne, Colorado College; Douglas McDowell, Pennsylvania State University, Ogontz Campus; James Owens, University of Idaho; and Robert E. Wood, Rutgers University; for the third edition by Peter Adler, University of Denver; C. Dan Danou, University of Wisconsin Center, Marshfield; Barbara Johnson, Augsburg College; George T. Martin Jr., Montclair State College; Harriet Miller, Framingham State College; and Kathryn Talley, North Central College; and for the fourth edition by Bradley Ebersole, Catonsville Community College; Norman Goodman, SUNY, Stony Brook; Dale Hoffman, California State University–Sacramento; Laura O'Toole, University of Delaware; Karen Rosenblum, George Mason University; and Ralph Wedeking, Iowa Central Community College.

The suggestions made in these reviews led, in large measure, to the overall shape of the final manuscript, and in many instances, to the specific material included in it.

Lastly, I would like to thank those friends and colleagues whose knowledge, advice, and support have been given so generously. I can't count the times I called on friends with questions about a reference, the date of a film, the name of a character in fiction, or one of a thousand other bits of hard fact or whimsical trivia. I am particularly grateful, then, for the friendship and assistance of Charles Angell, Arnie Arluke, Evelyn Brodkin, Walter Carroll, Ratna Chandrasekhar, Jackie Crews, Charles Fanning, Curtiss Hoffman, Don Johnson, Carol Kryzanek, Mike Kryzanek, Flea Levin, Jack Levin, Joyce Leung, Howard London, Betty Mandell, Joyce Marcus, Jim Scroggs, Janet Stubbs, Phil Sylvia, and Cynthia Webber. And to my friends and colleagues at Wadsworth, Andrew, Christy, Elliot, Jeanne, John, Marla, Sara, Serina, Sheryl, and Susan, thanks again.

William C. Levin

"We murder to dissect."

WORDSWORTH

Introduction

I was one very bored nine-year-old on that rainy Saturday morning, and my grandfather's pocket watch seemed to beg for investigation, so I pried it open. For as long as I could remember it had hung from a hook in the back of my father's bureau, its ticking so quiet I had to lean in to hear it. But now, with its intricate innards exposed, the magic of its operation would be made plain to me. With care, I placed the parts in a long row in the exact order in which I had removed them (the better to reconstruct the movement later). Tiny screws and pins so small they stuck to my skin, toothed wheels, and thin mounting plates all lifted out so cleanly they seemed to have agreed to fit together just so. As I pulled it apart, it all seemed so sensible. Watching it tick and whirr, I found the shape and interplay of the parts perfectly obvious, so very normal. It stopped running when the first screw was out.

It is still hanging in my parents' home, but in the living room now. My mother arranged the more attractive parts, the case, the front, the engraved back, some gears, and some jeweled pins, on a background of dark blue velvet and framed it all behind glass. Very pretty, I think, but certainly not a watch anymore.

You might think I would have learned my lesson, but as you will see in this book, I'm still at it. Only now, instead of a pocket watch, it's society that I'm pulling apart. And, like the watch, once dissected, it may seem hard to fit together again even mentally. It is true that "we murder to dissect." But, as students of anatomy, literature, economics, and a host of other fields can confirm, there seems to be no other way.

The aim of courses in introductory sociology, and of this book, is to introduce you to the component parts of society and how they fit together.

But there are two very important differences between learning how a watch operates and learning how societies operate. First, you can see and touch all the parts of a watch, but the elements of social order are not directly visible—they are abstract or conceptual. For example, you cannot look directly at the quality called *belonging.* You can see evidence of it in the way people behave, but, like gravity, it is a force that cannot be directly seen, touched, or heard. Second, the parts of a watch can fit together only one way, as I learned to my dismay many years ago, but sociologists disagree about how the parts of society operate and "fit together," and even the extent to which they do.

So there are two basic difficulties in learning about, and teaching, sociology. One is to define the abstract elements that together make social life. The other is to explain the many ways in which sociologists have suggested that these elements work.

The Plan of This Book for Learning Abstract Ideas

Except for the final chapter, each of the first twenty-one concepts, or chapters, in this book deals with a separate element of the study of society. Some years ago the American Sociological Association conducted a survey of sociology instructors and textbooks; it identified these concepts as "indispensable" in an introductory course. Each concept in the text consists of three parts: (1) a Definition section, (2) an Illustration (summarized reading), and (3) an Application (exercise), usually involving the collection of data.

The Definitions

The ideas being defined are usually quite abstract, so I have tried to use the clearest, plainest language possible. The special jargon of the field is often eliminated, especially if it has only made the ideas less clear. To make the abstract concepts more concrete, I illustrate them with real-life examples throughout the Definition sections. Key terms are printed in **boldface,** and their definitions are collected in a glossary at the end of the book.

The Illustrations

The concepts are further demonstrated by the Illustrations that follow the Definition sections. In most cases, these are adaptations of classic studies using the specific concept in research. Some are taken from more recent studies, selected because they are particularly interesting or illustrate the concept exceptionally well. Because so many sociological studies are very difficult to read without a great deal of experience with sociological terminology, I have summarized them all in the same writing style and language level as in the Definition sections. I also try to tie the findings presented in the illustrations to the core ideas in the definitions.

The Applications

It is one thing to read about the meanings of concepts or their use in research by sociologists and quite another to experience their use personally. I find the ideas of sociology exciting, important, and extremely *useful* in making sense of everyday life. These concepts do not have to be just ideas on paper to you. They have real application to your life in society. The Application exercises that follow the Illustration sections give you an opportunity to apply the concepts to the real world by collecting and analyzing data, usually on campus. In order to keep these exercises manageable, sample sizes are limited, and data analysis is fairly simple. The instructions are as plain and specific as I could make them.

The "Laboratory" Chapter

The last chapter in the book, the one on sex and gender in society, presents the only substantive topic in the text. It is the only one not focused on a sociological concept. New to the fourth edition, it is intended to serve as sort of a laboratory to demonstrate how sociological concepts can be used. Colleagues have told me that they use other topics, such as "medicine" and "education," and in my own classes I have used a range of topics as the basis for final papers in which the students propose how the concepts they have learned might be applied.

The Plan of This Book for Organizing the Ideas

Using a concepts approach does create a special problem for the organization of the ideas in the text. Even in the substantive chapter included to demonstrate the use of the concepts (Part Six), the problem remains. When the elements of a system are looked at individually, they lose some of their meaning. That is, one part of any system can only *partly* be understood out of its system context. For example, you can study a single gear from a watch and understand a good deal about its nature. But, for its *full* meaning to be grasped, you must study it in relation to all the other parts of the watch. How does each operate, and how do they relate to one another?

This problem is especially important in the study of society, because there are so many conflicting views among sociologists about how social systems work. Some stress the way people agree with one another about how to live in society. Others believe that conflicts among contending individuals and groups for favored positions create the relatively stable relationships of society. Still others assert that social order is constructed on the spot by individuals interacting every time they come into social contact. From each point of view, the components of social order have a different meaning and fit together in different ways. How is it possible, then, to organize the raw material of sociological concepts?

Part of the answer is that, in discussing each concept, each of these differing views must be expressed. I have tried to do this throughout the book. In addition, the differing perspectives are also discussed in some detail as separate concepts, those on function and dysfunction, conflict, and symbolic interaction and qualitative research.

This still leaves the problem of the order in which the concepts are to be presented. They could be arranged to proceed from the smallest unit of social behavior (*microanalysis,* it is sometimes called) to the largest (*macroanalysis*), that is, from the individual to the entire culture. Or the concepts could be divided into those useful to each of the major sociological perspectives, for example, conflict ideas, functionalist ideas, and symbolic interactionist ideas. It might even be possible to arrange the concepts alphabetically and let each instructor decide on an order of presentation.

I have chosen to organize the twenty-one concepts according to five categories into which they all fit logically. I developed this scheme over years of teaching and out of the ideas presented by many texts I have used in introductory sociology. The following list details the five parts of the book and the concepts in each, plus the sixth, and concluding, section—the "laboratory" chapter.

1. **Overview of the Sociological Perspective.** This part includes the sociological viewpoint, use of the "ideal type" as a device for understanding concepts in general, the role of science in sociology, and then each of the three major perspectives on social order—symbolic interaction, function and dysfunction, and conflict.
2. **Acquiring the Meanings of Social Membership.** The second part includes the concepts of culture, norms, and socialization.
3. **Basic Social Forces.** The concepts of solidarity, power, and ideology are included in the third part.
4. **Structural Components of Society.** This part includes status and role, groups, formal organizations, and institutions. (These concepts have been arranged in order of increasing complexity and size of the component.)
5. **Inequality, Change, and Social Disorder.** The concepts of stratification, discrimination and prejudice, social change, deviance, and alienation are included in the final part.
6. **Sex and Gender in Society: Using the Concepts.** This final chapter is the only substantive chapter in the book, intended to demonstrate how the concepts from the preceding chapters have application to an issue of great contemporary sociological importance.

Sociology teachers may prefer to assign chapters in a different order, as I often have done in my own classes. Since sociologists develop their own theoretical perspectives, each will not only illustrate each concept somewhat differently but also relate the concepts to one another in a particular way. That is what teachers are there for. I see the organizational scheme of this book as a beginning point, just one of many ways to see society.

PART ONE

Overview of the Sociological Perspective

Every person, every day, sees human social behavior but usually in a very casual, nonsystematic way. The sociological perspective provides a way of looking at human behavior that is as different from everyday observation as a powerful microscope is from a casual glance. Without the specific tools of inquiry and methods of analysis that the sociological perspective supplies, everyday observation misses a great deal.

Concept 1 of this part introduces the sociological perspective, comparing it with other fields of inquiry and explaining its unique way of viewing human behavior. The concept also discusses the concern of sociologists with both social order and social problems and their desire to develop theories to explain a wide range of human behaviors.

Concept 2 focuses on some of the basic conceptual tools of sociological inquiry. Sociologists search for stable patterns in human social behavior that suggest the existence of underlying social forces. Ideal types, models, and paradigms provide the sociologist with ways to describe such patterns.

Concept 3 discusses sociology as a scientific enterprise. It begins with a description of the scientific method, including the development and testing of hypotheses, and proceeds to a consideration of the major quantitative research methods used in sociology.

The last three concepts in this part present the major theoretical perspectives of sociology. These perspectives are so basic to the field that they are used in the analysis of many different issues. Concepts 4–6 introduce the perspectives and prepare you to understand their use as they are applied throughout this book. Concept 4 treats symbolic interaction theory and the special qualitative research methods it employs. Concept 5 introduces the idea of functionalism, focusing on two of the important ways it can be used in social analysis. Concept 6 deals with conflict theory and its application to many levels of social interaction.

CONCEPT 1

The Sociological Perspective

Definition

The Sociological Perspective A view of human behavior that focuses on the patterns of relationships among individuals rather than solely on the individuals themselves.

"What am I doing here?" A very popular question, often asked when you wish you were somewhere else. My friend asked it of me one summer. We were, at that moment, on a narrow path in the Montana Rockies. We were also frozen in our tracks by a rattlesnake that was coiled (and rattling) inches from our feet. "What am I doing here?" I ask it countless times while I am typing away indoors, knowing that outdoors there is hiking to be done or tennis to be played. You may be asking it of yourself at this very moment. How is it that you are enrolled in a sociology course? What forces have combined to place you in this particular spot at this scheduled time with all these other people? By developing an understanding of the sociological perspective, you can add to your ability to identify some important forces that shape your behavior, forces of which you have probably been quite unaware.

When we are being analytical we try to identify *all* the factors that contribute to our behavior. What a complex tangle of causes we can pick apart for even the simplest events! Take the rattlesnake encounter, for example. We were there because we wanted to fish for trout in the Madison River. But that oversimplifies it. Other factors contributing to our presence at that precise spot include that we chose that path, had good weather, were able to start the truck that morning, were in good enough shape to trek in five miles, grew up in families in which we learned to like fishing, had the free time to go fishing because of our job schedules, could afford the air fare to Montana, and were able to stay with a friend who not only put us up but also guided us to that particular canyon (bless him). The list of contributing causes could go on and on. The trick is to think of as many really important ones as we can. But how can we systematically do this? One possible answer is to ask some experts for help. Because everyone tends to see an issue from his or her own perspective, if we speak to people from enough fields, we may get the whole picture eventually.

Multiple Views

To illustrate how different perspectives offer different contributing causes and where the sociological perspective fits in, let's try to explain the fact that you are now enrolled in a sociology course. How did you get here? If a number of experts were asked to contribute explanations strictly from the perspectives of their own disciplines, a *physicist* might focus on the way gravity and friction allowed you to walk to registration or sit in a classroom seat; a *biologist* might talk about how your physical well-being influenced your ability to enroll in and attend classes (there is, you must admit, a strong relationship between being alive and ability to show up for class); a *biochemist* might discuss the biochemical properties of the brain that enabled you to think abstractly and learn, and how these properties evolved; an *architect* might talk about how the design of the building made access to the classroom more or less difficult; a *historian* might discuss how the school itself came into existence so that it could offer the course; an *economist* might focus on how you came to have the money to pay tuition and to forgo the earnings from a full-time job while you pursued your college degree; a *political scientist* might be interested in how the power of the teacher to distribute grades influenced your attendance (and acceptance of the teacher's ideas) or how your college education will increase your access to the power available to some members of society; and a *psychologist* would probably discuss how you became motivated to attend college, how attendance satisfies certain of your needs and desires, how rewards and punishments are associated with class performance, and whether your presence and performance in the class are influenced by factors such as your IQ, the authority of the teacher, and your personality.

Those are several interesting perspectives, and the more experts we import to help with the question, the more answers we get. Everyone will have something to say. But the problem is to find out which suggestions really help us answer the question and which ones, although true, either contribute minimally or are not central to the question. For example, although it may be true that class attendance is strongly related to the fact that you are alive, so is any other activity in which you are engaged. The contribution of the biologist is, in this case, true but not specific to the question asked.

Sociological View

We can see that certain approaches are more appropriate for certain questions. If I were interested in understanding the survival of a species of animals, I would undoubtedly talk to a naturalist first. But if I wanted to understand something about human social behavior, such as behavior in a college classroom, the **sociological perspective** would be most appropriate.

The sociological perspective focuses on the patterns of relationships among individuals rather than solely on the individuals themselves. It alerts us to the fact that a great deal of our behavior is shaped by our membership in groups with other people. Social forces, like physical forces, strongly influence our behavior. The analogy with physical forces is helpful, because in both categories the forces may normally be invisible to us, but their existence can be made obvious if we know how to look at behavior or know what kinds of questions to ask. Gravity is invisible until its effect on a dropped object is observed and measured. Air becomes "visible" when we see its influence on leaves or we turn an empty glass over and push it into a bowl of water—the water can't get in because there is something already there. In just this way social forces can be made visible by asking sociological questions, observing and measuring human social interactions, and then guessing about the nature of the underlying social forces that caused the observed behavior.

Just as there are physical facts that we learn when studying the workings of the physical world, so there are **social facts** that sociologists have discovered by studying human social behavior. Émile Durkheim, a French sociologist, coined the term *social fact* (Durkheim, [1893] 1958) to describe the forces that constrain (or control) human behavior and that result from our membership in groups rather than from what we are like as individuals. It is important to keep in mind that, when sociologists try to explain human behavior, the individual characteristics of the people involved are *not* of primary importance.

When I began studying sociology, I was interested in the things I read, but there seemed to be no common thread to it all. The issues of poverty, discrimination, divorce, and other social problems seemed worth understanding, but they also seemed disjointed. Was each really a separate problem, with separate causes? Then I read C. Wright Mills's articulate, passionate little book *The Sociological Imagination* (1959) (summarized in the illustration section of this concept), and I began to understand the benefits of a sociological view.

My problem in those first courses was that I had not developed a sociological perspective. Like most other people in American society, I had learned to think about the world in *individual* terms. It made sense that whatever happens to people must be a result of what they are like. This is a view that is built into our culture. Just as some people succeed because they are hard-working, clever, motivated, or lucky, so others fail because they lack these qualities.

Even when I read about widespread social problems, I focused narrowly, just as I had been taught. In American society we are taught to put great emphasis on individual responsibility and achievement; when we wish to control events in our everyday lives, we generally try to improve ourselves *as individuals.* Look at all the self-help books on the best-seller list at any time. This approach can be ineffective when social forces are at work. The fact is that a large percentage of our everyday actions are simply *not* the result of what we are like as individuals, mentally, emotionally, financially, or otherwise. These actions are due to our membership in groups and in the society at large.

For example, when you watch television at home you may not respond at all to what you see. The jokes don't make you laugh, or the game isn't exciting enough to merit cheers or even comment. But if you are part of the audience seeing the comedian in person or at the stadium watching the game, it is very likely that you will react much more strongly. We seem to be swept along with others around us as they react to what is going on. I can recall laughing loudly at a joke I didn't fully understand because everyone around me thought it was a riot. Laughing is a very social behavior. That is what we mean when we say that laughing is infectious.

Embarrassment, too, is a very social behavior. Consider the man who doesn't know he is being observed by three of his co-workers as he does his best Elvis imitation to his office mirror. He's pushed up both his hair and his collar, and he's singing into a water glass when he looks up to see these people he knows only professionally. When people feel "self-conscious," they are experiencing a heightened sense of themselves *as they imagine others to be seeing them.* Whenever this guy does the same Elvis imitation but goes undiscovered he is not embarrassed. People who accidentally reveal some private truth to others feel real, unpleasant physical symptoms that can redden their faces.

Or, if you have dealt with many bureaucracies, you may sometimes have felt that they hire only cold, uncaring, lazy people to deal with the public. The motor vehicle bureau clerk doesn't seem to care at all that you had to wait in line for two hours, only to find out that the bureau won't accept personal checks. Yet she probably is not basically nasty; after work hours she may be someone's sweet, doting grandmother. It is the way bureaucracies are organized that makes her act that way, not her individual nature. Bureaucracies are very large and complex social structures. They formalize the rules of conduct by which their workers operate and, in the name of efficiency, reduce the workers' jobs to repetitive, unchallenging routines. The very size and style of the organization of bureaucracies kills kindness and originality in its workers. Some may actually be nasty underneath, but workers in complex organizations don't have to start out nasty to be made to act that way after a while. So sociologists focus on the way social forces influence human behavior rather than on the individual characteristics of people.

Sociology and Social Order

Just as the natural sciences developed to deal with specific questions we have about the natural order, so sociology began when we started to ask questions about the forces underlying human social order. Although there have been social thinkers at various times throughout history, sociology is a relatively new discipline. Auguste Comte (1798–1857) did not coin the term *sociology* until the middle of the nineteenth century (Comte, [1848] 1957), and the thinkers who gave the new field its main theoretical direction, Karl Marx (1818–1883), Max Weber (1864–1920), and Durkheim (1858–1917), produced their most influential work between the mid-nineteenth and early twentieth centuries. Why did we suddenly become aware of social forces, even though they had certainly existed for as long as thinking humans have lived in organized groups?

The answer to this question illustrates something about the way social science proceeds. We generally do not notice orderliness. It is taken to be normal or everyday, the way things "ought to be." But occasional disruption of the normal order of events draws our attention and, strange as it may seem, points to the underlying rules of orderliness. The greater the disruption, the more attention it draws. Although the exception may not actually prove the rule, it certainly starts us thinking about what the rules might be. Attempts at proof come later.

For example, every day you walk around with your tongue and teeth in your mouth, and they do not bother you at all. You are unaware of their existence because they are part of the normal order of things. But should your tongue get swollen for some reason, or one of your teeth start to hurt, you certainly would wonder what happened and what you could do about it. In fact, you might be able to think of nothing else until you return to normal. If nothing ever went wrong with our health, the fields of medicine and dentistry would not exist. They developed, as almost any field develops, in response to instances of disorder that, in an attempt to set things right again, prompt the study of how things work in the first place. Sociology is no different in how it developed.

Marx, Weber, Durkheim, and others discussed in this book were reacting to the greatest disruption of all, world revolution. For more than a century the Western world had been experiencing severe and rapid changes in its basic structures. The feudal, aristocratic structure that had dominated for centuries in France, England, and the West in general was overthrown, replaced by national governments that were to be run, to varying degrees, according to the principles of representative republics. The populations became increasingly urbanized and industrialized. The Industrial Revolution was not just a matter of replacing hand work with machines or animal power with steam power; it was also a revolution in the way people evaluated their own worth. To the extent that their living came to depend upon their labor, they could now be judged on the basis of what they did, rather than on the basis of the village, religion, or family into which they had been born. The feudal order of things had been so stable for so long that questions about the forces that had created that order were unlikely to arise. But in the turbulence of social and economic revolution, and in its aftermath, such questions had to be asked. What was causing the changes? Would a new stable form of social organization develop, or would the upheaval continue forever? If a new order did arise, what would it be like? How long could it be expected to last? Would there be more monarchs and serfs in the new order? And the most important question of all: How can we discover basic laws of social order that will allow us not only to predict social change but also to influence or even control it in the future?

So sociology really developed in response to a problem in the social order, and our understanding of some rules of social order was a consequence (Nisbet, 1966). Sociology ever since has been a **problem-oriented** discipline. It is not the stable and orderly segments of society that attract our attention, but the areas we define as problems. Any catalog of the courses offered by a typical sociology department shows this orientation. Courses on social problems, such as racism and sexism, are obviously aimed at understanding problems, but even courses on urban sociology and the sociology of the family are problem-oriented. Urban sociology developed as a response to the apparent decay of social order in American cities around the 1920s. Courses on marriage and the family began as part of the effort to understand and control the weakening of the family structure on small family farms in the early twentieth century. Now family courses focus more on the problems of high divorce rates and the revolution in gender roles. The final chapter in this text, on sex and gender in society, represents an example of a topic that has become increasingly important in sociology as a reflection of struggles in many societies over power and gender inequality. Anyone can see that over the last few decades in America, things are changing with respect to gender roles. Changes in social order, the way we deal with one another in everyday life, can be unsettling, and even revolutionary.

As a consequence of this problem orientation, sociology has developed a reputation as a critical discipline. Sociologists tend to look beneath the official explanations of how things work to discover where problems may be hiding (Berger, 1971). Why are there so many poor people, and why do they tend to share so many characteristics—age, sex, race, ethnicity, educational level, geographical distribution, and so on? Such questions focus on problems. But they have the benefit of helping us try to make our social order live up to its highest goals. It is important to understand that by recognizing that sociology has a problem orientation, you should not conclude that sociologists are in the business of policing problems. It is not the job of sociologists to disapprove of, or to oppose, changes. Ideally, one goal of sociology is to understand why and how such changes come about. As individuals, sociologists may, like any other citizens, wish to influence the world around them in line with their beliefs. But as professionals, their task is analytical rather than political.

Sociologists are not just critics of society's problems. A major goal of the sociological perspective is to understand the basic principles of social order. Our constant attention to disruptions of order gives us the most direct route to understanding order itself. We want to test our guesses about how social order works. We want to build general **social theories** that can explain in the simplest possible terms a wide variety of behaviors.

To do this, we try to be as insightful and as objective as possible. Just as other disciplines—medicine, psychology, and political science—developed during and after the Industrial Revolution, sociology has tried, when appropriate, to emulate the methods of the natural sciences. (The details of the scientific method are discussed in Concept 3.) The objectivity of sociology depends on applying this method of inquiry. Basically, sociological inquiry becomes more objective when it is taken from the personal control of subjective individuals and conducted according to publicly agreed-on methods for the objective testing of ideas. By putting our guesses about the social world in the form of hypotheses that state clearly how everything is to be measured, we make it possible for anyone to verify (or contradict) our results by trying the same study (a process called *replication*).

Sociology as a scientific enterprise differs from other sciences in the types of forces on which it focuses and in the theories it attempts to build to explain them. Social behavior is very complex. Just trying to explain your presence in a sociology course clearly involves psychological, economic, and historical forces, to name just a few. So sociology often requires an awareness of a number of other perspectives. Sociology is *multicausal* in its approach; that is, it alerts us to the fact that virtually any social behavior has many contributing causes, all of which must be included in our explanation.

The nature of social order may be quite different from the nature of other systems of order. Humans, after all, are conscious actors who are capable of evaluating the meaning of their experiences and reacting in a variety of ways. In some circumstances, therefore, it is misleading to apply the methods of the natural sciences to human social behavior. (For a fuller treatment of this aspect of the sociological perspective, see Concept 4.)

Summary

The sociological perspective is an attempt to understand human behavior by focusing on the influence of interactions among individuals rather than on the characteristics of individuals. Sociology is multicausal in its attention to the many kinds of factors that influence human behavior, but it emphasizes the social forces. Social forces cannot be observed directly; they must be inferred from human behavior. Often questions about the operation of social forces are raised when disruptions in everyday social order occur. Because of this focus on social problems, sociology is seen as a critical discipline. Sociologists attempt to build theories to explain a wide variety of behaviors with a relatively small number of general statements. When possible, the ability of such theories to explain and predict behavior is tested in the real world using objective methods of research that allow others to confirm (or contradict) the results.

The illustration that follows discusses a work by C. Wright Mills, one of the most passionate and earnest statements yet made on the value of the sociological perspective. Mills not only explains what the sociological imagination is but also makes a plea for its usefulness in understanding the difficulties we face in modern life.

The Application for this concept is designed to give you practice in identifying many kinds of influences on social behavior and in recognizing those of particular interest to sociologists.

Illustration

C. Wright Mills, The Sociological Imagination *(London: Oxford University Press, 1959)*

At a time when American society stressed individual characteristics and achievements as the prime factors in a person's life, Mills provided an alternative way of thinking. He began by pointing out that people usually do not make a connection between their own troubles and the larger social forces surrounding them. Take, for example, the problem of divorce. Virtually all the divorced people I know talk about their divorces in personal terms. They ask, "Where did I go wrong?" Or they say, "What a rat he [or she] turned out to be." They blame their lack of money or the time spent apart, the demands of work or their lack of communication, the annoying habits of the other person or the need to "find my true self." Out of explanations like these come solutions that focus on individual adjustments: to put less emphasis on money, to work fewer hours, to talk more, to be more tolerant, and so on.

I almost never hear a divorced person talk about the fact that the divorce rate in the United States is about 50 percent. Something is happening to the society that is influencing *marriage itself as an institution*. Divorce is not just happening to this couple on the block and to that one in another town. Divorces are not unconnected events. But people get no satisfaction from looking at broader social trends, because they are taught to think in individual terms and because they experience problems as individuals. (No one can "see" problems happening to categories of people, except in tables or lists of data.)

Mills focused on this difficulty by distinguishing between *troubles* and *issues*. "*Troubles* occur within the character of the individual and within the range of his immediate relations with others: they have to do with his self and with those limited areas of social life of which he is directly and personally aware." Troubles raise private concerns. By contrast, "*Issues* have to do with matters that transcend these local environments of the individual and the range of his inner life." Issues raise public concerns. So your divorce or mine is *trouble*. The rate of divorce in America is an *issue*.

Having grasped this distinction, you may reasonably ask what good it does to recognize that divorce is an issue. How does it help a person who is trying to cope with the trouble of a divorce to realize that he or she is caught up in the issue of divorce as well? Part of the answer is that it is comforting to learn that not all our difficulties are the fault of our personal flaws, or those of others. But more than that, Mills pointed out, troubles are largely the manifestation of issues. To reduce troubles, we must deal with issues. If we ignore issues, then, we doom ourselves to continuing troubles.

For Mills, the key to understanding issues is adoption of the sociological imagination. It "enables its possessor to understand the larger historical scene in terms of its meaning for the inner life, and the external career of a variety of individuals." For example, the possessor of the sociological imagination can use an understanding of the causes of the high divorce rate in the United States to make sense of his or her own experiences in marriage. The broader understanding of social forces is the first step on the road to "doing something" about *both* troubles and issues.

The sociological imagination leads us to ask the following kinds of questions: (1) How is the overall society organized? What are its various elements, and how are they related to one another? How does one society differ in these qualities from others? (These types of questions focus on describing **social structure** at a specific time in its history.) (2) How does the current structure of the society compare with its past structure?

Is it changing? If so, how? In what segments of society are the rates of change greatest? (These are questions dealing with changes in the social structure over time.) (3) What are the characteristics of the members of the society? Which categories tend to be advantaged, and which disadvantaged? What is the process by which these advantages are distributed? How are the inequalities evaluated by the society? (These are questions focusing on the way the society operates.)

Not until questions such as these are asked can problems even be conceived of as issues. And not until they are acknowledged to be issues can we contemplate influencing broad social forces. Here is how Mills expressed the problem with respect to divorce. "In so far as the family *as an institution* turns women into darling little slaves and men into their chief providers and unweaned dependents, the problem of a satisfactory marriage remains incapable of purely private solution."

If it is so important to focus our sociological imaginations on the great issues of our time, why have we not already done so? Mills claimed that the difficulty lies in the relationship between our values and the extent to which we think they are threatened. Table 1-1 illustrates that relationship. When people hold values deeply and perceive no threat to them, they experience a feeling of *well-being*. When they hold values deeply but perceive them to be threatened, they experience *crisis* or even *panic*. When values are not deeply held and there is no perception of threat, the response is *indifference*. And when a threat is perceived, even though the values are not deeply held, a feeling that something is wrong, a response of *uneasiness*, or *anxiety*, sets in. Mills believed that "ours is a time of uneasiness and indifference—not yet formulated in such ways as to permit the work of reason and the play of sensibility." Lacking clear and deeply held values, we modern Americans are incapable of recognizing in concrete terms the difficulties we suffer. We can only struggle with a vague sense of unease or a maddening indifference. And while we seek help through self-improvement or seek escape in the selfish "good life,"

TABLE 1-1

Human response to the relationship between values and the perception of threat to them

		Are values deeply held?	
		Yes	No
Are these values felt to be threatened?	No	Response of well-being	Response of indifference
	Yes	Response of crisis or panic	Response of uneasiness or anxiety

the issues of social life go begging for attention. Mills's book is a call to social scientists to take up the task of focusing on issues and to pass on the awareness of the sociological imagination to others.

Application

The sociological perspective focuses on patterns of relationships among people. Sociologists want to understand what causes these relationships to take the shape they do and how these relationships influence human behavior. For example, what factors help create families, corporations, schools, friendships, and a host of other patterns of bonds among people? How does membership in any of these social groups cause people to behave differently than they would if they were not members?

Because social groupings are so varied and complex and touch our lives in so many ways, it may be difficult to see clearly how sociological ideas fit together. This application is designed to give you a technique for putting the ideas in perspective and some practice at it. I don't promise that this way of looking at sociological ideas will always make things completely clear for you, but I do find it helpful myself, and so do my students. To understand how this technique works, let me refer to the way it was used earlier in this concept.

Near the beginning of the Definition section, I used as an example the attempt to explain the social behavior of attending a sociology class. In the illustration, the viewpoints of a physicist, biologist, biochemist, architect, historian, economist, political scientist, and psychologist were imagined. How would each try to explain the presence of the students in the class on a particular day? Each discipline had something to contribute to the explanation from its point of view.

The sociological perspective makes use of explanations drawn from a wide variety of perspectives, including these. But it also uses ideas that are specific to sociology—that is, how *group memberships* might have contributed to the presence of the students in class that day. Perhaps the students were members of a social class in which college education is especially valued or even considered mandatory. Perhaps college education was a family tradition in their homes. Perhaps their high school classmates put social pressure on them to go to college. These particularly *sociological* factors might have contributed importantly to their presence in that sociology class.

Sociological analysis considers the influence of many factors, some from other perspectives and some particular to the sociological view. In this Application you are to list the factors that help cause some social phenomenon. Make your list as comprehensive as possible, and don't leave out a possible cause just because it may sound strange at first. The idea is to get some practice. It is not unusual for students to think of more than thirty contributing factors for a single phenomenon.

Here is a sample analysis of the factors that might influence the exam performance of a group of students:

1. Intelligence of students
2. Quality of notes used for study
3. Hours of study
4. Quality of study environments
5. Interest in subject
6. Personal motivation to do well on exams
7. Pressure from parents to do well in school
8. Confidence level
9. Anxiety level
10. Ability to concentrate under pressure
11. Verbal and reading skills
12. Whether students studied with others or alone
13. Whether students studied with good or poor students
14. Amount of sleep before exam
15. Whether students cheated
16. From whom students cheated
17. Extent of desire to outperform friends
18. Degree of class attendance
19. Degree of attention in class
20. Quality of precollege education
21. Degree of career ambition
22. Perceived consequences of doing badly
23. Length of time available to take exam
24. Number of other exams that week
25. Weather conditions that day
26. Lighting conditions in exam room
27. Skill at taking tests (aside from actual knowledge)
28. Grading techniques of teacher
29. Level of distracting financial worries
30. Health conditions, such as having headache

In this sample list, which factors do you think would be of particular interest to sociologists? That is, which ones have something to do with the group membership of students? Which ones seem to be of special interest to other disciplines? Anxiety, confidence level, intelligence, motivation, and ability to concentrate under pressure are likely to be of particular interest to psychologists, for example. Economists might be concerned with the financial worries of students. Meteorologists would study weather conditions. Biologists or physicians might focus on students' health and sleep habits. Educators would study the grading techniques of teachers and the students' verbal and reading skills, attention paid in class, cheating, and quality of precollege education. All of these are of interest to sociologists to the extent that they contribute to an understanding of the social phenomenon under study. But sociologists also focus on factors influenced by group membership. For example, intelligence tests have been found to favor the skills and knowledge of certain segments of society (races, age groups, income groups) over others. So group membership influences IQ scores. Also, certain segments of the society receive better precollege training than others, are taught better verbal and reading skills, and learn better skills at taking tests. Some segments of society value education more highly than others; so they tend to motivate their members more to do well, to have high career ambitions, to take the consequences of exams more seriously, and to attend classes more faithfully and pay stricter attention in them. Finally, the group memberships of some students enable them to study with others in a quiet place, share notes, get adequate sleep before the exam, and cheat from one another (if they want to).

You may have noticed that some of these factors seem easier to label than others. For example, confidence and anxiety levels are pretty much the concern of psychologists, and weather conditions of meteorologists. But others, such as intelligence, quality of precollege education, sleep, motivation, cheating, attention in class, and financial worries, have multiple labels. They are of interest to a number of disciplines, even though each discipline studies it in a different way and for different purposes. Biologists may look for the biochemical origins of intelligence, psychologists for its operation as a mental process, statisticians for its stability as a measure, and *sociologists for its unequal distribution among various social groups.*

Doing an Analysis of Contributing Factors

Using the analysis sheet provided, list in the left column factors that help cause a specific social phenomenon. Think of as many factors as you can, up to thirty. After you have completed your list, identify, in the right column, the discipline most appropriate for each contributing factor in the left column. Begin with the sociological factors, those that are influenced by the group membership of individuals. Then move on to the factors that are of particular concern to other disciplines. You will have a relatively easy time identifying disciplines for some factors, such as biological or economic factors. But other factors will be more difficult, either because they are the concern of obscure disciplines or because they are the concern of a number of disciplines. When in doubt, take a reasoned guess and move on. Here are a few suggested social phenomena to analyze:

1. The stability of marriages—that is, factors that might influence the likelihood of divorce
2. The social class of individuals, including their wealth, their power over others, and the esteem in which they are held
3. The likelihood of getting a promotion at work
4. The likelihood of committing serious crimes
5. Membership in a protest group, such as the peace movement or an environmental protection group
6. The decision to run for political office
7. The decision to become a physician
8. The decision to go to graduate school
9. Taste in music, literature, or television programming
10. The degree of prejudice against any category of people

Analysis Sheet

Social phenomenon analyzed ____________________

	Possible factors that have influence on the phenomenon	Discipline(s) concerned
1.	____________________	____________________ ____________________
2.	____________________	____________________ ____________________
3.	____________________	____________________ ____________________
4.	____________________	____________________ ____________________
5.	____________________	____________________ ____________________
6.	____________________	____________________ ____________________
7.	____________________	____________________ ____________________
8.	____________________	____________________ ____________________
9.	____________________	____________________ ____________________
10.	____________________	____________________ ____________________
11.	____________________	____________________ ____________________
12.	____________________	____________________ ____________________
13.	____________________	____________________ ____________________
14.	____________________	____________________ ____________________

15.

16.

17.

18.

19.

20.

21.

22.

23.

24.

25.

26.

27.

28.

29.

30.

CONCEPT 2

Ideal Type, Model, and Paradigm

Definition

Ideal Type An abstract definition of some phenomenon in the real world, focusing on its typical characteristics.

In high school biology we studied diagrams of microscopic plants and animals and their internal structures. These pictures were marvels of complex draftsmanship—all those little cell walls and nuclei, all those different colors showing where one part ended and another began. We copied everything in our notebooks just as clearly. Then we went to the laboratory microscopes to identify the same things the diagrams had shown us, only "for real" this time. But they never looked the same. In fact, sometimes they were barely recognizable from the pictures we had been shown. Under my microscope, when I could focus it at all, was something with no apparent nucleus, or something with lots of little dots where there should have been only a few, or something not round enough. My biggest problem was that I wanted the structures I saw in the lab to match perfectly with the picture drawn on the board. I didn't understand the concept of the **ideal type.**

When we wish to understand the character of some object, we rarely describe one unique object. Instead, we speak about a category of such objects that have characteristics in common. Red blood cells, for example, are generally disk-shaped with dished (concave) centers and a yellowish color. But all the cells are not exactly alike in reality. Categories of objects have characteristics in common but also differ from one another. To explain, or define, such a grouping to a person who has had no experience with it, we must ignore the differences that occur and develop an ideal type, a description focusing on characteristics typically held in common among otherwise differing cases. The dictionary is filled with such ideal types. My dictionary has wonderful illustrations and photographs for selected definitions. The drawings of the flowers are of perfect, idealized flowers, and the photographs of the dogs are of breed champions who represent the ideal for the specific animal. Every concrete example of an ideal type will differ from the theoretical definition, but having the definition in mind makes the common features of the real-world examples stand out clearly.

Ideal Types in Sociology

The concept of ideal types has existed for as long as people have been able to group things that have some qualities in common. It seems likely that this concept was applied first to physical objects such as flowers, dogs, and (when microscopes extended our vision) red blood cells. By the end of the nineteenth century, the developing field of sociology had focused our attention on patterns of behavior and ways of thinking that, like physical objects, could also be seen as having characteristics in common. The German sociologist Max Weber (1864–1920) coined the term *ideal type* (Weber, [1925] 1946) as an aid in describing such regularities in patterns of social behavior.

As an example let's take the concept of a minority group. The American sociologist Louis Wirth (1897–1952) developed a famous definition of a minority as a group of people who "because of physical or cultural characteristics are singled out from others in the society in which they live for differential and unequal treatment, and who therefore regard themselves as objects of collective discrimination" (Wirth, 1945:347). As an ideal type, Wirth's definition should fit the experiences of all minority groups to some degree, but we should not expect it to fit any single minority group perfectly. Not all minorities need be physically or culturally distinctive in the exact same ways to fit the definition. They need not have experienced the same kind of discrimination, nor do they have to express the same kind of collective consciousness of discrimination. For example, blacks in the United States have experienced a very different history of discrimination than women, yet both are good examples of minority groups. The cultural characteristics by which Jews are identified are entirely different from those of Vietnamese-Americans, but both are examples of minority groups as well. The elderly have only recently begun to see themselves as objects of collective discrimination. Before this self-awareness developed, the elderly had only a kind of negative group consciousness that led them to avoid association with the status of old people.

Yet this negative group consciousness is a form of awareness of collective discrimination, so the elderly should also be defined as a minority group. The ideal type fits all these groups, but in different ways. Although they are different from one another in some ways, they all still share the essential characteristics of minority groups as spelled out in the definition.

Because ideal types in sociology reflect social reality, they can change just as society does. For example, think of the ideal type of the American family. What characteristics do all families have in common? Just after World War II, George Murdock (1949) defined the family as "a social group characterized by common residence, economic cooperation, and reproduction, . . . including adults of both sexes, at least two of whom maintain a socially approved sexual relationship, and one or more children." But are there people you know whom you think of as a family but who do not fit this definition? What about a divorced woman living with her children, or an unmarried woman and man who have no children and no plans to have children.

When we begin to pay attention to patterns in social life, a wide variety of such ideal types appears. Family, friendship, social class, bureaucracy, even major types of societies (such as hunting-gathering, horticultural, agrarian, and industrial) are all examples of ideal types. By observation, you could develop an ideal type for some common experience of your own. For example, what are the minimum, abstract characteristics of a college class? Are there certain necessary participants? How many are needed? Can there be too many? What essential relationships must be present? What forms of behavior, exhibited at what times, in what order, and in what kind of setting are required? When you begin to think of such questions, the existence of some underlying order can be discovered and eventually described.

Ideal Types and Models: "What If?"

The attempt to discover order in the social world involves roughly equal parts of observation and speculation. We watch, listen, and make guesses from moment to moment about whether what we have seen is a genuine pattern or a chance event. For example, is it just in my classes that a certain thing happens, or is it true of other college classes also? That kind of speculation is what leads us to test our ideas against reality and eventually to develop stable ideal types for describing social order.

Another type of speculation that helps us in this process is "what-if" speculation, or **model** building. Assume that we have come to some agreement about how a specific part of the social world operates. For example, we have agreed that the family is primarily a unit of kinship in which a number of tasks are accomplished, such as procreation; care, socialization, and education of the young; sexual access between adults; and consumption of goods. Based on this ideal type, we can speculate what shape the family would take in American society under a variety of conditions.

Clearly, things have changed in America, and so has our ideal type of the family. Its definition generally still includes common residence and economic cooperation. But the extent to which the definition should include the elements of reproduction or sexually approved relationships is now a matter of heated debate. To some extent the debate is framed by the struggles of various groups for recognition of their interests, such as divorced people and same-sex partners. And to some extent the debate responds to changes in the way life is lived in America. According to the U.S. Bureau of the Census, since 1960 the percentage of American families headed by one parent has more than tripled, from about 8 percent to nearly 28 percent. At the same time the divorce rate has more than doubled, from 2 divorces per 1,000 American citizens in 1960 to 4.7 per 1,000 in 1988. Just since 1980 the number of unmarried couples has approximately doubled, to over 3 million as of 1991.

Suppose an absolutely safe, effective, and extremely inexpensive birth control technique were invented. What impact might this have on the American family? Or what if the economy suddenly developed a greatly increased demand for the work skills that women traditionally have provided and simultaneously diminished the need for traditionally male work skills? What if housing became much more difficult to obtain and several generations of parents and children were required to live together in extended families? What changes in the family could we expect under any of these conditions? Model building allows us to guess, based on our assumptions about how the social world currently operates, how a variety of circumstances would influence that social order. Not only does model building have the practical consequence of helping us plan, but it also clarifies the ideal types on which we depend for our normal examination of social order.

I want to make it clear that this definition of the term *model* has been chosen for two specific reasons. One is that in many texts the terms *ideal type, model,* and *paradigm* (which is discussed next) are used interchangeably. They have lost their usefulness as separate terms. I want the distinctions among them to be clear. The second reason is that *model* is used in a similar way in other fields. Engineers build models of physical objects to test their behavior under a variety of conditions (such as car tests in wind tunnels). Urban planners build computer models of vehicle and pedestrian movements to test the effect of changes in street patterns on mobility. Planners in various segments of the federal government build models of population demands for food, housing, fuel, and so on. (The military even has models for the various outcomes of war under a variety of, we hope, theoretical conditions.) So I chose this definition of *model* to fit with those in other fields.

Paradigms

Ideal types and models are fairly concrete statements about the way things work and, therefore, lend themselves to testing against reality. For example, we can observe how well various groups fit our ideal type for a minority group. Or, using a model, we can speculate about the influence various conditions might have on the structure of the American family. But underlying both ideal types and models are sets of very broad assumptions that often go unexamined, even though these assumptions shape the ideal types and models we are capable of developing. We can call this collection of broad assumptions a **paradigm** (Gouldner, 1970).

Two opposing paradigms that have shaped much of our thinking about social order deal with the question of how order itself develops. The *conflict* paradigm argues that social order is the result of a balance of contending forces, each struggling against the other for some valued good. The result of such a struggle is either domination of one group by the other or a balance of contention between them. In either case, a sort of orderliness is the consequence of the conflict. By contrast, the *consensus* paradigm argues that social order is the result of agreement among groups and individuals about what the distribution of valued goods should be. Accordingly, order is seen as the consequence of a consensus (whether conscious or not) about what is in everyone's best interests.

Very often my students are surprised to discover that they have held one of these paradigms in preference to the other without awareness that they were doing so. The kind of paradigm a person holds can be traced, in turn, to even more basic assumptions about the nature of human beings, such as their capacity for cooperation and generosity or the degree of their natural competitiveness. Whatever the origins of paradigms, they inevitably shape how we think about the character of social order. A person holding the consensus paradigm would be likely to develop an ideal type for the family that stresses the agreements among its members about what each is to do as a family member. By contrast, a person holding the conflict paradigm would probably see the family as a social unit in which the differing aims of members produce a relatively stable balance of contending forces. Does one of these views seem more likely than the other to you? If not, then you may hold a different paradigm, which would lead to the development of alternative ideal types and models. Or perhaps the consensus and conflict paradigms are not in complete disagreement. At times consensus may be possible, whereas at other times conflicts may contribute to stable social order. They may be complementary processes.

The usefulness of concepts such as paradigm, ideal type, and model is that they cause us to examine the way we study and think about the social world. As you read material in sociology (or any other discipline), keep in mind how broad assumptions, especially unexamined ones, shape the ideal types and models we can produce.

Summary

Ideal types are abstractions of reality. They focus on the typical characteristics of the phenomenon being defined, so they cannot be expected to fit any single case in the real world exactly. All dictionary definitions, then, are ideal types, as are sociological definitions. In an effort to develop useful ideal types, we often speculate about the extent to which the ideal type fits reality and what effects specific conditions might have on the social world as the ideal type describes it. Such mental pictures and speculations about the social world are called *models*. Underlying both ideal types and models are broader, often unstated, assumptions about how order itself is created. Such assumptions are called *paradigms*, and they influence the kind of ideal types and models that sociologists create.

The Illustration for this concept examines the role of paradigms in the world of science. According to Thomas Kuhn, all science is limited and shaped by the dominant paradigm of a given time in a specific field. Only when the research findings of scientists consistently contradict the assumptions of the dominant paradigm can a new, more adequate paradigm take its place.

In the Application for this concept, you will see how models allow us to speculate about the ways in which specific conditions might change our social world. You will be asked to choose from among a list of such specific conditions and then model what American society might be like if the condition were to occur.

Illustration

Thomas Kuhn, The Structure of Scientific Revolutions *(Chicago: University of Chicago Press, 1970)*

A paradigm is a set of assumptions that shapes how we think about the world. In our everyday lives paradigms are almost always unexamined assumptions. In the physical and social sciences, however, the role of paradigms has been closely examined, thanks largely to the work of Thomas Kuhn.

How Paradigms Work in Science

The information we can gather from our environment is infinite. The impressions we can gather just through our senses are more than we can deal with if they are not organized in some way. Paradigms are essentially tools for organizing information, usually by focusing attention on some category of information while ignoring all others. For example, much modern medical research uses a paradigm that rests on the assumption that germs cause most disease. Kuhn recognized that when paradigms are used to organize information, they greatly limit the kinds of research questions a discipline can ask. Researchers who use the germ paradigm, for example, inevitably search for specific germs. If a disease were caused by something else (a vitamin deficiency, for example), the very assumptions of the germ paradigm would rule out the discovery of an alternative cause. Even the research tools within a given paradigm limit the focus of inquiry. Microscopes are used to look for germs. They cannot be used to search for vitamin deficiency. Kuhn noted that certain paradigms come to dominate within specific sciences, as the germ paradigm dominates medical research.

The germ paradigm was not always dominant in medicine, however. Well into the nineteenth century (long after germs had been identified) the dominant medical paradigm attributed disease to "evil humours," and leeches were therefore prescribed to bleed off the "bad blood." Once a paradigm is in place within a given field, its hold is tenacious. Students and novice practitioners are taught its beliefs and methods. Powerful and famous leaders of the discipline lecture on its uses and distribute career opportunities only to those who share the dominant paradigm. Careers become so invested in the paradigm that any threat to its stability becomes a threat to its leading practitioners.

Once a paradigm is in place, according to Kuhn, it leads to the practice of what he called "normal science," that is, the application of the paradigm to the explanation of the real world. Thus, efforts are made to test the extent to which disease is, in fact, caused by germs. Every time a paradigm does a good job of prediction or explanation, it seems confirmed and becomes even more entrenched. When it fails to explain some information or events, efforts may be made to adjust the paradigm. Events that are not predicted or explained by a paradigm are called *anomalies* (a departure from the general rules of a paradigm). And when a paradigm is totally dominant in a field, anomalies are routinely dismissed or ignored altogether. For example, when Louis Pasteur in the nineteenth century confronted the medical establishment with concrete evidence of the existence of germs, he was fighting the evil humours paradigm. He was ridiculed and dismissed as a crackpot. But in spite of this resistance, the stage had already been set for a "scientific revolution" in medicine.

According to Kuhn, even firmly entrenched paradigms can deal with only a limited number of serious anomalies. When a paradigm becomes overburdened by an inability to explain events or when another paradigm proves able to explain those events in a much more satisfactory way, a "scientific revolution" takes place. This is simply the replacement of one paradigm by another. Thus, the germ paradigm replaced the evil humours paradigm. Once this has occurred, the process of "normal science" begins again, now guided by the new paradigm.

The common belief is that science progresses in an orderly, cumulative way, that one individual discovery builds on another along a single line of advancing knowledge. How different Kuhn's view is: (1) A given paradigm explains the world as best it can and becomes the basis for the conduct of "normal science." (2) By investing belief in the paradigm and training new scientists in its assumptions and methods, adherents cause it to become deeply entrenched. (3) Such a dominant paradigm resists numerous anomalies, but, as increasingly serious anomalies accumulate, they cause the paradigm to be replaced by a new one that seems to do a better job of prediction and explanation. (4) The cycle begins again after the "scientific revolution."

Applying Kuhn's Work to Sociology

Most discussions of Kuhn's work focus on paradigms in the natural sciences, as I have done here. This is because Kuhn was trained as a physicist before he began working as a historian of science, so he used the natural sciences to illustrate the concepts in his book. But these ideas also apply to sociology. I did not apply them to sociology before now for two reasons. First, people are generally more familiar with the paradigms of the physical sciences than those of the behavioral sciences, so I thought a physical science paradigm would illustrate some of Kuhn's ideas more easily. Second, sociology does not now have, and may never have had, a *dominant* paradigm. In fact, sociology is conducted according to the assumptions of three main paradigms: (1) conflict, (2) consensus (sometimes called *functionalism*), and (3) symbolic interaction. These paradigms are so different from one another and there are so many adherents of each, that you will find them discussed throughout most sociology books (including this one). Often a single topic will be treated from the point of view of each paradigm, one right after the other.

The conflict paradigm rests on the assumption that the members of society are constantly in competition with one another for valued resources (such as jobs and power). Out of their struggles social order develops, either because one side wins a conflict and becomes the dominant force or because contending parties reach a "balance of forces" in which neither side is able to gain further advantage over the other.

The consensus, or functionalist, paradigm rests on the assumption that the members of society are integrated into a basically cooperative, organized system. The primary mechanism by which this integration occurs is **socialization,** the process by which members of the society adopt its beliefs and values as their own. Thus, social order is the result of a shared, deeply held desire to follow society's rules.

The symbolic-interactionist paradigm rests on the assumption that in every situation in which individuals interact, social order is created anew. Participants create order in each social situation by agreeing on the meaning of the symbols expressed in their interaction. From this point of view, social order can be understood only from the perspective of the actors in each setting.

The problem in applying Kuhn's notions of how paradigms operate is that in sociology no paradigm dominates (although adherents of each paradigm might argue that theirs does). In fact, the three do not share much with one another. For a conflict sociologist and a symbolic interactionist to communicate, they must keep each other's paradigms in mind. They have to do some "translation." With no paradigm in a dominant position in sociology and data not well shared among them, the conditions for a "scientific revolution" as Kuhn described it seem unlikely to occur. Each sociological paradigm has carved out an area for itself and pretty well ignores the others. The fact that all three might be studying the same phenomenon is often accommodated by the idea that each can contribute something valuable to our understanding.

Without having to decide whether this is a good or bad thing, we can explain the coexistence of these sociological paradigms in two ways. First, sociology may be in a "pre-paradigmatic" stage. That is, sociology has not yet formulated ways of organizing and measuring the social world that are compelling enough to become dominant. Thus, no single paradigm has emerged that comprehensively predicts or explains social behavior. Second, the information that sociology studies (human social behavior) is so different from the data of the physical sciences that no satisfactory paradigm could exist. That is, human interaction is not stable enough, or objectively observable enough, to be measured or predicted.

The fact that sociology lacks a dominant paradigm is not necessarily a failing. I, for one, am perfectly comfortable with the coexistence of several social paradigms. The comparison between the behavioral and the physical sciences can easily be overdone. Disciplines need not study the world in the same way or undergo the same sort of "scientific revolutions."

Application

A sociological model is a mental picture designed to promote speculation about what effect specific conditions might have on the social world as we understand it. Some of the most fanciful, and often revealing, sociological models are seen in science fiction. In fact, I think these books are badly named. They should be called *social* fiction, because they almost always focus on the changes in our social lives brought about by imagined technical changes, usually in some distant future. For example, what would a world be like if people could read one another's minds, could communicate in emotions rather than ideas, could travel across time, or (most classically) were visited by powerful aliens from another planet? Once the premise of the work is introduced, we may like the special effects and glittering machinery created by the author or filmmaker, but what really makes the story, what interests us, is the type of society that would result.

This is precisely how films like *Total Recall, Blade Runner, Terminator,* and *Aliens* get written. In each of them a technical premise is posed, and the writers speculate about the changes that would be likely in the operation of (roughly) modern American society. If a teenager were to go back in time to when his parents were in high school, how would he fit into his own mother's family? How would the American military deal with a visit by peaceful aliens? What would happen to our own social order if slightly strange-looking aliens arrived who were eager to do any kind of job for any kind of pay?

Developing Your Own Model of Society

For this application you will get a chance to speculate about one of a number of possible sociological models based on a premise that you can choose from a list I will provide. They vary widely in terms of how fanciful they are, but whichever premise you choose, you will have to focus on how our current social order might be changed by different conditions. The idea of the assignment is both to make you familiar with the way models work and to get you thinking more about the sociological perspective, as it was introduced in the first concept of the book.

To give you a bit more guidance, here is an example. In 1920, about 5 percent of the population was over the age of sixty-five; by 1950 the figure was 8 percent; and by 1984 it was almost 12 percent. It has been predicted that approximately thirty years after the turn of the century the percentage of Americans over sixty-five will have increased to approximately 20 percent. What are the possible social consequences of this dramatic demographic change? Here are a few I can think of:

1. The cost of medical care for the elderly may become so high and the need for services so great that money will have to be taken from other areas of the economy to pay the bills, and production of physicians, nurses, and other medical personnel will have to increase proportionately. As a consequence the status (and pay) of physicians may decline as they increasingly have to care for patients whom they cannot "cure."

2. The leisure industry will expand along with the changes in geographic settlement patterns, especially around the Sun Belt.
3. Conflict between generations over resources may replace the current tendency for aid to flow from elderly parents to their children.
4. The elderly may increasingly form a special-interest voting block as their entitlements (such as Social Security and medical benefits) become threatened. The extreme consequence would be a society controlled by older Americans (a "gerontocracy").
5. Retirement may become less available to individuals. The age of retirement may be raised to above seventy, and the benefits reduced. This would change the nature of the workplace as the talents and capacities of the elderly are matched to jobs, work schedules, and so on. Also, salaries may not increase steadily with seniority but may rise until a certain age and then decrease so as to encourage retirement and make places for younger workers.
6. The society may decide to allow certain forms of euthanasia (mercy killing) by techniques such as the rationing of emergency medical service, kidney dialysis, and even hospital bed space.
7. There may be large communities of institutionalized elderly, and new careers may be introduced in association with those institutions. (Fewer than 5 percent of Americans over the age of sixty-five are today in nursing homes.) There may be an increase in family size as longer-lived parents are compelled to live with their children's families by lack of money or available housing.
8. If people continue to live healthier and longer lives, there may be a great expansion in education for the elderly. Colleges are already experiencing a significant increase in attendance by retired people.
9. Current minority groups, such as blacks and Hispanics, may be victimized less by prejudice if the elderly become a major target themselves.
10. If the elderly manage to keep control over enough resources, many products may be directed at them. Perhaps the current youth-oriented advertising and programming would become less dominant in the culture.

That should give you the idea. Pick one of the "what-if" ideas from the following list, and use it to develop a social model for American society. It may help, if you run out of ideas, to think of some of the major elements of the social order and what might happen to them. These elements are the family, marriage, education, work, the economy, government and power, housing, leisure industries, medicine, social welfare, the military, transportation, and so on. In each of these areas we have certain ways of doing things, rules and habits that would probably be influenced by any important change on the list. Use your imagination.

Conditions for Sociological Modeling

1. Global warming floods 30 percent of the American land mass.
2. A technique is developed by which couples can safely, though at great cost, genetically determine the characteristics of their babies.
3. Food in the United States becomes essentially free and limitless.
4. The U.S. population becomes so great that the density makes privacy impossible and food in extremely short supply.
5. Technology becomes so advanced that home computers can, by verbal instructions, do everything from listing and showing entertainment and information programs to shopping, banking, and allowing a person to work from home.
6. Production techniques become so efficient that 75 percent of the population needs to work only one day a week, and the remaining 25 percent never needs to work.
7. A technique is developed giving men the ability to bear children exactly as women do.
8. The divorce rate increases from the current rate of 50 percent (approximately) to over 80 percent.
9. All fossil fuels run out.
10. Travel is made very cheap and very efficient by every means, including by air. For example, it costs $10 to fly anywhere in the United States.
11. At age fifteen, Americans are able to discover with accuracy how long they will live, as long as no physical accident intervenes.
12. The epidemic of acquired immune deficiency syndrome (AIDS) grows suddenly and greatly, infecting the heterosexual population at the same rate as the homosexual population and afflicting 50 million Americans by the year 2000.

Analysis Sheet

Condition chosen for sociological modeling: ____________________

____

____

List of social consequences: ____________________

CONCEPT 3

The Scientific Perspective and Quantitative Research

Definition

The Scientific Perspective An approach to studying the observable world that stresses systematic, objective measurement aimed at the discovery and explanation of stable order in that world.

Quantitative Research A research method that attaches numbers to the qualities of objects, behavior, or relationships.

I have finally come to enjoy poetry, having disliked it intensely most of my life. I suspect this change in taste arose because a friend convinced me that poetry understands the world so much differently than sociology that it can give me welcome relief from my work. Take, for example, light and darkness. What are they? My first instinct is to think about light waves and to discover what they are and how scientists measure them. But, instead, here are two ways that poet Don Johnson (not the actor) has talked about darkness (1984: 27):

I lay there, hearing the loons on Lake Aroo,
waiting for the beaver slaps to start,
waiting for the night to crawl
into folds and crevices,
haired cracks,
to stay.

If you have ever been camping you probably recognize the feeling of this scene, though it takes a poet to capture how darkness seems to arrive, in this case like some liquid animal crawling into the folds, crevices, and cracks of the woods by a lake.

Or consider another of Johnson's poems (1984:39):

Not falling
but curling up
out of oak woods
at the lawn's edge,
summer darkness
is flooding our yard
like the tide.

This time the darkness really is a liquid, a tide flooding into his yard like water, first filling the low spots of the land, then in time rising. For Johnson, darkness and night have meaning because of his specific experiences, such as camping on the shores of a lake and watching night rise up from the ground around his home. The poet's view of the nature of things is so personal, so imaginative and individual. It startles me to experience things through another person's senses. How different from the "truths" for which scientists search. A scientist would never describe the growing darkness in these ways, at least not during working hours.

What poets and scientists have in common is that both make guesses about the nature of the real world. Scientists call these guesses **hypotheses.** The appeal of poetry, at least in part, lies in its lack of restrictions and the total subjectivity of the poet's descriptions. Scientists, however, test their hypotheses in such a way that their observations can be checked by others who use the same methods of investigation. The basic idea is not to have to take someone's word for things. Until the development of the **scientific perspective** and methods of **quantitative research,** belief in the truth of some explanation of how things worked depended solely on either the power of someone to enforce that belief (it was dangerous to contradict the king or priest) or the beauty or persuasiveness with which the idea was expressed. Of course we are still subject to such influences today. But the scientific perspective is our main way of removing the process of discovery from the control of powerful or persuasive individuals and making it more of an objective, communal search.

Science and Verification

The first step in this process of discovery is objective observation. Objectivity requires that we focus on only those characteristics in the world that can be sensed, rather than on our beliefs or feelings about them. The best check on whether we have observed objectively is whether another individual, observing the same thing and using the same measurement instruments, will see the same thing. Can an observation be verified (or contradicted)? Can you verify that the night is like the tide or like an animal? I think not. But you *can* verify the amount of light in a given location with a light meter. So an important step in science is the careful recording of observations, and of the exact methods used to obtain them, so that others can test them.

When we speak in normal conversation we use **concepts,** abstractions drawn from observed events that make up our everyday vocabulary. The word *darkness,* for example, names a category of events that share some characteristics, although they are not all exactly alike. By grouping these characteristics, we can abstract our experiences (such as dusk, full night, turning off the light in a room, or closing one's eyes) and name them with a conceptual definition. The examples I just cited are some of the different experiences whose common quality we will call *darkness.* The dictionary is loaded with such definitions. To increase the accuracy of observation however, and to make it possible for all researchers to observe in exactly the same way, scientists use what are called **operational definitions,** which specify precisely how a defined concept is to be measured. By operational definition, for example, darkness is indicated by the amount of deflection of the needle of a light-sensitive photoelectric cell. While such operational definitions are not romantic or evocative, they are accurate and reproducible.

You can see how operational definitions help make observations comparable when we are studying something concrete, such as spiders. They are even more important when we are studying a more abstract concept, such as ambition or social class membership. It is much easier for us to be mistakenly talking in differing terms when we are discussing more abstract concepts. Do we always mean the same thing when we are talking about social class? You may think of it only in terms of yearly income, whereas I may want to include level of education and quality of home. An operational definition of social class must specify exactly what elements are included in a measure and exactly how they will be calculated.

Observations made using clear operational definitions not only make it possible to check the studies that others have done but also help organize the way we see the world. Once we begin to observe things in this way, patterns quickly emerge. The patterns of nature, whether in form or in behavior, surround us in mind-boggling abundance. The types of questions facing naturalists, psychologists, or sociologists can also be clearly differentiated. For the naturalist: "Why do certain birds migrate to specific places on specific dates, and how do they do it?" For the psychologist: "Why do some individuals develop such inability to confront authority that they greatly exaggerate its importance and identify strongly with it?" For the sociologist: "Why do Protestants commit suicide at a greater rate than Catholics, single and divorced people at greater rates than married people, and army officers at a greater rate than enlisted personnel?" Each question resulted from the observations of naturalists, psychologists, or sociologists, and each forms a hypothesis worth testing in the real world. The process of testing such hypotheses against reality is called **empiricism,** and it is a primary characteristic of the scientific perspective. It is sometimes contrasted with the subjective process called "armchair" theory, in which we make guesses but make no attempt to test them against reality.

When scientists test hypotheses using operational definitions, they generally refer to the measured concepts as **variables,** things that vary and take on differing values. The patterns for which scientists search express themselves in *consistent* variations in the qualities being observed. (Such patterns are called **distributions,** because a quality being measured is often distributed unevenly throughout a population.) For example, if the suicide rates of officers and enlisted personnel were always about equal or if sometimes one were higher and at other times the other were higher, then no clear pattern would have emerged that was worth studying. But when variations in suicide rates for the two groups are observed to be consistent, the scientist feels justified in trying to confirm (or contradict) the pattern empirically and explain it.

Descriptive and Causal Hypotheses

Using variables as our building blocks, we can create three types of hypotheses: (1) A **simple descriptive hypothesis** is a guess about the distribution of a single variable; (2) a **correlational descriptive hypothesis** is a guess about the simultaneous distributions of two or more variables; and (3) a **causal hypothesis** is a guess not only that variables change together (are *correlated*) but that a change in the value of one variable actually *causes* a change in the value of another variable.

Simple descriptive hypotheses are very common in everyday life. We may wonder how much it will rain (the distribution of precipitation in an area) or what other people think about the cafeteria chili (the distribution of attitudes toward a food). In election years, political pollsters are employed by candidates and news organizations to test simple descriptive hypotheses about the distribution of votes for the various candidates "if the election were to be held today." These are called simple descriptive hypotheses because they make guesses about the distribution of one variable at a time.

The results of measurements of individual variables can be displayed in a number of ways, depending on the level at which the variable is measured. The levels of measurement common in social research are *nominal, ordinal,* and *interval.* **Nominal measurement** names information; that is, it categorizes the information into mutually exclusive groupings between which there are no distinctions of degree. For example, religious group membership is nominal data. In America it is common for a person to be categorized as Protestant, Catholic, Jewish, other, or no

FIGURE 3-1

Histogram (also called a bar graph) for the distribution of sexes in a college class

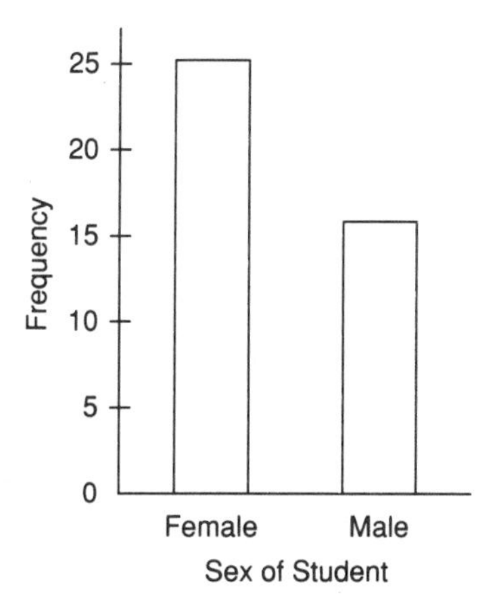

religion. The categories are mutually exclusive since a person can fit into only one of them, and there is no distinction of degree between them. That is, the measure is of religious group membership, not degree of religiousness, for which distinctions of degree are very possible. Other examples of nominal measurement are sex (male or female), occupation (teacher, truck driver, acupuncturist, etc.), and major in college (sociology, math, and so on).

A common way of graphically displaying the distribution of a nominal variable is the bar graph, or **histogram,** as it is often called by social scientists. Figure 3-1 shows a histogram for the distribution of sexes in one of my classes. The categories for the nominal measure are spread across the bottom of the chart, and the frequencies for each category are indicated against the scale on the left. As you can see, it would not matter if the bar for females were on the left or the right of the histogram, since there is no particular order in which the nominal categories have to be arranged. There are no distinctions of degree between categories. There may be more males than females in the class, but we are not comparing "maleness" and "femaleness" here. Also, it is easy to display the distribution nongraphically, as in the **frequency table** for the same data, shown in Figure 3-2.

Ordinal measurement ranks or orders information into categories in terms of the extent to which a given characteristic is present, but still with no precise knowledge of any distinctions of degree. Ranks typically are arranged from low to high. For example, *wealth* often is expressed at the ordinal level of measurement, ranging from poor, to moderate, to high income. We know that a person with a moderate income has more than a poor person, but we do not know by exactly how much. Other examples of ordinal measures that people often use are *intelligence*

FIGURE 3-2

Frequency table for the distribution of sexes in a college class (This table summarizes the same data as used in Figure 3-1)

Sex	Frequency	Percent	
Females	25	62.5%	(25/40)
Males	15	37.5%	(15/40)
Total =	37	100.0%	

(slow, average, smart, brilliant), *year in college* (freshman, sophomore, junior, senior), *quality of housing* (substandard, average, luxury), and *height* (short, average, tall).

As with nominal data, ordinal distributions can be displayed in histograms or frequency tables, as in shown in Figure 3-3. The one differ-

FIGURE 3-3

Histogram and frequency tables for the same set of ordinal data (year in college of students in a college class)

Histogram for Ordinal Data

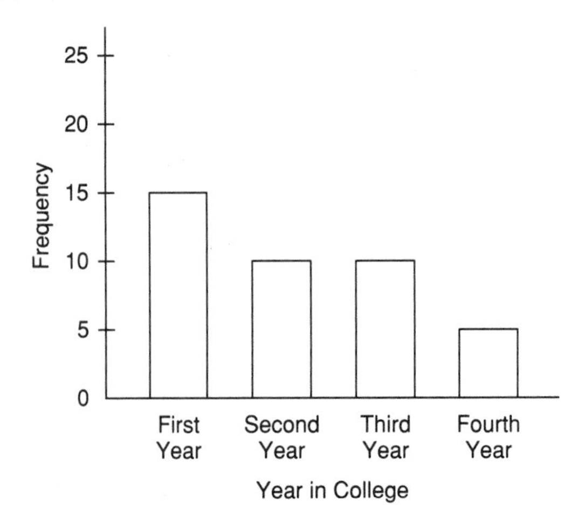

Frequency Table for Ordinal Data

Year	Frequency	Percent	
First Year	15	37.5%	(15/40)
Second Year	10	25.0%	(10/40)
Third Year	10	25.0%	(10/40)
Fourth Year	5	12.5%	(5/40)
Total =	40	100.0%	

ence is that the order of categories as shown must be the same as in the ordinal measure. Thus, for example, it would be misleading, and confusing, to arrange the bars of a histogram for year in college in any other order than from first year to fourth.

Interval measurement has all the qualities of ordinal measurement, plus it allows for the measurement to reflect the exact degree to which a given characteristic is present. For example, height measured in inches improves on the ordinal measure of height because we can determine not just that one person is taller than another, but by precisely how much. The key is the use of a unit of measure that is the same at all points of the scale. The inch that distinguishes between 50 inches and 51 inches is the same as the inch that distinguishes 78 inches from 79. This is why it is called interval measurement: the intervals remain equal throughout the scale. Other examples of interval variables are income in dollars per year after taxes (the equal interval is the dollar), credits completed toward a degree (the equal interval is the credit, not the course), and religiousness (perhaps measured by the number of times one attends religious services in a year).

Interval data, too, can be displayed graphically or nongraphically. One common graphic technique is the **frequency polygon,** an example of which is shown in Figure 3-4. It is a sort of smoothed-out version of a bar graph, with frequency along the side, and the scale for the interval measure spread across the bottom. The nongraphic techniques for describing interval-level distributions are called **descriptive statistics,** which include measures of (1) **central tendency,** such as the *mean* (the arithmetic average), the *median* (the point on a scale of

FIGURE 3-4

Frequency polygon for the distribution of degree credits completed by the members of a college class (120 needed to graduate).

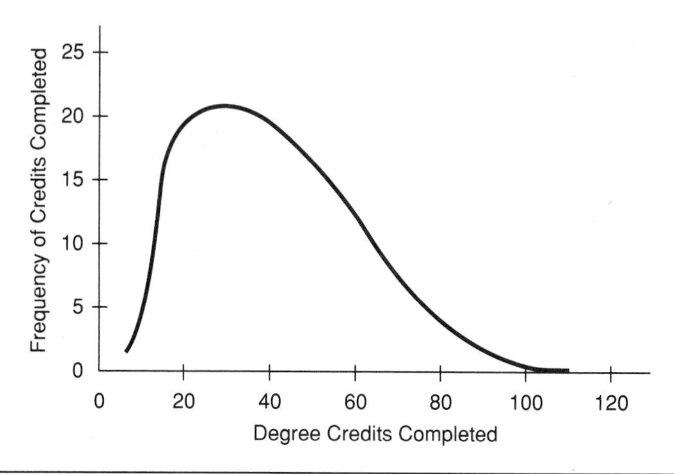

FIGURE 3-5

Interval level scores for the grade on a ten-point exam of ten individuals and some descriptive statistics for those scores (mean, median, mode, range and mean deviation).

	Grade on the Exam		The Mean Minus Each Grade
	5		3
	6		2
	7		1
	8		0
	8		0
	8		0
	9		1
	9		1
	10		2
	10		2
Sum of Grades =	80	Sum of Deviation Scores =	12

Descriptive Statistics:

Measures of central tendency

1 The mean = 8.0 (The arithmetic average, it is the sum of the scores, 80, divided by the number of scores, 10.)

2 The median = 8 (The median divides the scores in half. It is the point on the list of scores below which half the scores fall.)

3 The mode = 8 (The mode is the most frequently appearing score.)

Measures of variability

4 The range = 5 (The range is the highest score minus the lowest score.)

5 The mean deviation = 1.2 (The mean deviation is the average difference between all the scores on the test and the mean for that list. To calculate the mean deviation, subtract the mean for the list from each score ignoring whether the difference is negative or positive. I have provided a list of these deviation scores above. Then add the deviation scores, "12" in the example, and divide this by the number of scores, "10" in the example.)

scores below which half the scores fall), and the *mode* (the most commonly appearing score in a list of scores); and measures of (2) **variability,** such as the *range* (the difference between the highest and lowest scores in a list) and the *mean deviation* (the average difference between all the scores in a list and the mean for that list). Figure 3-5 shows a short list of scores and each of these descriptive statistics.

The other type of descriptive hypothesis is a guess about the **association** (or **correlation**) between two or more variables. For example,

FIGURE 3-6

Scattergram of association between students' grades and their seating position

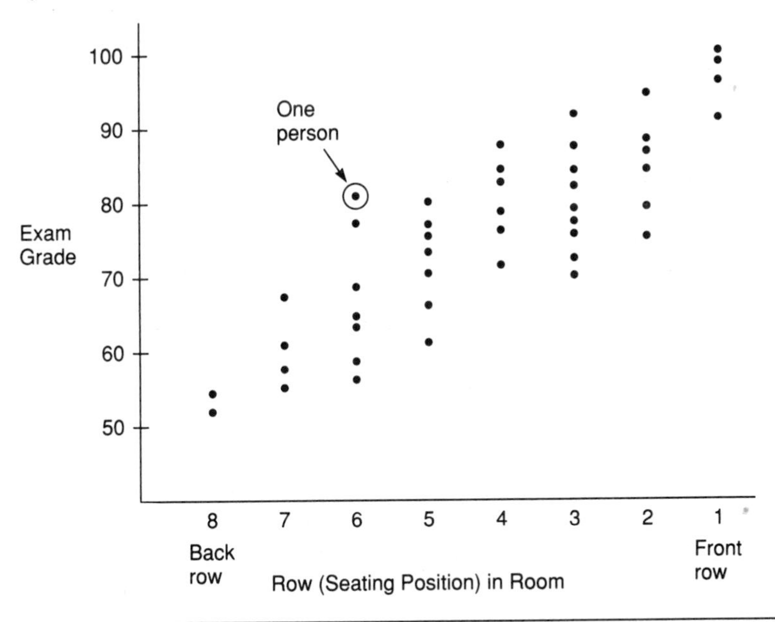

you may wonder whether the farther forward in the classroom a person sits, the higher his or her exam grades will be (the correlation between exam grade and position in the room). Or you may wonder whether the more weight people put on, the jollier they get (the correlation between weight and good humor).

If you want to test whether these relationships between variables actually exist, all you have to do is measure each variable carefully and then plot the data on a scattergram such as in Figure 3-6. Each dot on the scattergram represents the grade and seating position of one person. So the person whose position on the chart is circled sat in row 6 and got a grade of 81. The scattergram reveals an association between *exam grade* and *position in the room,* two variables that change together (that is, systematically). Perhaps you can see that the scattergram is merely two simple (single-variable) distributions stuck together. The simultaneous charting of the scores for grade and seating position shows the relationship that could not be revealed if you drew the histograms of each distribution separately. The pattern of dots shows that there is a strong tendency for grades to be higher for people who sit more toward the front of the class. The more tightly packed the dots are from lower left to upper right, the greater the strength of the association between the variables. The more widely scattered the dots, the weaker the association.

The examples I chose to illustrate the idea of correlation may seem a little strange, but I have heard them suggested many times. The reason I used them is that, although it may seem possible for each to be true, in neither hypothesis is it very likely that a true *causal* relationship exists. That is, a change in one variable is unlikely to *cause* a change in the other variable. It is possible for variables to be strongly correlated without being causally related. For example, for many years there was a strong relationship between the distance of Halley's comet from the earth and the cost of living in the United States. For several decades, as the comet flew toward us, the cost of living steadily increased. But clearly the comet did not cause the inflation. This became evident in the mid-1980s when the comet circled our sun and started away from us but the cost of living continued to climb.

It is extremely important not to confuse correlational and causal relationships. This is especially true because most research is aimed at eventually doing something concrete about problems. If we try to influence events by manipulating purely correlational variables, we will have no effect and will just waste our resources. So, to establish causal relationships, we develop causal hypotheses. One example of a causal hypothesis is: "The more hours you study, the higher the test grade you will get." Notice that the causal hypothesis sounds just like the correlational version of a descriptive hypothesis. The difference is the way it is used to conduct research. We begin by trying to establish that as the value of one variable changes, so does the value of the other. This is all that is necessary to show a correlation. But in addition, to establish a causal relationship, we have to show that changes in the causal variable must occur *before* changes in the other variable and that it is *only* the changes in the first variable that bring about changes in the second. In causal hypotheses, the variable that we believe to be *causing* changes is called the **independent variable.** The variable whose value changes because the value of the independent variable changed is called the **dependent variable.** It is called this because its value depends on the value of the independent variable.

In the hypothesis "The more hours you study, the higher the test grade you will get," *hours of study* is the independent variable and *test grade* is the dependent variable. To test this causal hypothesis, a researcher would have to demonstrate the three conditions mentioned above: (1) that the two variables are correlated (changes in one are always accompanied by changes in the other), (2) that changes in the independent variable (hours of study) occur before changes in the dependent variable (test grade), and (3) that changes in the dependent variable are due *only* to changes in the independent variable and to nothing else. For example, the researcher must show that higher grades did not result from better notes, higher intelligence, or better attendance among some people.

When we state causal hypotheses, factors like these, whose influence on the value of the dependent variable must be controlled (prevented from influencing the value of the dependent variable), are called **moderating variables.** Here is an example of a causal hypothesis that includes a moderating variable: "The more you study, the better your test grade will be, if your notes are good." Once again, *hours of study* is the independent variable (the assumed cause), *grade* is the dependent variable (assumed effect), and *quality of notes* is the moderating variable. If we are testing whether studying more hours raises grades, we do not want differences in grades to be due to anything other than hours of study. We cannot show a causal relationship between study and grades if the quality of notes influences grades (or, as researchers would say it, if the quality of notes moderates the relationship). In conducting research, sociologists are constantly aware of the many variables that can influence any single dependent variable. So sometimes hypotheses get complex. For example: "Assuming your notes are good and you have had enough sleep, the more you study, the higher your grade will be, if you read the test instructions carefully." This may at first look like a very confusing hypothesis. It is really just proposing that more study raises grades, but it includes three moderating variables, two at the beginning and one at the end. They are (1) quality of notes, (2) amount of sleep, and (3) attention to the exam instructions.

A more sociological hypothesis, taken from the literature on the sociology of aging, is: "The greater the level of one's social activity, the greater one's life satisfaction, as long as one is healthy and enjoys social involvement." This hypothesis is focused on understanding what influences *life satisfaction,* the dependent variable. The independent variable hypothesized to influence life satisfaction is *level of social activity.* Lastly, there are two *moderating variables,* that is, variables held constant, or removed from the relationship between activity and life satisfaction. They are (1) *the health of the individual,* and (2) *the degree to which the person enjoys social involvements,* which not everyone does. The hypothesis suggests that social activity contributes to life satisfaction, but not for people who are ill or who prefer privacy and isolation.

Methods of Testing Hypotheses

Once you have mastered the language of hypotheses, you will find it easier to think clearly about how the social world operates and how to conduct research to test your ideas. A number of research methods have been developed in sociology, each appropriate for one type of question being investigated.

Descriptive hypotheses are usually tested by conducting **survey research.** It is the kind of social research familiar to most people and emphasizes accuracy and reliability of measurement. Election-night polling and marketing research ("Which soap do you prefer?") are two common examples. Carefully worded questions, administered either in person by an interviewer (interviews) or on a written form filled out by a respondent (questionnaires), are developed to measure the variables under study.

By contrast, causal hypotheses are often tested by **experimental**

research. In this method, a researcher manipulates one or more independent variables in order to measure the effect of such manipulation on the value of some dependent variable. Once again, accuracy of measurement is extremely important, but because the experimenter wants to account for the influence of a variety of moderating variables, the experiment also aims at control of the environment of the study. An experimental study of the relationship between hours of study and test grades might work as follows: Divide a class of students into two equal groups, by random assignment, to ensure that the groups do not differ in important characteristics, such as intelligence or motivation. Control the moderating variables by giving all students the same quality of notes from which to study under the same conditions (light, noise, and so on). Then manipulate the independent variable by allowing one group to study for only one hour and having the other group study for three hours. Give all the students the same test, making sure that both groups take it under the same conditions (control of the experimental environment). If hours of study (the independent variable) is causally related to test grade (the dependent variable), the group that was made to study more hours should have higher grades than the group that studied less.

In this discussion on the scientific perspective, I have focused on the quantitative methods of sociology and other sciences. Quantitative methods are aimed at trying to evaluate numerically the qualities being investigated. It usually seems reasonable to attach numbers to variables such as hours of study or test grades, but it is somewhat more difficult to evaluate concepts such as group cohesion or social class membership numerically. This is the major problem sociology has in using the methods of the natural sciences. All quantitative scientific research proceeds under the assumption (called **positivism**) that what is being studied has a stable reality that can be measured from the outside by an objective observer. But there are a considerable number of questions in sociology that do not lend themselves to this kind of approach.

Another Approach to Social Research

One of the most interesting ideas in sociology is **symbolic interaction.** (There is a separate discussion of this topic in Concept 4.) Basically, it suggests that the behavior of humans in social situations can be understood only from the point of view of the actor, because that individual constructs the meaning of the behavior for each specific situation in which it occurs. Symbolic interactionists generally argue that the quantitative methods on which I have focused here cannot fully represent the meaning of social behavior. The problem is that attaching the same numerical evaluation to two different people who behave the same way can misrepresent the meanings of those actions to them. The main method of research that symbolic interactionists have developed is called **participant observation.** It is designed to deal with this dilemma, and, although it is generally discussed in sociology texts along with the other research methods described here, I have included it in the concept on symbolic interactionism because it is so different from the quantitative methods in this concept.

Summary

Science is a set of methods for testing guesses (hypotheses) about how systems of order work. Hypotheses may be descriptive (guesses about the distribution of some characteristic), correlational (guesses about the association of the distributions of two or more characteristics), or causal (guesses about the extent to which a change in the value of one characteristic actually causes a change in the value of another). The process of testing hypotheses against observable events in the real world is called *empiricism.* The scientific perspective requires that measurements be as objective and verifiable as possible. That is, other investigators must be able to reproduce the findings using the same tools for measurement and analysis. Scientific methods usually require the evaluation of reality in numerical terms. It should be kept in mind that human social order is not exactly like other systems of order, so the methods of the natural sciences do not always apply as well in sociology. In fact, a significant proportion of sociologists, the symbolic interactionists, argue that the methods of the natural sciences have virtually no application to the study of human behavior.

The Illustration for this concept shows how science can be applied to the study of humans either poorly or well. The naturalist Stephen J. Gould has examined the way scientists from a variety of disciplines have studied human intelligence, and he exposes the misuse of science in the pursuit of personal interests. Yet, by his own method of investigation, Gould illustrates the care and objectivity of good scientific analysis.

The Application for this concept provides an opportunity to practice random sampling, one of the most frequently used scientific methods in social research. A small illustration "population" is provided, and the method for drawing a simple random sample (including a table of random numbers) is presented.

Illustration

Stephen J. Gould, The Mismeasure of Man *(New York: W. W. Norton, 1981)*

The word *science* often is used imprecisely. Somehow it has come to refer to the trappings of science rather than to its methods. This is quite understandable, because the trappings are exciting—test tubes, machinery with flashing lights and banks of switches, computers, severe-looking lab coats, and thick printouts of data. Who could reveal all the mysterious secrets of those data except a scientist? It gives the scientist an air of power and magic. The fact is that science is none of these trappings. It is only the method behind them—careful, orderly, and even tedious. The problem is that we often take the shortcut of substituting the appearance for the process. As a consequence, the findings of people who look or act like scientists may be taken at face value. Checking on how their results were obtained is as demanding as generating the results in the first place and often requires more training than the average person has. Gould has proved himself a master at exposing some "science" as nothing more than personal bias dressed up in the impressive trappings of science.

By the second half of the nineteenth century science had become something of a craze. In matters of intellectual authority it had virtually replaced the aristocracy and the church. Scientists investigated subjects ranging from the movement of the planets to the nature and capacities of the human being. While Charles Darwin was carefully documenting his observations of the natural world, other "scientists" were making their own sorts of measurements.

Led by Paul Broca, an eminent French professor of surgery, a number of men concerned themselves with the relationship between brain size and intelligence (see Haller, 1971). They generated a great amount of data that they said "proved" that larger brains were associated with greater intelligence and that various groups could be ranked in order for both qualities. At the top of the list were white, male Europeans. Below them, in decreasing order, were Asians, Africans, Mongols, Mexicans, Negroes, Javans, and Australians. Males were reported to have, on average, larger brains than females and therefore to be more intelligent.

The data the brain-weighers presented to support their theories were extremely impressive. Broca, for example, wrote long studies of the proper way to measure brain size. Some favored pouring mustard seed into the vacated skull, then measuring the weight of the seed that would fit. Broca argued for lead shot, since it packed more evenly. According to Gould, Broca even discussed techniques for tapping the shot to a settled position. Broca also liked to weigh the actual brain immediately after autopsy. The data that resulted looked quite convincing, and the people who generated the information were considered extremely authoritative. Their proof was almost universally accepted in spite of the fact that it was the result of purely subjective bias.

Gould looked carefully at the techniques used to measure cranial capacity and at the original data from which the conclusions were drawn. He found that often the techniques were extremely inaccurate and that, when the data were accurate, they were either selectively collected or badly interpreted.

Measurement Techniques

Before Broca's group began measuring cranial capacity, phrenologists had tried to determine human mental and behavioral qualities by studying the bumps and shapes of the outside of the skull. Broca's more impressive "science of craniometry" was thought to be a significant advance. But as Gould points out, when a brain is weighed, a number of problems in standardizing techniques arise. At what point below the brain is the brain stem to be cut? Less stem means less brain weight. Are the tissues that cover the brain to be removed or not? These can weigh enough to wipe out the measured differences between two categories of people. Is the brain preserved in any fluids before weighing, and at what temperature? How long after death is the brain weighed? Brains lose weight over time. Even if the brain-weighers had agreed on standard techniques (which they did not), other problems existed.

Selective Data Collection

In order for samples of data to represent the larger populations from which they are drawn, rigorous methods of selection must be used. The brain-weighers did not use these methods. They measured whatever brains or skulls fortune gave them to examine. There was no possibility of widespread sampling among groups who did not normally live in Europe or of people wealthy enough to keep themselves from being subjected to such indignities. The Broca group also chose, whenever possible, the brains of eminent white males of the day, who were expected to have large brains—the German mathematician K. F. Gauss, the French naturalist Georges Cuvier, and the Russian novelist Ivan Turgenev, to name a few.

In a most blatant example of sampling bias, E. A. Spitzka (whom Gould calls "the most prominent practitioner" among the brain-weighers) compared "one of his largest brains from an eminent white male, a bushwoman from Africa, and a gorilla"—one brain from each group! There are such wide variations in brain size *within* any group that had he chosen the smallest white male brain on hand and the largest African bushwoman brain, his results would have been reversed. Of course, given the sample he selected, he felt justified in claiming that white men were as superior to the bushwomen as bushwomen were to gorillas.

Even if such obviously intentional sampling bias had not occurred, some bias would have unintentionally resulted from the differences in the groups being considered. For example, the finding that women had, on average, smaller brains than men could have been a result of the life expectancy of each. The average age of the female brains used in Broca's studies was much higher than that of the male brains. This is not surprising, given the longer life expectancy of women. Because a high percentage of these women were likely to have died of long-term degenerative diseases and because brain weight diminishes with both age and such diseases, the women's brain weights would certainly be lower.

Even if the methods of brain measurement had been accurate and the sampling techniques fair, the data would not have escaped the most biasing of all the brain-weighers' techniques. As Gould documents, these "scientists" started with a point they wanted to prove and then went about doing what was necessary to prove it. This a priori approach (proceeding from an assumed cause to the events that "confirm" it) is found throughout the work of the brain-weighers. The ominous thing about a priori research is that, even when all the data collection is accurate, the results can be invalidated entirely by the way the data are interpreted.

Biased Data Interpretation

In its extreme, biased data interpretation involves simply ignoring data that do not fit our ideas about what is true. For example, among the collected brains of eminent white men, a number were of only average size and some were smaller than average. No problem: Broca simply attributed their small size to the person's advanced age or small physical stature. The trick is to ignore data that do not confirm the hypothesis.

The brain-weighers made little attempt to account for the influence of physical stature in the comparison of male and female brains. An elephant, for example, has a larger brain than a human simply because of its great size (but no one attributes greater intelligence to elephants than to humans). So it is reasonable to assume that the larger average physical stature of men would contribute to gender differences in brain weight. When Gould reanalyzed the male-female differences after removing the effect of body size, he virtually wiped out group differences in brain weight.

Gould also recounts another instance of ignoring inconvenient data (this time more recent). H. T. Epstein (1978) reported data from the 1930s purporting to rank professions by average head circumference. At the top of the list were professionals; then below them in descending order of head circumference were semiprofessionals, clerical workers, trades workers, public service workers, skilled trades workers, personal services workers, and laborers. Epstein concluded that head size (and therefore intelligence) was related to vocational status. Yet clerical workers (third on the list) are not generally considered higher in vocational status than public service or skilled trades workers (fifth and sixth on the list). Gould also discovered that Epstein (or somebody from whom Epstein had copied the data) had removed from the original list of occupations three groups—factory workers, transportation employees, and farmers and miners. The reason? They all had higher average head circumferences than the most prestigious group, professionals, although the vocational status of the omitted groups was relatively low.

In sum, Gould has given us dramatic illustrations that the *appearance* of scientific rigor is no substitute for the rigor itself. Broca's "scientists" were merely men who wished to prove the superiority of their own group and cloaked their "evidence" in scientific trappings. They wrote more in justification of their assumptions than in examination of them. One brief quote from Broca will illustrate: "The relatively small size of the female brain depends in part upon her physical inferiority and in part upon her intellectual inferiority" (Gould, 1981:104). Broca had argued throughout that women's smaller brains caused them to have lower intelligence. Then, amazingly, he concludes that their lower intelligence causes (in part) their smaller brain size! Quite clearly, just because something looks like science doesn't make it so.

Application

A **population,** in research terms, consists of all the people who conform to a set of characteristics. For example, all the citizens of the United States form a population, as do all the registered voters, all the women at a given university, or all the left-handed pitchers in the National League. Populations are generally too large to be studied. There are so many people in each that the time and costs of questioning all members would be prohibitive. So social research focuses on **samples,** people selected from a population, who, it is hoped, represent the characteristics of the larger group.

One problem that social researchers face is finding selection techniques that give reason to believe that the sample will be like the population. The purpose of this application is to illustrate how one of these techniques, simple **random sampling,** is done. To begin with, let's dispel some of the common misunderstandings about random sampling.

The word *random* is often misused by reporters on television and in print. Many times I have heard them say, "We conducted a strictly unscientific study of attitudes toward the oil price hike by selecting shoppers *at random* from supermarkets around the country." The selection they really made is called *accidental sampling.* They just grabbed whoever came out of the store. Random sampling does not mean "without purpose or control." It is a *very* scientific method of sampling in which every individual in a population has an equal chance of being selected for the sample.

The word *random* refers to the fact that members of a population (called *elements*) can be selected with no pattern or favoritism. Picking names from a hat is not random, because there is a tendency for names at the top and the edges of the hat to be excluded from selection. Interviewing shoppers as they leave supermarkets is not random, since it systematically excludes from selection people who shop in neighborhood stores or have groceries delivered.

TABLE 3-1

Table of random numbers

48663	91245	85828	14346	09172	30168	90229	04734	59193	22178	30421	61666	99904	32812
54164	58492	22421	74103	47070	25306	76468	26384	58151	06646	21524	15227	96909	44592
32639	32363	05597	24200	13363	38005	94342	28728	35806	06912	17012	64161	18296	22851
29334	27001	87637	87308	58731	00256	45834	15398	46557	41135	10367	07684	36188	18510
02488	33062	28834	07351	19731	92420	60952	61280	50001	67658	32586	86679	50720	94953
81525	72295	04839	96423	24878	82651	66566	14778	76797	14780	13300	87074	79666	95725
29676	20591	68086	26432	46901	20849	89768	81536	86645	12659	92259	57102	80428	25280
00742	57392	39064	66432	84673	40027	32832	61362	98947	96067	64760	64584	96096	98253
05366	04213	25669	26422	44407	44048	37937	63904	45766	66134	75470	66520	34693	90449
91921	26418	64117	94305	26766	25940	39972	22209	71500	64568	91402	42416	07844	69618
00582	04711	87917	77341	42206	35126	74087	99547	81817	42607	43808	76655	62028	76630
00725	69884	62797	56170	86324	88072	76222	36086	84637	93161	76038	65855	77919	88006
69011	65795	95876	55293	18988	27354	26575	08625	40801	59920	29841	80150	12777	48501
25976	57948	29888	88604	67917	48708	18912	82271	65424	69774	33611	54262	85963	03547
09763	83473	73577	12908	30883	18317	28290	35797	05998	41688	34952	37888	38917	88050
91567	42595	27958	30134	04024	86385	29880	99730	55536	84855	29080	09250	79656	73211
17955	56349	90999	49127	20044	59931	06115	20542	18059	02008	73708	83517	36103	42791
46503	18584	18845	49618	02304	51038	20655	58727	28168	15475	56942	53389	20562	87338
92157	89634	94824	78171	84610	82834	09922	25417	44137	48413	25555	21246	35509	20468
14577	62765	35605	81263	39667	47358	56873	56307	61607	49518	89656	20103	77490	18062
98427	07523	33362	64270	01638	92477	66969	98420	04880	45585	46565	04102	46880	45709
34914	63976	88720	82765	34476	17032	87589	40836	32427	70002	70663	88863	77775	69348
70060	28277	39475	46473	23219	53416	94970	25832	69975	94884	19661	72828	00102	66794
53976	54914	06990	67245	68350	82948	11398	42878	80287	88267	47363	46634	06541	97809
76072	29515	40980	07391	58745	25774	22987	80059	39911	96189	41151	14222	60697	59583
90725	52210	83974	29992	65831	38857	50490	83765	55657	14361	31720	57375	56228	41546
64364	67412	33339	31926	14883	24413	59744	92351	97473	89286	35931	04110	23726	51900
08962	00358	31662	25388	61642	34072	81249	35648	56891	69352	48373	45578	78547	81788
95012	68379	93526	70765	10592	04542	76463	54328	02349	17247	28865	14777	62730	92277
15664	10493	20492	38391	91132	21999	59516	81652	27195	48223	46751	22923	32261	85653

Source: Abridged from *Handbook of Tables for Probability and Statistics*, second edition, edited by William H. Beyer (Cleveland: The Chemical Rubber Company, 1968).

Drawing a Random Sample

The most commonly used method of random selection of elements from a population involves the use of a table of random numbers. This is a series of numbers arranged in rows and that have no numerical pattern. (See Table 3-1; you are going to use it in this application.)

For the application I have invented a "population" of ninety-nine people from whom you will draw three simple random samples using Table 3-1: the first sample will be of just ten names, the second of twenty names, and the last of thirty names. The reason for drawing these three samples is that it will illustrate the "law of large numbers." The law works like this:

Let's say you are flipping a completely fair coin; that is, it has no tendency to come up heads rather than tails. The likelihood of tossing the coin and having it come up heads is 50 percent on any trial. Even though the coin is totally fair, the likelihood is quite high that in just ten tosses the distribution will differ from five heads and five tails. In fact, it would not be very surprising to get seven heads and three tails, or the reverse (a 70-30 split), in just ten tosses. But you would be *very* unlikely to get a 70-30 split in thirty tosses (twenty-one heads and nine tails, or the reverse). And it is virtually impossible to get a 70-30 split in one hundred tosses. Of course, another law, the principle of random sampling of events, is also operating if the coin has a tendency to come up on one side more than the other. If it has a tendency to come up heads 68 percent of the time, then after fifty tosses the underlying tendency to come up 68 percent heads will be represented quite closely in the sample results. The point is that, as the number of random samplings of events increases, the underlying distribution of events being sampled is more closely approximated. Just a few selections can easily differ from the underlying distribution, but more selections will cause the sample to approach the "real" population's characteristics. This is true whether the underlying distribution is very simple (like the 50-50 underlying distribution of the fair coin) or very complex (like the distribution of characteristics in a population of people).

In order to illustrate the principle of random selection and the law of large numbers, you will draw three samples of increasing size using the table of random numbers (Table 3-1). The population I have invented is numbered from 01 to 99. (See Table 3-2.) Each person in the population is also identified by sex and age. To conduct the selection, enter Table 3-1 at any point. (I like to close my eyes and lower my hand holding a pencil, point down, from a foot above the table.) Then, moving from left to right and down the page, underline ten two-digit numbers, ignoring the spaces between groups of numbers. If your entry point is very near the end of the table, go to the beginning of the table once you reach the end. If the same two-digit number comes up twice, select an

TABLE 3-2

"Population" of students listed by identification number, sex, and age

ID no.	Sex	Age	ID no.	Sex	Age	ID no.	Sex	Age	ID no.	Sex	Age
01	F	21	26	F	44	51	F	33	76	F	23
02	F	26	27	F	31	52	F	27	77	F	27
03	M	32	28	M	25	53	F	43	78	F	33
04	F	19	29	F	24	54	F	30	79	M	17
05	M	29	30	M	21	55	M	22	80	F	24
06	M	22	31	F	21	56	M	18	81	F	28
07	F	22	32	F	34	57	M	21	82	M	21
08	F	31	33	F	28	58	F	29	83	M	25
09	F	40	34	F	30	59	F	28	84	F	31
10	M	23	35	M	17	60	F	27	85	F	35
11	M	29	36	F	29	61	M	25	86	F	25
12	F	18	37	M	22	62	M	21	87	M	21
13	F	33	38	F	27	63	F	28	88	M	24
14	F	24	39	F	37	64	F	27	89	F	25
15	M	30	40	F	24	65	M	22	90	F	48
16	M	17	41	M	26	66	F	33	91	M	19
17	F	28	42	F	33	67	F	35	92	F	27
18	F	20	43	F	32	68	F	27	93	F	26
19	M	19	44	F	27	69	M	21	94	M	22
20	F	29	45	M	25	70	M	17	95	M	27
21	F	27	46	M	31	71	F	21	96	F	33
22	M	22	47	F	22	72	F	28	97	F	27
23	M	21	48	F	38	73	F	40	98	F	26
24	F	29	49	M	23	74	M	28	99	M	18
25	M	25	50	M	23	75	M	22			

extra number to make ten different ones. Each of the two-digit numbers chosen represents a person in the population selected for the first simple random sample. Because there is no pattern in the table of numbers, there can be no pattern in the selection of elements from the population. For example, say the first two-digit number you underlined was 67. The person number 67 on the population list (Table 3-2) is in your sample. The next might be person number 47, the next person number 58, and so on. Here is how your first random sample selection of ten people *might* look:

96063 95138 61087 92046 25674 75855 33428 45166 76774 75600 06471
24085 09365 78163 78500 57301 22175 78452 70504 56370 46175 78352
55352 56829 97037 27561 11950 08573 35427 91930 57776 17564 88308

Each of the underlined two-digit numbers is a person included in the first sample.

In the first ten spaces on the recording sheet for this application, list the people who were selected for the first sample. Continue selecting ten more people to make up the second random sample of twenty names, then select ten more to complete the third sample of thirty, recording each selection as you proceed.

Next, beginning with the first sample of ten people, calculate the percentage of the sample that is male (the number of men selected divided by the sample size of ten) and the percentage of the sample that is female. Also calculate the average age of the entire sample of ten (the total of the ten ages divided by ten, carried to one decimal place). Fill in this information in the spaces provided. Then calculate the same statistics for Samples 2 (twenty people) and 3 (thirty people). (Remember that the percentage of males in Sample 2 is the total number of males in Sample 1 plus the next ten people selected divided by the new sample size of twenty.)

Analysis of the Results

After you have recorded and calculated the information for all thirty people, you can figure out how closely the characteristics of each sample approximate those of the population from which they were chosen. The percentages of men and women and the average age for the entire population are:

1. Percentage of men in the population: 39%
2. Percentage of women in the population: 61%
3. Average age of the population: 26.6

How closely did your samples approximate the underlying characteristics of the population? Were larger samples more accurate than smaller ones? They should have been. If you want to do one more bit of analysis, here are the average ages of males and females in the population:

4. Average age of men in the population: 22.9
5. Average age of women in the population: 29.1

You can use your sample data to calculate these data for each of your samples and compare the averages to the total population averages.

Recording Sheet Simple random sample

In the spaces provided, list the identification number, sex, and age of the people chosen by random sampling.

No. of person in your sample	Person's original ID no.	Sex	Age
01	______	______	______
02	______	______	______
03	______	______	______
04	______	______	______
05	______	______	______
06	______	______	______
07	______	______	______
08	______	______	______
09	______	______	______
10	______	______	______
11	______	______	______
12	______	______	______
13	______	______	______
14	______	______	______
15	______	______	______
16	______	______	______
17	______	______	______
18	______	______	______
19	______	______	______
20	______	______	______
21	______	______	______
22	______	______	______
23	______	______	______
24	______	______	______
25	______	______	______

26	______	______	______
27	______	______	______
28	______	______	______
29	______	______	______
30	______	______	______

Using the data from the first sample of ten people (Sample 1), calculate the following:

1. Percentage of men in sample: ________%

2. Percentage of women in sample: ________%

3. Average age of entire sample: ________%

Using the data from the first twenty people (Sample 2), calculate the statistics again:

1. Percentage of men in sample: ________%

2. Percentage of women in sample: ________%

3. Average age of entire sample: ________%

Using the data from all thirty people chosen (Sample 1), calculate the statistics one last time:

1. Percentage of men in sample: ________%

2. Percentage of women in sample: ________%

3. Average age of entire sample: ________%

CONCEPT 4

Symbolic Interaction and Qualitative Research

Definition

Symbolic Interaction The view that social reality is constructed in each human interaction at the level of symbols.

Qualitative Research A research method in which the qualities of objects, behavior, or relationships are evaluated in textual terms (words) rather than quantitative terms (numbers).

One of my favorite films is *Rashomon* (1951), by the great Japanese filmmaker Akira Kurosawa. It tells the story of four people who are involved in a rape and murder—or perhaps not. We see the events unfold as each survivor tells the story of what happened, but each tells a different tale. At first it seems to be a mystery in which we are being asked to discover the liars. But the genius of the film is that Kurosawa compels us to realize that each of the conflicting stories is actually true. It is not just that each participant remembers events differently but also that, from the point of view of each person, the events *actually were* experienced differently. (An American version of *Rashomon,* called *The Outrage,* is almost as good as the original and easier to find on late-night television.) There is a very good sociological explanation for what these films say about how humans construct reality.

Symbolic Interaction and the Construction of Social Reality

Human beings communicate and interact with one another at the symbolic level. That is, we have the ability to abstract the qualities of objects and experiences and to assign **symbols** such as words or gestures to them. Imagine observing a prehistoric man developing this ability (though it is equally likely to have been a woman). He sees several different kinds of animals kill their prey. One quality they all have in common is that they are dangerous. The abstracting process requires the man to ignore the differences among these animals (such as size, coloration, markings) while making the important quality they have in common (the fact that they can kill) central in the naming process. The prehistoric man can then create a symbolic label or representation of such dangerous animals. It may be a gesture (such as an arm thrown across the eyes) or a word (such as "big-trouble"). Once the symbol has been taught to others, it can be used to warn of danger, because it calls up the image of danger without the actual threat having to be present. This is communication and interaction at the symbolic level.

The ability to communicate by abstract symbols means a number of things for human conduct. For one thing, it is not necessary to have the object or experience present in order to react to it. I can tell you about a prehistoric being without having to show you one or even without knowing that just such a prehistoric scene ever occurred. We humans, therefore, can speculate; we can re-create reality or create our own versions of it. Without symbolic systems, such as language, none of this would be possible. Reality would be limited to what we could sense immediately and what had been implanted genetically into our system of "knowledge" by instinctual patterns. This is the way reality is for animals other than humans.

Mead and Symbolic Interaction

G. H. Mead, an extremely important figure in the development of symbolic interactionism, proposed in *Mind, Self and Society* (1934) that the unique ability of humans to symbolize led to the construction of a special kind of social order. According to Mead, our ability to communicate by symbols means that we can contemplate possible actions, guess about their outcomes, and compare the qualities of different objects; in short, we can think. Mead referred to this ability as **mind.**

Just as we are able to evaluate our own potential actions, we can also evaluate the actions of others. (Mead called this "taking the role of a particular other.") Mead concluded that by accumulating such evaluations we develop a **self,** that is, a cumulative idea of who we are that is constructed from the actions of others toward us. He spent a good deal of time discussing the way the development of the self operates in the process of socialization. (It is covered at length in Concept 9, on socialization.) It is enough to say here that Mead believed that there was no inborn, or unvarying, human character (or self). Rather, since each of us experiences a unique set of interactions with people and the environment, each necessarily develops a unique self. Ultimately, humans develop the ability to evaluate and internalize the expectations of a group of people simultaneously. (Mead called this "taking the role of the generalized other.") Accordingly, society is the sum total of the ways in which people agree to act in given situations and that are internalized by its members.

Social Reality and the Definition of the Situation

There is nothing in social order that is real unless we participants in society agree that it is real. Or, to use the now famous words of W. I. Thomas, "If [people] define situations as real, they are real in their consequences" (Thomas & Thomas, 1928:572). Shaking hands is a friendly greeting only because we define it as such. In another time or culture it might mean hostility (for example, "What are you hiding in your hand?").

In everyday interaction, having agreed rules for behavior makes things predictable and, therefore, relatively stable. But everyone is taught the rules of everyday behavior under different circumstances, and everyone lives a unique life. Even if we were all taught pretty much the same rules of behavior, we would inevitably learn them differently. Also, we sometimes face contradictory rules or ambiguous situations in which the proper rule is not apparent. In short, there is a great deal of room for individual evaluations of the meaning of a social situation to differ. At a party, a casual comment taken as a compliment by a woman is later recalled as an insult by her husband. They may discuss it, and one may convince the other; but, if they don't, the "real" meaning of the original comment will remain undisturbed in the mind of each person forever. In the film *Rashomon,* what was a rape to one character was a seduction to another, because that was the meaning applied to the situation by each.

Symbolic interactionists emphasize how humans decide on the meaning of every situation in which they interact. Thus, the agreements that are internalized in the process of socialization are only guidelines for social behavior, not unvarying, narrow commands. How critical is it to our interaction that we interpret the meaning of each other's actions and statements? Consider that humans are capable of saying something in such a way as to mean its opposite. For instance, my friend asks his teenage daughter if she wants to speak to the boy who has just telephoned. My friend knows the boy has been pestering her to go out with him, and has no trouble interpreting the meaning of his daughter's response: "Sure, Dad. Right!" Her *sarcasm* ("a cutting, often ironic remark intended to wound," according to the *American Heritage Dictionary,* 3d ed.) turns the meaning of the words upside down. If we cannot interpret meaning subtly, we cannot communicate or interact with confidence.

Herbert Blumer (1969), whose work has been important in shaping modern symbolic interactionism, described social order as a process of continual creation. Every time we interact, we create a new version of social reality. When we find ourselves in a specific situation, we do not act automatically, as if we were born with instincts or had been conditioned to respond a certain way. Rather, we evaluate the stimulus and decide what it means before we act. Two people can obviously interpret the same stimulus differently. When the boss told me, "Please sit down," it sounded like a command to me, but my co-worker thought the boss was just being courteous. The symbolic-interactionist view of social reality has serious consequences for the ways in which we study human social behavior.

Participant Observation and the *Verstehen* Method

In the natural sciences, the measurement of objects, including their behavior and relationship to one another, is overwhelmingly quantitative. That is, researchers use numerical evaluations such as weight in grams, velocity in meters per second, length in centimeters, and energy in joules. This is an effective approach, because there seems to be a stable, physical reality in the natural world that yields consistent measurement results (when we know how to measure it), and that does not seem to mind the process at all. Of course, rocks, electrons, and red blood cells have no consciousness with which to respond to the measurement they undergo. But people do.

Weigh a rock and it does not react, but a human might try to weigh less or more (depending on his or her self-image) or might refuse to be weighed at all. The problem becomes even more acute when the quality being measured is more abstract, such as an opinion about welfare or prayer in schools. If the person being interviewed does have an opinion (we cannot assume that the quality being measured is present in everyone), she or he might lie about that opinion, distort it, or even refuse to reveal it. One possible solution is to avoid asking people to respond to questions. That way they cannot distort answers. Instead we might just observe their behavior and evaluate it. But as the symbolic interactionists point out, the same situation can mean entirely different things to different participants. So it can be extremely misleading to apply numerical evaluations to human statements or behaviors. In many situations the objective, quantitative methods of the natural sciences are completely inappropriate for studying human social behavior. How, then, can it be studied?

The answer of the symbolic interactionists is that, to understand the meaning of a particular social situation, we must both observe and take part. The idea is to put ourselves in a position to construct the meaning of the situation just as the participants do. Max Weber ([1925] 1946) referred to this approach by the term *verstehen* (German for "to understand"). It is the process through which subjective reality can be discovered. The researcher must see a situation from the point of view of the participants, which makes the fact that the researcher is a subjective interpreter of events actually an advantage rather than a detriment to the research process. This is exactly opposite to the quantitative approach of the natural sciences, in which every effort is made to remove the influence of the researcher's evaluations from the measurement process. The difference in the two approaches results from their different views of the nature of social reality. Positivists (who use the methods of the natural sciences) see social order as relatively fixed and measurable, like the order of the rest of the natural world (see Concept 3). By contrast, symbolic interactionists see social order as constantly being created, subject to the meanings and evaluations of the participants in each interaction. Thus, social order cannot be measured quantitatively, because it is the *quality* of meaning that lies at its core. The qualitative approach of symbolic interactionists requires close, careful, and systematic observation and thus is, in an important sense, scientific. But the methods employed by symbolic interactionists are quite consciously not those of the positivists.

Qualitative Approaches

Many of the tools used by qualitative researchers are the result of common sense. When observing social interaction, for example, it is a good idea to describe and record the setting, participants, actions, and verbal exchanges carefully. But when people know they are being observed, they are likely to act differently than they normally would. So the researcher may not reveal that he or she is observing.

Of course, this raises important ethical problems of deception that are a continuing issue for participant observation. Consider the case of a sociologist who wants to discover the inside workings of a local restaurant. If she were to reveal that she is doing research, she might never gain the confidence of the people whose cooperation she needs. So she takes a job as dishwasher in the restaurant and is eventually taken into the confidences of the staff and, to a lesser degree, the management. Along with all the information she gathers about the hierarchy of the place and the way relationships and work are negotiated, she also discovers that when they are angry, staff members sometimes sabotage food in imaginative and grotesque ways, or steal food and equipment. This information is clearly important as part of the overall picture of the operation of the place as a social system. However, revealing it would also endanger certain staff members, even if they were not named. What is the right thing to do? Is there some way you can think of to report the findings that would not harm the subjects of the study? Do you think the findings of the research could justify the potential harm their publication might cause? Is any of this the researcher's problem in the first place? These are some of the ethical dilemmas that are the subject of debate within the field.

Beyond the mechanics of participant observation there are a number of theoretical models within symbolic interactionism that have led to important insights into the way social reality is constructed.

Goffman's Dramaturgy

Erving Goffman developed an approach he called **dramaturgy,** in which social interaction is viewed as a series of small plays, or dramas. Each interaction is, in a sense, scripted by the roles the participants play. The lines are not as specific as in scripts, but once the roles are chosen for the situation, the participants enact lines appropriate for the characters being portrayed. We have these roles available because we are socialized to know what others expect of our behavior in given situations. For example, if I decide that I am expected to play the role of teacher in a situation, I will say lines appropriate for a teacher. We become quite skilled at playing a variety of roles and at deciding when to play each role.

Goffman's numerous studies (for example, *The Presentation of Self in Everyday Life* [1959], *Asylums* [1961a], *Stigma* [1963], and *Relations in Public* [1971]) uncovered how individuals present themselves to others within the framework of social roles. His work revealed previously unrecognized rules that we use every day to make sense of our relations with one another. Among these are the rules used by the physically handicapped and mentally ill to cope with their stigma and the rules that people use to maintain socially appropriate distance between themselves and those with whom they interact (this set of rules is the focus of the Illustration in Concept 8 from Goffman's work). As a methodological key for the participant observer, the idea of dramaturgy can be extremely helpful. Imagine observing the interaction of several people in a bar. Without some organizing principle, the interaction might seem to be a disorganized, random mass. But if you ask yourself what role each person is playing in each separate interaction, a number of patterns are likely to unfold. You might identify the role of "old hand," or "protector," or "shark," or "lost lamb," or "no-nonsense drinker," or "good-time Charlie" (or Charlene). Once the various roles are specified, the way each person deals with another becomes clearer. For example, what kind of conversation would you expect between a "shark" and a "lost lamb"? It would most certainly be different from the encounter of the "no-nonsense drinker" and the "shark," who probably would have nothing at all to say to each other.

Garfinkel's Ethnomethodology

Another approach to the understanding of social interaction is provided by the work of Harold Garfinkel (1967). **Ethnomethodology** is an approach to understanding interaction based on the assumption that social reality is the result of our agreement to agree with one another. That is, we negotiate reality by exchanging accounts of what is going on between us, with the unstated assumption that we will reach an agreement eventually. A methodological key used by ethnomethodologists is intentional disruption of the process of reality negotiation or meaning construction. When a researcher upsets the process by which the meaning of a situation is negotiated, the normally hidden methods of the negotiation can be starkly highlighted. For example, suppose you have just been introduced to a stranger and you stand with your face only inches from this person and stare unflinchingly into his or her eyes. The normal process by which two newly introduced people negotiate the meaning of an interaction would definitely be disrupted by your behavior. There are likely to be some awkward moments during which the "target" person scrambles for a way to construct a new meaning for the situation. ("What is this, are you crazy?" or "Are you from another country?" or "Did Ted put you up to this?") The normal rules that have temporarily been abandoned may become clearer by their very absence. On the other hand, some people seem to use ethnomethodology as an excuse to be rude. They are not actually trying to carefully reconstruct the techniques by which people negotiate reality. This sort of careless prodding and poking of social order should absolutely be avoided.

Summary

One of the major perspectives in sociology is symbolic interactionism. It is based on the belief that social reality is constructed by people every time they interact. Through a process of negotiation, people come to agreement about the meaning of the situation in which they are involved. Once this definition of the situation is in place, further interaction can occur according to its specific, agreed-on rules. Accordingly, human social interaction takes place primarily at the level of symbols, which are ways of expressing the meaning of an experience. G. H. Mead, one of the most important figures in the development of this perspective, saw the self as the sense of who we are, which is constructed from our understandings of the actions of others toward us. Because symbolic interactionists believe that reality can be understood only from the point of view of the actors in social situations (who define the meaning of the situation), special methods are needed for the study of human interaction. Unlike the methods of the natural sciences, which are based on the idea that the nature of a phenomenon can be measured accurately from an objective, external point of view, the methods of the symbolic interactionists are based on the idea that no such objectivity is possible. Instead, they use participant observation, taking part in the interaction so that it can be understood from the point of view of the participants.

The Illustration for this concept begins with a summary of one of the best known participant observations in sociology. William Whyte lived in Boston's Italian North End for three and a half years, then published a book in which he described the way people lived their lives there. His study has served as a model for the conduct of participant observation, a central concept of this chapter. Recently, Whyte's study was criticized by a researcher who revisited the North End and concluded that he had gotten it pretty much all wrong. The Illustration also includes a summary of some of that controversy.

The Application for this concept provides an opportunity to conduct a participant-observation study of a social situation. Some specific tools of the observer are discussed, and their use in the research situation is made clear.

Illustration

William Foote Whyte, Street Corner Society: The Social Structure of an Italian Slum, *3d ed. (Chicago: University of Chicago Press, 1981)*

Patricia A. Adler, Peter Adler, and John M. Johnson (eds.), Journal of Contemporary Ethnography *21(1) (1992) (Entire issue devoted to reexamining William Whyte's* Street Corner Society*)*

Beginning in 1936, William Whyte lived in Boston's Italian North End for three and a half years; then he published a book in which he described the way people lived their lives there. On the face of it, Whyte's *Street Corner Society* is merely a somewhat dated, though interesting, participant observation study of life in an ethnic neighborhood. However, Whyte's book is also the subject of a recent argument among scholars that has been, at times, not so gentle. The book and the flap about it tell many stories that reveal the most human, most compelling side of what sociologists do and what they are like as people. Certainly, like the other illustrations in this book, this one is intended to show how concepts from the chapter are applied in social research. However, this illustration consists of equal parts science, sociology, and soap opera, and I urge you to read at least some of the original sources summarized here. They may bring the field alive for you. In the case of this story, I think it best to start near the end.

In 1992 an entire issue of the *Journal of Contemporary Ethnography* was devoted to a reexamination of the methods and conclusions of Whyte's study. The editors of the journal listed five reasons why, after 50 years, *Street Corner Society* deserved such close examination. The first three reasons have to do with the methods Whyte used to conduct his study, and the remaining two with the substance of what he found.

First, the study has served as a model for the conduct of participant observation, a central concept of this chapter. In an appendix to his book,

Whyte published a detailed summary of the way he conducted his research. It has been a valuable teaching tool ever since. Second, in the same appendix, Whyte proposed that researchers check the validity of their conclusions by showing their work to some of the key people they have studied. This procedure has become a common practice for social researchers. Third, Whyte's lively, descriptive, narrative writing style broke with the traditionally dry and abstract style that had been common before *Street Corner Society.* Fourth, Whyte's conclusions about the relationship between social structure in the neighborhood and the behavior of individual residents provided important information for further study of small groups. And last, Whyte's examination of this poor, immigrant neighborhood led him to contradict assumptions of his time that such "slums" were highly disorganized places. He found the neighborhood to have stable and clear patterns of social and political organization, though they differed from the model held up by middle-class America of the time.

But the editors of this issue of the *Journal of Contemporary Ethnography* had another compelling reason to look again at *Street Corner Society.* It seems someone had revisited Boston's North End a number of times between 1970 and 1989 and concluded that Whyte had gotten the story of its people all wrong in the first place. In an article entitled "Street Corner Society: Cornerville Revisited," W. A. Marianne Boelen accused Whyte of (1) getting important facts wrong, (2) interpreting data incorrectly and with bias, and (3) being unethical and disloyal in his research practices. Clearly there is something interesting going on here. These are not cool and detached scholars making gentle suggestions about possible flaws in one another's work. Boelen confronts directly the worth of a classic study and accuses its author of bad research and bad faith. In order to make sure the issues are understandable to you, let me begin with a summary of part of the original study. I'll give you enough of it so we can move on to Boelen and make sense of her criticisms.

William Whyte's Original Study

In 1936 William Whyte, recent college graduate, arrived at Harvard University, having been offered a fellowship to study just about whatever he wished for a period of three years. Motivated by an uneasy sense that his middle-class upbringing and life had shielded him from much of the excitement and challenge of the world, Whyte consciously set out to understand how people lived in a slum, specifically, Boston's Italian North End. There was something of the reformer in him, but his ideas about how such a study would "make things better" for anyone were not concrete. After some months of trying to figure out how to gain access to the people there, he was introduced to "Doc," a neighborhood insider then in his late twenties who was eager to serve as Whyte's guide, protector, and contact. A short time later Whyte moved out of his room at Harvard and moved in with the Martini family, North End residents who owned a restaurant. He was on his way.

Doc took "Bill" all over the neighborhood. Whyte was conducting a participant observation in which he tried to position himself carefully within the community he was studying. He wanted to be close enough to the people he was observing to learn what they did, who was involved, and how each interpreted the meaning of their actions. At the same time, however, he wanted to remain removed enough from the center of these activities that he would never influence them directly. He met the street corner gang members and, by keeping his mouth shut and observing closely over a period of several years, Whyte was able to record masses of observations, which only much later began to form a coherent picture.

The picture that emerged was of a community whose social structure was very different from the upper-middle-class one in which Whyte had been raised. According to Whyte, the street corner society on which he focused consisted of three major components: (1) groups of young men affiliated with Doc's more street-oriented gang and/or Chick Morelli's Italian Community Club, (2) the Settlement House, which was the community social work organization of the time, and (3) the local racketeers and politicians, whom Whyte called the "big shots." To provide you with an adequate sense of the research Whyte did, I would like to focus on the young men's organizations. You should read the original study to get the rest of the story.

Whyte had most of his early contact with Doc's street corner gang, called the Nortons. According to Whyte, these men spent little time at home, preferring to hang out on the streets with one another, play cards, and go for beers or bowl. Doc was the leader, and the hierarchy within the group became clear in the ways members spoke and acted toward one another. For example, prestige rankings among Nortons were reflected in their patterns of dating with members of the Aphrodite Club, a group of females in the community, with the lower-status Nortons dating the matching Aphrodites. Whyte also concluded that an individual's status within the group was revealed and reinforced by his performance at bowling, but *not* because status in the group was based on bowling skills. Higher-status group members were expected to win even if they were not such good bowlers. Whyte cited instances in which lower-status gang members who were better bowlers lost to higher-status gang members, thus reinforcing social distinctions. But why did this happen, unless better bowlers lost intentionally in order to "stay in their places" (which Whyte says was not the case)? The answer reveals why participant observation is the best (or even only) method for studying certain aspects of social behavior.

According to Whyte, to lose in this way was not a conscious decision on anyone's part but rather a natural consequence of the social hierarchy within the gang. To explain this phenomenon, Whyte recalled the night of the most important bowling match of the year. Largely because of his friendship with Doc, Whyte was a fairly high-ranking group member, but he was also a bad bowler. Listen to the researcher's voice as he explains (p. 319, 2d ed.): "As my turn came and I stepped up to bowl, I felt supremely confident that I was going to hit the pins that I was aiming at. I have never felt quite that way before—or since." Whyte's bowling scores that night closely reflected his status in the Norton group. No outsider to the community could hope to discover such a facet of the lives of the North End residents through questionnaires or interviews, or by observation from a distance. The observer would have to take part in order to "know" from the point of view of the participants what was going on among them.

Whyte also got to know the members of another group of young men, Chick's Italian Community Club. (Whyte was proposed for membership but, being non-Italian, was voted down and given a guest membership.) More educated and upwardly mobile than the Nortons, this group was organized for "the social betterment of the members and the improvement of Cornerville." The club held meetings with formal rules and organization, put on a play, and dated members of the Italian Junior League. Notice the social class difference in the activities of the two groups. A number of the Nortons, including Doc, were voted membership in the Italian Community Club, but there was a clear and persistent split within the group. Whyte described the hierarchy as consisting of three layers. At the bottom were the street corner boys, who focused on social activities in the local community. At the top were the college boys, who were interested in social advancement for themselves and Cornerville. Between these two layers were people like Doc, who served as intermediaries between those at the top of the hierarchy and those at the bottom. Eventually, despite the efforts of the intermediaries, the friction between the street corner guys and the college men weakened the club. Its membership declined, and it died as an organization. Many of the college men moved on to join the district Republican Club and became active in politics.

W. A. Marianne Boelen Revisits Cornerville

Boelen was born and raised in Holland and lived in Italy for a number of years. In the late 1960s she was a sociology student at Columbia University, where she read and discussed Whyte's book, finding something in it that rang untrue. She recalled from her years in Italy that the men there, like young men of the North End, were also in the habit of hanging out on street corners. Boelen wondered whether Whyte was wrong to conclude that street corner behavior was part of gang membership, and whether, instead, these men were merely exhibiting a cultural habit imported with immigration. In short, she questioned whether his entire book was based on a flawed interpretation of the meaning of this behavior. As with Whyte's original study, I don't have the room here to summarize all of Boelen's work, but I can give you some sense of her study, "Street Corner Society: Cornerville Revisited" (Adler, Adler, & Johnson, 1992: 11–51).

In 1970 Boelen went to the North End and began reinterviewing members of the community who had been part of Whyte's study thirty years earlier. Over the next twenty years she went back to the community "25 times, usually for 3 or 4 days, a few times 10 days, 2 weeks, or a month, and the last time for 3 months in order to have sufficient time to discuss the draft of this article with most of the characters of Street Corner Society" (Boelen, 1992: 13). What she heard convinced her that Whyte had made serious errors in his study.

For example, Boelen was told that people felt hurt that Whyte had characterized the community as a "slum" and the street corner men as "gangs." She concluded that Whyte was biased by his upper-middle-class upbringing and was determined to make the North End seem like something it was not—chaotic, criminal, and dangerous. She accused Whyte of incorrectly characterizing informal street socializing in the North End as gang behavior, for she believed it to be merely the transfer of the normal Italian style of community interaction to American streets. Boelen also concluded from her interviews that Whyte had exaggerated the importance of a "handful of isolated racketeers in the area and had overlooked the role of the family." Lastly, a number of individuals who were part of Whyte's original study claimed that they had not been informed that they were being studied, and were not given a chance to evaluate the accuracy of the manuscript before its publication. One member of the Nortons told her that "if we had known he [Whyte] was writing a book, we would have helped him in getting a real picture of the neighborhood" (Adler, Adler, & Johnson, 1992: 29). Comments like this led her to charge that Whyte had been unethical in his methods and inaccurate in his conclusions.

Whyte's Response to Boelen

William Whyte was given the opportunity to respond to Boelen's criticisms in the same edition of the *Journal of Contemporary Ethnography* in which her article was published. Whyte said that the terms *slum* and *gang* were used in his study with their meaning at the time he did his research. In the 1930s, *slum* referred to any place densely populated and with a good deal of poverty. Apparently, sixty years ago the term did not carry quite the sense of violence and hopelessness that it does today. As with *slum,* the term *gang* has changed in meaning. At the time, Whyte claimed, it meant any group of people who hang out together. It, too, lacked the menacing connotation it has today.

What about the exaggerated emphasis on the criminal element in the North End? Whyte claimed that Boelen was the one who knew nothing about the people involved in rackets (at one point she showed living members of the old Norton group the wrong lists of names for identification purposes). He admitted that he did not penetrate the racket organization as deeply as he had hoped he would but that there was plenty of evidence in his original study of the connection between the rackets and street corner social structure in the community.

Lastly, Whyte defended himself against the charge of unethical methods by stating that he had checked his work with the subjects of his study to a much greater extent than was common at the time. In addition, he argued, he had made it plain to Doc and others that he wanted to write about Cornerville and noted that a participant observer who reveals too blatantly or too often that he is studying his subjects runs the risk of leading them to perform for the researcher in ways they normally would not behave.

Evaluating the Controversy

What are we to make of this debate? On the surface it is merely a difference between academics about the best way to study a community of people. But I think if you read the original works you will sense the hostility underlying Boelen's critical article and Whyte's response to it in the same issue of the *Journal of Contemporary Ethnography.* There is a great deal at stake here. Even the classics in our field are always open to criticism. The data on which we base our understandings of how people relate to one another, what various subgroups are like, and how our communities should be organized are not fixed. This is good, because it helps protect against incorrect ideas being cemented too firmly in our thinking.

It also subjects our research methods to constant reevaluation. Consider for a moment whose version of the truth about Cornerville you tend to believe from my summaries. Is it possible that Whyte was biased by his upbringing and ill informed about informal public behavior in Italy? Was Boelen capable of understanding thirty years after the fact and in twenty-five visits what Whyte could discover only after three and a half years of living with the people in the North End? Given his upper-middle-class upbringing, could Whyte really have been accepted into the community enough that its residents would act naturally in his presence? Could he have understood the meaning of their behaviors from their points of view? For example, what about that section on the relationship between group position and bowling skill? Whyte described it as one of his most exciting research experiences in Cornerville. When I first read this section of the book I was impressed with his insights. However, I also wondered if he was justified in extending his personal responses to the members of the Norton gang.

On Doing Participant Observation Versus Doing Nothing

I have chosen to summarize Whyte's study and Boelen's criticism of it to point out a few of the payoffs and the pitfalls of doing participant observation. I believe there will never be clear resolutions to the issues raised here. But I also believe that the debate is healthy and that there are aspects of human social behavior that cannot be studied nearly as well by other methods as they can by participant observation. In this case I think Whyte's study stands up much better than Boelen's criticism of it. But my saying so should not suggest that criticisms are always, or even usually, weak. It is clearly a weakness of participant observation that studies cannot be replicated, so conclusions are more subject to the biases of researchers than is the case with, for example, experimental studies. However, participant observation is uniquely suited to the task of discovering what meanings people attach to their social behaviors. We cannot count on their telling us outright, if only because such meanings often are privately held. So participant observation will continue to be our best method for understanding the "other," the person into whose point of view the most insightful of sociologists may enter, though it will be impossible for others to follow to see if she or he got it right.

A Personal Note on This Illustration

I have a personal reason for revisiting *Street Corner Society* through this edition of the *Journal of Contemporary Ethnography.* It has to do with the place of Whyte's book in my own education. In order to protect the privacy of the subjects of his study, Whyte disguised the location of his research by referring to the place in his book as "Cornerville" in "Eastern City." But when I read the book in graduate school in Boston in 1970 it was pretty easy to figure out where "Cornerville" was. Whyte conducted the study while on a fellowship at Harvard, and, at least at the beginning of his research, he commuted daily from his Harvard room to the research site. The book had been wonderful to read, filled with evocative descriptions of the people and the street life they lived, so I took the MTA to the Italian North End and tried to see for myself what Whyte had seen.

The problem was that I couldn't find it. The North End was there, all right, with its crowded streets, popular restaurants, and people hanging out on the streets late into the night. What was not there was the community Whyte had described. Had the corner gangs gone out of style in the thirty years since he had been there? Was he wrong about them in the first place, either exaggerating them or even making them up in the interest of a good tale? Or, would the community and its relationships remain invisible to me that night merely because I was a rank amateur observer and a total outsider who had spent less than an hour looking for the soul of a people? Ah! I began to understand.

Application

Symbolic interactionism is a school of sociological thought holding that social reality is constructed by humans every time they interact. People basically negotiate a reality that they can agree on by exchanging sets of ideas or symbols that give meaning to the social situation. For example, when two people meet, they must decide whether they should define the situation as business or social. Are they to see each other as colleagues, acquaintances, friends, or close friends? Once questions such as these have been settled, the social situation takes on a shared meaning that gives it stability.

Because of the nature of such interactions, symbolic interactionists claim that a social situation can be understood only from the point of view of the actors. As a result, a research method is required that allows the student of social behavior to experience the social world as the subjects being studied experience it. The method is called *participant observation,* and it is the focus of this assignment.

It is important to understand that good participant observation consists of more than just taking part in a social activity and reporting on what you see and hear. Every day we engage in social interactions as *interested* members of groups, but, because we have the narrow, subjective perspective of a single individual, our observations are necessarily biased and lack broad insight into the overall meaning of the interaction. For example, a member of a church congregation experiences religious services from his or her own point of view and does not look for the larger meaning available to a more systematic observer.

The great benefit of participant observation is that the researcher is capable of understanding the meaning of a given social situation from the point of view of a participant in it. According to the symbolic interactionists, we each take part in a range of social situations every day. If we are to interact in any useful way with others, we must come to share with them a definition of the situation in which we interact. The researcher who conducts participant observation simply uses this skill intentionally to adopt the perspectives of the people being observed.

Conducting a Participant Observation

In this application you will take part in a social situation as a participant observer. It is best if you choose a social setting that is accessible but relatively unfamiliar to you, so that you can bring an open mind and a flexible view to the interaction. (It is all too easy to settle into learned assumptions about the meaning of situations with which we are familiar.) Here is a list of some social situations that students of mine have observed with good results over the years:

1. An Alcoholics Anonymous meeting
2. A Weight Watchers meeting
3. A small neighborhood beauty parlor or barber shop
4. An aerobics class
5. A community meal at a senior citizens' center
6. Bingo night in the basement of a church
7. An afternoon in the television room of a college dorm during the Super Bowl or soap opera time
8. An evening in a bar, such as one that has a popular weekly darts tournament
9. A busy airport, bus, or train terminal waiting area during a storm
10. The emergency room of a hospital
11. A religious meeting of some kind, including a retreat or a Pentecostal meeting

These are just some possibilities. Avoid situations that are spread over a wide area and therefore too large to observe effectively, situations in which little happens, and situations that are so compartmentalized that they lack a central focus (such as restaurants divided into booths and tables).

You should choose a situation that you can observe without getting into trouble (don't, for example, crash the executive board meeting of the local Hell's Angels). You will have to take notes in such a way that others will not notice you are observing. Sometimes that is easy; in a doctor's waiting room or a hospital emergency room, you can simply take notes on a pad. But in a bar or at a meeting you will have to rely on your memory more and slip out of the room now and then to make a few notes. A number of observational techniques will help focus your study and give it structure:

1. *Draw a diagram* of the setting for the interaction. Include the physical elements of the setting—doors, seating, tables, lighting, paths of travel (such as aisles), and so on. Mark on your diagram where each participant is. In a bar you would mark the positions of the bartenders, customers, people playing electronic games, people dancing, and so on.
2. *Describe the nature of the interaction.* How much talking is going on? Who is talking to whom? With what intensity and for how long?

Is there only one conversation at a time or many simultaneously? What kind of movement is taking place? Are most people sitting still? Are a few people moving? When do people move, and for what purposes?

3. After you have noted the physical setting and some of the interactions going on, shift your focus to the structure of the interactions. *Look for "central" people.* There are actors who seem to be at the center of things. They command the attention of others and may have some control over what takes place and at what rate. At a church service, for example, a priest or minister is obviously a central figure. But there are others as well, people who sit in the front, center seats and those who seem to know what is going on before others do. Note who these central figures are and why they seem to have such positions. What do they do that makes them central? (Most social interactions have central figures, but they are not always so obvious.)
4. After a while, *shift your focus to "marginals," "transitionals," "deviants," and "loners."* In every social situation of any size or complexity, there will be people at the fringes of the interaction. Often they are physically at the edges, but sometimes they merely seem not to know what is going on or how to act. By looking for people who do not fit in well, you can clarify the essential meaning of the interaction. That is, there is something central to the interaction that the outsiders do not share. That is what makes them marginal. Are they new to the group? In transition in some way? Outcasts actually being shunned by others? Just as observing central figures can give you a clue to the meaning of the interaction, so can observing people on the edges.

Depending on the kind of interaction you have chosen, it probably will take you a few hours of observation to get some insight into the essential meaning of the social interaction. Remember that the less familiar the setting is to you, the better the techniques of participant observation will work. In some cases you will be surprised by your discoveries. For example, in one report of observations made at a Weight Watchers meeting, a student found that losing weight was not the focus of the interaction. The meeting was really designed to provide support for members so that they would not feel guilty or alone about being overweight. The central characters were not the leaders or those who were slimmest but, rather, those who still felt "good about themselves" in spite of not losing weight.

Report your observations on the sample observation sheet, which has sections on the setting, behavior, central characters, and marginal characters. In the last section of the report, write a paragraph in which you speculate about the essential meaning of the social interaction from the point of view of the participants. Please realize that most published participant observation studies take much more time than this assignment requires. They are aimed at much more detailed insights than you can hope to develop quickly. Just keep in mind the need to see the situation from the point of view of the participants. Abandon your own point of view and ask, "What does this situation mean to each of them?" The answer will almost certainly differ for each person. You may develop the ability to see social interaction as having as many "social realities" as there are participants.

Sample Observation Sheet

A. Physical setting. On a separate sheet of paper, draw a diagram, marking the positions of the participants.

B. Interaction. What behaviors took place? Between which participants? With what frequency and intensity?

1.
2.
3.
4.
5.
6.
7.
8.
9.
10.

C. Central participants. Describe them, and tell what made them central.

1.
2.
3.
4.
5.

D. Marginal people. Describe them, and tell what made them marginal.

1.
2.
3.
4.
5.

E. Summary statement. Describe the essential meaning of the social interaction from the point of view of the actors.

CONCEPT 5

Function and Dysfunction

Definition

Function A consequence that aids the stability of a system.

Dysfunction A consequence that detracts from the stability of a system.

Almost five hundred years ago, around the time Christopher Columbus got lost trying to find India, the artist and scientist Leonardo da Vinci was dissecting the bodies of humans to learn how muscles and bones looked and moved beneath the skin. Leonardo's curiosity led him beyond questions about what things looked like; he also wanted to know what they did. What were the functions of the organs? What part did each play in the operation of the human organism?

Questions like these assume that, if a structure is found in the body, it must have some consequence for the operation of the overall system. Leonardo made this assumption when he guessed (incorrectly) that the appendix must function to collect gases. Nothing is in the body "just because," so a function must be identified. This type of reasoning forms the basis of the functionalist perspective. In its various forms, **functionalism** has been used by all the sciences, including sociology. In this concept I discuss two different forms of functionalism as they have developed within sociology. To keep them separate, I'll refer to them by the names of their principal authors, Talcott Parsons and Robert Merton. Before they can be distinguished, some of the basic ideas of functionalism need to be set out.

Function and System

Functionalism begins with the idea that any stable **system** (such as the human body) consists of a number of different, but interrelated, parts that operate together to create an overall order. Systems are all around us. A table has a top and legs that work together to make a stable system. The human body has a heart, lungs, skeleton, blood vessels, kidneys, and so on that interact to create the overall **system stability** we refer to as good health. A car has an engine, transmission, steering, and so on that (usually) interact to operate smoothly.

For each of these systems, any single part can be understood in terms of the way it contributes to the operation of the system to which it belongs. Any contribution to the stable operation of a system is called a **function,** or **positive function.** So the leg of a table is functional for the table, because it contributes to the stability of the system. Similarly, the heart is functional for the human organism, the engine is functional for the car, and the sparkplugs are functional for the engine. Notice that one system can also be part of a larger system. The sparkplug is an element of the engine, a system; but the engine is also an element of the car, another system. Systems, then, can be analyzed at many levels.

Elements of a system vary in their **functional importance.** For example, the heart and the outer ears are both positively functional for the human organism. The heart pumps blood and the outer ears help collect sound. But the heart is functionally *vital* to the stability of the system (without it the person dies), whereas the outer ears contribute much less critically to the operation of the system. It is possible to lose the outer ears with only *some* reduction in system efficiency.

Functionalism in Social Systems

Talcott Parsons's Functionalism

So far I have used examples only of physical systems, because everyone is so familiar with them. But just as tables, cars, and human organisms can be seen as systems, so can human social behavior. The functional model was first applied extensively to human social order by Parsons in the late 1930s. He developed an elaborate description of society as a vast, complex system with interrelated parts at many levels. The levels he identified were (1) the organism; (2) the personality; (3) the social structure, ranging from small social structures such as friendships to large ones such as the economy or the political structure; and (4) the culture, consisting primarily of society's most deeply held values. Parsons (1937, 1954, 1966, 1971) argued that the overall system and the subsystems of which it is composed work together to form a balanced, stable whole and that the system naturally tends toward stability rather than toward disorder. Because he believed that systems of social order are self-balancing, Parsonian functionalism has been called *conservative* (social order is "conserved"). His functionalism was concerned with the search for the positive functions of a wide variety of human ideas and behaviors in much the same way that biologists have used functionalism to discover how a wide variety of structures are positively functional for the systems to which they belong.

For example, the fact that people work can be seen as positively functional (1) at the level of the organism, because work puts food on the table; (2) at the level of the personality, because it contributes to a person's self-esteem; (3) at the level of the social structure, because the economy depends on work being done; and (4) at the level of the culture, because it reinforces the cultural value for achievement.

Behaviors such as work, raising families, and getting an education obviously seem to be functional at the levels of the organism, personality, social structure and culture. But this sort of search for the functions served by social actions has also been extended to things like inequality in society and disengagement in later life. For example, in 1961 it was proposed by Elaine Cumming and William Henry (1961) that retirement and withdrawal from social involvements were mutually functional for the disengaging individual and for society. They argued that the individual benefited because disengagement from social roles in later life allowed for time and energy to face ill health and the tasks of putting one's life in order. It also prevented the feelings of guilt a person would experience if declining abilities made her or him inefficient while still "on the job." For society, disengagement was thought to be functional because it allowed for the replacement of critical older workers with younger ones before the older workers became inefficient. This "disengagement theory" is one example of the attempt made by the sort of functionalism developed by Parsons to account for many aspects of our social system. The assumption was that any aspect of the social order that was long-standing must, in some way, be functional for society and its members. (For another example of the application of Parsonian functionalism, see the discussion, on page 304, in Concept 17 of Kingsley Davis and Robert Moore's argument that inequality in America is functional.)

Parsonian functionalism (sometimes called *structural functionalism*) is like the functionalism used by naturalists, in that both assume that the behaviors and structures they identify must in some way be positively functional for a larger system. But there is an important difference. For naturalists, the assumption that structures or behaviors are positively functional is supported by the theory of natural selection. Elements of biological systems (such as the coloration of animals) are generally believed to have developed as adaptations for survival within specific environments. So the force generally thought to underly the stability of biological systems is natural selection. What force might underly the stability of human social systems? Parsons claimed it was the strong agreement among the members of a group about how to think and act, or **consensus.** According to him, society's members are taught (socialized) to want to conform to society's standards for accepted behavior. Because societies encourage beliefs and behaviors that help maintain the stability of the social system, conformity with those standards inevitably is positively functional. By contrast, any belief or behavior that is seen as disruptive to the social order (such as crime) must, therefore, come from outside the normal, accepted rules of the group. Parsonian functionalism treats the apparently disruptive behaviors of some people as "strains" that come from outside the system, strains that must be "managed."

Given this point of view, it is no surprise that Parsonian functionalism has trouble explaining why disruptive human behaviors such as crime or conditions such as poverty are so persistent in society. Or if they were acknowledged to be part of the basic structure of society, how could they be seen as positively functional? These problems in Parsonian functionalism are addressed by Merton ([1949] 1968) in his reformulation of the theory.

Robert Merton's Functionalism

Merton recognized that not all human behaviors or ideas are positively functional, even if they are generally seen as such. For example, some kinds of work (handling radioactive materials, let us say) are apparently harmful for workers at the level of the organism (leading to sickness or death) and at the level of the personality (causing anxiety). Merton termed such system-disrupting consequences **dysfunctions,** (or **negative functions**). Notice that a single action, object, or idea can be functional for some systems but dysfunctional for others. For example, the work that diminishes the health and mental well-being of the worker handling radioactive materials still might contribute to the physical well-being of people who use the product of those labors, might benefit the economic well-being of the worker and of the employer, and might simultaneously reinforce the cultural value for productivity. Mertonian functionalism challenges us to identify all the functions *and* dysfunctions of a particular action, idea, or object and to identify the various systems to which each consequence applies. (The Illustration for this concept gives an example of how this is done, and the Application gives you an opportunity to try it yourself.)

Merton made another major addition to functionalist analysis when he distinguished between **manifest functions** and **latent functions,** (or **manifest** and **latent dysfunctions**). As we try to identify all the functions and dysfunctions of a behavior (work, for example), some consequences are obvious. Work is both normally recognized and normally intended to be positively functional for various systems. Insulting or yelling at someone is normally recognized and intended to be dysfunctional for the target person. Merton called such intended or normally recognized functions or dysfunctions *manifest*.

Some consequences are not so obvious. *Latent* functions or dysfunctions are those whose consequences are generally unintended or unrecognized. For example, whereas colleges have manifest positive functions such as educating and training students, they also have less obvious, unintended functions such as training students for careers in professional sports, providing opportunities for students to meet potential mates, and supporting textbook publishers and authors. These generally unrecognized or unintended positive functions of colleges are, in Merton's terms, *latent positive functions*. Colleges may also have latent dysfunctions, such as causing some students to suffer great anxiety and even mental illness, overtraining some students, supplying the economy with too many trained in one skill and too few in others, and putting a heavy burden of taxes on states or tuition costs on families. Since these dysfunctions are normally unintended or unrecognized, they are also called *latent*.

Table 5-1 illustrates the various categories of Mertonian functional analysis. These categories encourage us to think of the many kinds of consequences an action, idea, or object might have for systems at many levels. To show how this works, Tables 5-2 and 5-3 present brief analyses of two very different social phenomena. One, a dinner party given to benefit charity, generally is thought to be quite praiseworthy; the other, war, generally is thought to be evil.

The Balance of Functions and Dysfunctions The analyses of a charity dinner and of war in Tables 5-2 and 5-3 illustrate an important characteristic of Mertonian functionalism. Any one idea, action, or object can have both functions and dysfunctions. By adding up the functions and weighing them against the dysfunctions, we can calculate what Merton called a **net aggregate of consequences.** If an analysis reveals many more

TABLE 5-1

Functional consequences, manifest and latent

	Manifest	Latent
Functional	Contributes to the stability of some system and is intended or recognized as doing so.	Contributes to the stability of some system but is not intended or recognized as doing so.
Dysfunctional	Diminishes the stability of a system and is intended or recognized as doing so.	Diminishes the stability of some system but is not intended or recognized as doing so.

TABLE 5-2

Manifest and latent functional and dysfunctional consequences of a charity dinner

	Manifest	Latent
Functional	1. Provides aid for the sick or poor in the form of money, medicine, research funds, and so on. This consequence operates primarily at the level of the organism (helps the poor and sick remain alive).	1. Provides work for caterers, jewelers, tailors, limousine services, and so on. (Since this is not an officially recognized reason for or consequence of a charity event, it is a *latent* functional consequence. The same is true of all the other consequences in this box.) This consequence operates at the level of (a) the organism (feeds the families of caterers and so on), (b) the personality (increases the sense of security of the caterers and others), and (c) the social structure (supports the catering and other industries in general). 2. Provides a setting in which the rich can display their wealth (jewelry, for example). This consequence operates primarily at the level of the personality (enhances the egos of the rich). 3. Reinforces the norms of hierarchy (showing who is of higher status than whom) both among the wealthy participants and between the rich and the poor (who are being helped). This comparison also highlights for the rest of society the shame of poverty and dependence, thus reinforcing societal norms of hard work and saving. These consequences operate at the levels of the personality, the social structure, and the culture. 4. Provides a tax benefit for wealthy participants, since charitable donations are tax deductible. This consequence operates at the level of the personality and the social structure. 5. Provides tax benefits and free or inexpensive advertising through news coverage for corporations listed as sponsors of or donors of goods to the event. These consequences operate at the level of the social structure (advertising the products and deducting the value of the goods from taxes).

TABLE 5-2
(continued)

	Manifest	Latent
Dysfunctional	[There would be manifest dysfunctions only if the charity event were held expressly to diminish the well-being of some system or were officially recognized as doing so. Because this is never the case, to my knowledge, this box should be empty.]	1. Impairs the prestige of wealthy people not invited to participate or relegated to the bottom of the status hierarchy if they do make it to the event. This consequence operates primarily at the level of the personality. 2. Reduces the tax bills of the wealthy through tax deductions for charity, thus decreasing the tax income available for expenditures on other programs and increasing the tax bills of others. This consequence operates at the level of the social structure.

TABLE 5-3
Manifest and latent functional and dysfunctional consequences of war

	Manifest	Latent
Functional	1. Defends territory. 2. Takes land. 3. Achieves independence. 4. Protects allies. 5. Roots out evils (such as Hitler). [These are manifest functions only when they are officially stated purposes of going to war.]	1. Strengthens the economy, for example, by increasing employment. 2. Stimulates inventions and technical innovations. 3. Controls population growth. 4. Increases internal national cohesion. 5. Expands and strengthens the military structure (a function from the military viewpoint). 6. Expands territory. [This is a latent function only when not an officially stated goal of going to war.] 7. Increases profits for some industries.
Dysfunctional	1. Kills people. 2. Causes suffering from injuries and from sacrifices made by certain members of the society—especially the poor. 3. Imposes financial costs on much of the society—especially the poor and lower middle classes.	1. Forces people to kill in violation of their religious and moral beliefs. 2. May deteriorate relations with some nations, such as allies. 3. Alienates certain subgroups in society (such as Japanese-Americans during World War II). 4. Decreases profits for some industries.

or stronger functions than dysfunctions, it is likely that the action, idea, or object being studied will persist. Using the charity dinner as an example, it appears that many more people will benefit from such an event than lose. Therefore, all the people for whom charity dinners are functional can be expected to work to have them take place, and, if these people are powerful, they can expect to prevail easily over the objections of those for whom charity events are dysfunctional.

When the phenomenon being studied is generally considered desirable (such as a charity dinner), this outcome is not worrisome. Unfortunately, the same kind of analysis also applies to morally offensive social phenomena such as war. If Mertonian analysis reveals that the net aggregate of consequences of war is positive (that is, if war is shown to have many more system-maintaining functions for powerful interests than system-detracting dysfunctions), it too can be expected to persist. This is true even though war continues to have horrible dysfunctions for many, less powerful people.

Mertonian Functionalism as a Conflict Theory By introducing the concepts of dysfunction, latency, and the net aggregate of consequences, Mertonian functionalism is not tied to the exclusive search for system-maintaining consequences, as Parsonian functionalism is. Merton's approach, then, is not as conservative as Parsons's in its view of social order. It can alert analysts to the conflicts that are built into the social system. Conflict no longer has to be seen as coming from outside social systems or from "errors" in the order of society. In a sense, then, Mertonian functionalism is a variety of conflict theory, an approach to social behavior that sees a balance of contending forces as a normal facet of the social order.

Functional Versus Moral Evaluations It is important to keep in mind the special, sociological meaning of functions and dysfunctions. They refer only to the consequences that some action, idea, or object has for the stability of a variety of systems. To conclude that something is functional for a system is not to conclude that it is therefore good in a moral sense. To the extent that the heart contributes to the operation of the body, it is positively functional, even if the body in question is that of a mass murderer. The fact that the person is evil in a moral sense does not make his or her heart any less functional. And to conclude that war has some positive functions is not to conclude that war is good. Our analysis of functions and dysfunctions must be kept separate from our moral judgments about whether the systems involved are good or evil. We can then continue to have moral standards and can use the evidence that functional analysis supplies about how things can be specifically changed to reflect our best moral judgments. It is one thing to oppose war or bigotry and quite another to be aware of how they benefit certain segments of society. It may not be possible to change the order of things by force of moral disapproval alone. A knowledge of how the systems operate adds a critical element to our ability to improve the world.

Summary

Functionalism is one of the major perspectives in sociology. It is based on the assumption that society operates as a system with many interrelated parts. Each of the parts can be best understood in terms of its consequences for larger systems, at the level of the organism, the personality, the social structure, or the culture. A function (or positive function) is a consequence that aids the stability of some system, while a dysfunction (or negative function) is a consequence that detracts from the stability of some system. One important version of functionalism, associated with the work of Talcott Parsons, focuses on the identification of positive functions in society, assuming that society naturally tends toward equilibrium or balance. Parsonian functionalism is often considered conservative because of its emphasis on the existence of system-maintaining positive functions. It views ideas, actions, or objects that disrupt systems as deviations from normal social order.

By contrast, Robert Merton's ideas about social order allow for a broader, less conservative use of functionalism. Merton recognized that ideas, actions, or objects can have dysfunctions as well as functions and that each of these can be generated from within operating social systems. Further, he noted that functions (whether positive or negative) may be manifest or latent. Manifest functions are those whose consequences are intended or recognized, and latent functions are those whose consequences are unrecognized or unintended. Mertonian functionalism allows the calculation of what he termed a *net aggregate of consequences,* a total evaluation of the functions and dysfunctions of a single action, idea, or object for a variety of systems. The net aggregate of consequences tells us whether the action, idea, or object is likely to persist. It does not reflect whether it is morally good or bad.

The Illustration for this concept is an example of the use of Mertonian functionalism. Herbert Gans's analysis of the functions of poverty reveals some surprising ways in which poverty has latent positive functions for the operation of society. If a net aggregate of consequences of poverty were calculated, Gans's work suggests, poverty would persist as long as it contributed to the maintenance of certain powerful segments of society, even though it was severely dysfunctional for the poor.

The Application for this concept reviews some of the techniques of Mertonian functional analysis and has you conduct such an analysis of your own. A number of possible subjects are suggested, although you might choose one not on the list.

Illustration

Herbert J. Gans, "The Positive Functions of Poverty," American Journal of Sociology *78 (September 1972): 275–89*

In 1972 Gans, a sociologist at Columbia University, published an article entitled "The Positive Functions of Poverty." What a title! To anyone unfamiliar with Merton's model of functional analysis, it must have seemed that Gans was praising poverty. But Gans probably chose the title for just that effect. He wanted to list the latent positive functions that contribute to the persistence of poverty. Our society has spent a great deal of time identifying and being distressed about the manifest dysfunctions of poverty, war, discrimination, and prejudice. Yet knowing about these does not help us understand their persistence. If poverty were only dysfunctional, wouldn't we eliminate it from American society? As Gans realized, to know why a system persists, we must figure out how it contributes to the well-being of various powerful segments of society.

If you look back at our analyses of war (generally considered a bad thing by society) and of a charity dinner (generally considered good), you can see that the latent functions are always the most interesting. This is because we sociologists are at our best when we are fulfilling our role as "debunkers." We spend a great deal of time trying to poke holes in the official and widely accepted explanations for the ways things work. That is what Gans's work does. As you read about each of the positive functions of poverty that he identified, keep in mind the question of power. Is the group for whom poverty is functional in a position to influence its persistence in any way? On the other hand, could a group for whom it is dysfunctional do anything to eradicate poverty?

Economic Functions

Gans noted that in every society there is "dirty work" that must be done. Jobs that Americans think of as unpleasant, dangerous, or low in prestige are obviously not in demand. But society could not keep operating if garbage were not picked up, public bathrooms not cleaned, or crops not harvested. To get people to do these jobs, we could pay them extra high salaries. Or we could fill these jobs with people who must take them because they qualify for no other jobs. The poor provide a labor market for the least desirable jobs in society and at the lowest cost. If they didn't do this kind of work, they would starve.

Because the poor work for low wages in a variety of industries, they raise the profit margins of those industries. This consequence allows for the investment of those profits in other areas of the economy or in a better quality of life for the rest of the society.

The poor also act as "guinea pigs" for medical research and training. They are trapped in the inner cities and must go to hospitals in which universities train medical students and conduct research. The rest of society benefits from the training the doctors bring with them when they go to practice in more affluent areas outside the cities (which they disproportionately do) and from the findings of the research conducted.

A number of industries have developed to deal with the activities, needs, and desires of the poor. Welfare workers and social service professionals would have no jobs without poverty. Police work focuses primarily on crimes by and against the poor. Social researchers, in both universities and private companies, depend on government and foundation grants to study poverty and the poor. (During my graduate school training, one such researcher put an ironic poster on the door of his office stating, "There's money in poverty.") Gans also listed a number of less respected activities that depend on the patronage of the poor—gambling, drug traffic, cheap liquor sales, storefront religions, quasi-medical practices (such as faith healing), loan-sharking, and so on.

The value of goods is extended by forcing the poor to buy used and day-old products (such as bread), because they cannot afford new or fresh goods. In addition, because the poor cannot afford to travel around, they must shop in their own neighborhoods, and the same goods may actually cost more in poorer neighborhoods than in wealthier suburbs, creating more profit for store owners.

Social Functions

The dominant norms of our society stress hard work, saving and planning for the future, and certain ways of talking and acting toward others that together characterize the *middle class*. The poor, by not fulfilling these norms, become deviants in the eyes of the rest of society. By comparison they provide a way of emphasizing the value of the dominant norms of the society. They give middle-class Americans a "negative comparison group" and so legitimize the dominant norms.

Not all of the poor are seen as lazy or incapable of planning for the future. Some are considered to be "deserving poor" who are disabled, old, or otherwise suffering poverty because of bad luck. These people provide the rest of the society with objects of pity, compassion, and charity. They allow us to feel good about how sorry we feel for them and how much we would do for them if we could.

A somewhat different emotional function is served by the poor who are believed to indulge in uninhibited, unlimited sensual activities such as sex, drug use, and drinking. The belief that the poor are on some sort of "permanent sex-filled vacation" allows some members of society to enjoy vicariously the "depravities" of the poor. Notice that this function is served even if there is no truth at all to the belief about the poor.

Just as the poor provide a negative comparison group for the norms of the middle class, so they also reinforce the status of the rest of the society by being of such low status themselves. A status hierarchy is purely relative, and without people on the bottom rungs the upper rungs would not exist.

The poor provide the wealthy with an object for their charity, thus helping to justify the existence of wealth itself. After all, if the rich didn't exist, who could hold charity events for the poor?

Cultural Functions

The poor have always supplied the labor for the construction of those monuments that are the marks of "advanced" civilizations. There are no pyramids without slaves or welfare states without the poor.

The poor help shape the culture itself by providing some of its richest and most original material, such as jazz, street language, art, and other examples of folk culture.

Political Functions

The poor serve as an important political constituency for parties and candidates. They also serve as the targets for politicians who campaign against welfare chiselers and other social misfits.

The poor bear much of the cost of social change. They are employed in the least stable jobs and so are the first fired during economic down cycles. They are available for immediate employment when economic changes require a greater labor supply.

By listing these functions of poverty, Gans showed that poverty survives "in part because it is useful to a number of groups in society." He has *not* claimed that poverty should exist, only that it will continue to exist as long as its functions outweigh its dysfunctions for those who are in a position to do something about it. We cannot leave the eradication of poverty to the poor or to the good will of the rest of society. We must make the net aggregate of consequences of poverty negative.

Application

In the Definition section of this concept I discussed Merton's version of functionalism. I believe this approach to studying social systems can be one of the most useful analytic tools in sociology. Once its techniques are understood, it can be used to understand a very broad range of social behavior.

Merton distinguished between a function (an action, idea, or object that *aids* the stability of some system) and a dysfunction (an action, idea, or object that *diminishes* the stability of some system). In addition, he recognized that functions and dysfunctions may be manifest or latent. Manifest functions and dysfunctions have consequences (whether aiding or detracting from system stability) that are intended or recognized. By contrast, latent functions and dysfunctions have consequences that are *not* recognized or intended. Table 5-1 earlier illustrated these categories of Mertonian functional analysis. (You might like to look back at it.) In addition, ideas, actions, or objects can have consequences for human systems at four levels: the organism, the personality, the social structure, and the culture.

Here are some examples of the consequences of having a party that fit into each of the four cells in Table 5-1:

1. **Manifest Function.** Having a party is manifestly functional for the people who go to it, because they have fun and relax and because they create new friendships and strengthen or renew old ones. Having fun and relaxing are positively functional at the level of the personality. Creating and strengthening relationships are positively functional at the level of the social structure. Both are examples of manifest functions, because these are the stated and intended consequences of having a party.
2. **Latent Function.** Having a party is latently functional for caterers and people who sell liquor. It aids them at the level of the personality, because they feel better when they sell more, and at the level of the social structure, because it helps their businesses to do better. These are latent functions because they are neither intended nor generally recognized as reasons for having a party.
3. **Manifest Dysfunction.** Having a party is manifestly dysfunctional for neighbors, because the noise keeps them up late at night (dysfunctional at the level of the organism) and makes them upset (dysfunctional at the level of the personality). It is dysfunctional for the owner of the building in which it is held, because she or he receives noise complaints (dysfunctional at the level of the personality) and because wine is spilled on the carpeting and stains it, which costs money to clean or replace (dysfunctional at the level of the social structure). Notice that for such dysfunctions to be manifest, they must have been among the stated purposes of having the party or recognized as likely consequences of the party. If a party is held with the express purpose of *not* inviting someone, that is another example of a manifest dysfunction at the level of the personality for the individual who was snubbed.
4. **Latent Dysfunction.** At a party, some people drink or eat so much that they get sick (dysfunctional for them at the level of the organism). Others make fools of themselves (dysfunctional at the level of the

personality, because they feel embarrassment or shame) or get into arguments and make enemies (dysfunctional at the level of the social structure, because relationships are weakened or lost). These are latent dysfunctions, because people normally do not hold parties or attend them with these purposes in mind or with the expectation that such things will happen.

Conducting a Functional Analysis

For this application you will conduct a functional analysis using Merton's scheme. To show you how the result will look, Table 5-4 presents the analysis of having a party, with a few consequences in addition to those I have mentioned.

TABLE 5-4

Mertonian functional analysis of having or going to a party

	Manifest		Latent	
	Consequence	Level at which it operates	Consequence	Level at which it operates
Functional	1. Relaxing, having fun, "getting away" from cares	Personality	1. Buying food and liquor from merchants	Personality Social structure
Functional	2. Making friends, renewing old friendships, meeting people who can help you	Social structure	2. Getting new ideas about decorating or giving parties	Personality Social structure
Functional	3. Eating in moderation	Organism	3. Providing income for baby-sitters	Social structure
	Consequence	Level at which it operates	Consequence	Level at which it operates
Dysfunctional	1. Making noise that disturbs the neighbors, who complain to the landlord	Organism Personality Social structure	1. Eating and drinking so much you get sick	Organism Personality
Dysfunctional	2. Damaging the rug with wine stains	Personality Social structure	2. Making a fool of yourself	Personality
Dysfunctional	3. Intentionally snubbing people not invited	Personality Social structure	3. Arguing and making enemies	Personality Social structure
Dysfunctional	[Note: these are manifest dysfunctions only if they are intended or recognized consequences.]			

Now conduct your own Mertonian functional analysis of some fact of American social life. Here are a few possible topics for analysis:

1. Religious services
2. Parades on holidays such as Armistice Day and St. Patrick's Day
3. The Super Bowl
4. Orientation week at a college or university
5. Final exams
6. Basic training in the army
7. Presidential inauguration speeches
8. The stockholders' meeting of a corporation
9. Graduation ceremonies at a college
10. Social welfare agencies
11. Traditional sex roles in American family structure
12. Groups concerned with saving the environment
13. College or university governments
14. Newspapers in the United States
15. Profit-making businesses in the United States

Obviously, there are hundreds of topics you could analyze. You will find not only that it gets easier to use Merton's scheme after a while but also that other topics will occur to you. A blank copy of the analysis format is provided for you to use. You might use several copies, so you will have room to identify a wide variety of functions and dysfunctions. At the beginning, don't rule out anything just because it sounds a bit strange to you. Leave it for a while, and come back to it. You may decide to move it to another category, or it may suggest further functions to you.

Mertonian Functional Analysis Sheet

Mertonian functional analysis of: ______________________________

	Manifest		Manifest	
	Consequence	Level at which it operates	Consequence	Level at which it operates
Functional	1.		1.	
	2.		2.	
	3.		3.	
	Consequence	Level at which it operates	Consequence	Level at which it operates
Dysfunctional	1.		1.	
	2.		2.	
	3.		3.	

CONCEPT 6

Conflict: Individual, Group, Class, and Societal

Definition

Conflict Any condition of disagreement between individuals, groups, or categories of people that can be expressed in attitudes or behaviors.

Conflict Theory The sociological view of conflict as a source of both social order and social change.

A friend of mine is an expert on bird calls. He records them and can distinguish one kind of bird from another. I'm amazed by this, since they all sound like chirping to me. He tells me that in an open field he can tell how each bird is marking out its territory with its personal message. This pattern of territorial behavior is repeated by other animals. Many leave scent markers to establish boundaries. Even humans maintain territories. My neighbor, for example, clips only halfway across the top of the hedge that runs between our houses. (My side of the hedge is usually more ragged than his.)

Sometimes boundary markings are directly disputed, and conflict breaks out. Birds squawk at one another; neighbors exchange hostile glances, or argue about who allowed those nasty leaves to blow onto whose nice, neat driveway. If we look beyond the unpleasantness and cruelty of conflicts, it is possible to see a pattern emerging from them. The fact is that conflicts can create order by establishing a balance of contending forces. The balance may not always be what we wish or may not even seem fair, but it is a balance nonetheless.

In its simplest sense, social order is nothing more than the predictability of social behavior. We have expectations about what will occur in a situation, and if it does occur, things are seen as orderly. There are two basic perspectives in sociology about the origins of orderly behavior and about the nature and likelihood of changes in social order. One view focuses on consensus and the other on conflict. The consensus view (discussed in Concept 5) argues that social order is a consequence of agreements among people about how to act, and these agreements are passed from one generation to another through socialization. The consensus view also contends that changes, or strains, in a particular order are caused by individuals or groups of people who fail to share the beliefs on which agreements are built. By contrast, the conflict view argues that social order, and changes or strains in it, result from disagreements between people as they contend for the valued resources of society (Dahrendorf, 1959). First I am going to discuss the nature of social conflict and its contribution to social order and social change. Then I'll move on to the various levels at which this process occurs.

Conflict, Social Order, and Social Change

In every society the members learn to value certain things. In American society we value among other things material possessions and the esteem of others. We therefore value the means to gain these assets, such as jobs, money, and power to make decisions for ourselves and others. When these valued goods are thought to be in limited supply, they can become the objects of competition and conflict.

Conflict as Behavior and Attitude

Individuals use a variety of tools to win conflicts, and the tools available depend on the person's position in the society. Those who are already in advantaged positions have access to the mass media, systems of education, the police or army, and other powerful aids in the struggle for control over resources. They can hire others to help confront the competition, or they may influence the legislative process so that control over the scarce good is officially given them. Poorer and less powerful people are limited to personal threats of violence and to public protest. Their access to societal institutions that might help, such as the court system, is very limited. The poor are rarely made aware of how to gain such access.

In addition to specific actions the competitors may take, either side is likely to develop a fierce dislike of its opponents. This negative attitude serves to justify and strengthen the actions taken against an opponent. During war, for example, basic training includes the dehumanization of the enemy (calling them "gooks" or "monkeys," for example), which enables a soldier to fight without concern for the well-being of a fellow human on the opposite side. Conflict, then, exists at both the behavioral and the attitudinal levels.

The Outcomes of Conflict: Dominance and the Balance of Forces

There are only two possible outcomes of a conflict: **dominance** or a **balance of forces.** Either one side defeats the other and gains the dominant position, or neither side defeats the other, and a balance of forces is established. In either case, a form of social stability develops. In the situation in which one side dominates, the stability of the situation comes from the fact that the loser is kept under control by the winner. Slavery is an example. It is not the style of social order that the losers would choose, but it *is* a form of order, because it makes the social relations between people predictable. Both slaves and slaveholders know the roles they are to play.

Once dominant and subordinate positions have been clearly established, it is common for the winners to become convinced that the way things have worked out is not only satisfactory but also fair and just. They can sometimes even convince the losers to adopt these beliefs, further stabilizing the order of things. One of the most common justifications for slavery given by slaveholders was the story that blacks would be much worse off without whites to run things. To maintain this story, whites denied freed slaves the opportunities to borrow money, buy land, get jobs at fair wages, or in any way compete with whites. The freed slaves failed outside the "protections" of slavery because they were caused to fail. The lesson about the appropriateness of white domination was convincing to some blacks, especially those with the less hateful plantation lives who might actually have lived more deprived lives out of slavery.

When neither side of a conflict is a clear winner, the result is a balance of forces. This is a form of social stability only to the extent that each side can keep the other from gaining a further advantage, although each is unable to advance its own interests. Think of two playing cards leaning against each other. Each "wishes" to fall to one side but is prevented from doing so by the other card. The balance of forces (in this case dependent on gravity and friction) creates a stable structure. But the resulting stability is not as reliable as that in the dominance situation, because a delicate balance must be maintained. If anything changes in that balance, the character of the structure must also change. This is why conflict theorists believe that both social order and social change result from conflict. The order that comes from conflict between contending forces is constantly threatened, because the underlying forces continue to struggle to gain access to the valued goods of the society. Conflict theory argues, then, that when changes do occur, they are the result of shifting conflicts.

Levels of Conflict

Individual Conflict

It is very common to see people arguing—a son disagreeing with his father over household chores, two drivers yelling about a contested parking space, even friends disagreeing about the point of a movie. The negotiations can range from logical points backed by subtle insults to physical threats and actual violence. Typically, such conflict is seen as disruptive and harmful, to the individuals who engage in it and to the social order within which it occurs. Psychologists, courts, and the police are concerned with the management or control of conflict. But conflict between individuals, especially when it occurs at the level of everyday interaction, can be seen as an important source of social order as well. One important aspect of the sociological perspective is the view that social order is constantly being created by individuals as they interact. Two people who meet in a social situation must share the same definition of the situation before they can interact meaningfully. For example, if a student and a teacher meet outside the classroom, one of them might define the situation as social and the other as part of the classroom experience. The confusion and embarrassment resulting from such mismatched definitions of the situation show how much meaning is lost in the interaction. Even though we are taught widely shared rules for everyday behavior, we learn them under different circumstances, live unique lives, and face contradictory rules or ambiguous situations. In short, there is a great deal of room for differences to arise in individual evaluations of social situations. Using the conflict theory from sociology, we can study the process of creating order when individuals negotiate their conflicting views of the meaning of a specific situation. (For a more complete discussion of this process, see Concept 4, on symbolic interaction.)

Group Conflict

Conflict can occur in a group setting, with consequences for the relationships between and within groups. Group conflict can be divided into two main categories: (1) within-group conflict (conflict between two or more members of the same group), and (2) between-group conflict (conflict between two or more groups).

Within-Group Conflict Lewis Coser (1956), building on the earlier work of Georg Simmel ([1908] 1955), proposed that conflict within a group has the potential to *reduce* the group's solidarity, because group members have inside information that enables them to harm one another more than outsiders can. But Coser also proposed that internal conflict can *increase* group solidarity by leading to the resolution of minor disagreements that might otherwise accumulate and eventually become too large and explosive to handle. Marriage counselors have used this last idea in their advice that couples "clear the air" through scheduled, controlled arguments.

Between-Group Conflict Coser also hypothesized that between-group conflict tends to *increase* the solidarity within each group by providing an external "enemy" against which each group must struggle. Any differences within a group must be forgotten, or overcome, in the interest of presenting a strong front to the opposition group. Thus, internal cohesion is increased by external conflict. This principle provides one explanation why wars are sometimes started by the governments of countries having severe internal problems. It may not be legitimate to suppress a rebellious or divided population in peacetime, but during war such measures are legal and acceptable if national survival can be held up as a goal.

Class Conflict

Perhaps the most widely studied form of social conflict exists at the level of social class. Karl Marx (1818–83) saw history in terms of the struggle between two classes—the capitalists (or bourgeoisie) and the workers (or proletariat). According to Marx ([1844] 1964, [1867] 1967), profit comes from the ability to pay workers less than their labor is worth when measured in terms of raising the value of the raw materials that their labor transforms. For example, if a worker takes one dollar's worth of wood, nails, and paint and makes a product worth five dollars (after other overhead has been deducted), then the work is worth four dollars. But if the worker is paid only one dollar, a three-dollar profit can be made. In terms of conflict theory, capitalists must control the marketplace of labor in order to keep wages down and profits up. Workers, because they are taught the conventional values and norms of the society (a process controlled by the dominant members of the society), are unaware of their exploited position. In Marxist terms they have a *false consciousness.* Marx predicted that this false consciousness would eventually be replaced by an awareness among workers that they have to take control of the means of production to end their exploitation. Marx called this awareness *worker consciousness.* The result, he said, is inevitable: worker revolution.

Class conflict in American society has not taken the precise form predicted by Marx. It has been suggested that the failure of American workers to conform to Marx's model of class conflict does not mean that class conflict does not exist. Beth Rubin (1986) uses data from the period 1902–76 to show that worker-capitalist conflict in America has merely become increasingly institutionalized in the form of trade unionism. The major medium for the expression of conflict is now the labor strike, or the threat of one, which drives the bargaining process. Though the institutionalization of labor-capitalist conflict has stabilized the relationship between these contending forces, it also has created a split in the labor market (Bonacich, 1972), in which lower-tier laborers who are not unionized are pitted against both management and unionized workers. And other forms of class conflict exist. Poor people protest and riot, even though they have less effective means available to express their dissatisfaction than organized workers do. But the proletariat has not taken control of the means of production away from the bourgeoisie. Although Marx may not have predicted the progress of American class conflict, his ideas have been very influential in revolutions in the former Soviet Union and the People's Republic of China and in continuing conflicts in South America, Africa, and other countries around the globe.

Societal Conflict

War is the ultimate social conflict, especially since the most powerful tools of societal conflict, nuclear weapons, now have the capacity to end all life on earth. But not all societies have nuclear weapons, and some wars are fought for limited, specific aims—to gain territory, to control natural resources, or to reduce internal problems by diverting attention to an external enemy. In some cases wars are not actually fought but only threatened, to gain advantages in trade or political struggles. Whatever the character or scale of war, it is useful to ask who benefits from it. Who gains land, or money, or time?

Let's return to the first level of conflict, that between individuals. As strange as it may sound, war between countries can be viewed in much the same way one views conflict between individuals. Recent events in the relationship between the United States and what once was the Soviet Union illustrate the operation of the principles of balance of forces, dominance, and conflict in the creation of order and disorder at the level of societies. For more than forty years the relationship between the United States and the Soviet Union was stable only to the extent that a balance of force existed between them. Whatever else was happening in the economy of the Soviet republics, the balance was ensured by the capacity of each country to annihilate the other with nuclear warheads mounted on intercontinental ballistic missiles. The calculations of the balance were made largely within the secret councils of the military organizations of each side, and slight shifts were enough to shake the world. As precarious and costly as the balance was, it was maintained.

Events since the collapse of the Soviet Union have revealed that it was only Russia's and the Soviet military's dominance over the member republics that kept them from breaking away or from warring with one another. Formerly composed of fifteen republics, the USSR has now split into fifteen separate states (including Russia, Georgia, Ukraine, Belarus, and Armenia), twelve of them currently realigned in a precarious confederation. Yugoslavia, which was not one of the fifteen but was economically and politically tied to the communist, Eastern bloc, not only separated from its alliance with the Soviet Union but fell apart as a state. It fragmented into warring ethnic groups, including the Croats (now the independent nation of Croatia), the Slovenes (also an independent nation, Slovenia), and the Serbs (the majority ethnic group that has been documented by the United Nations to be committing genocide against the Muslims in Bosnia-Herzegovina, a former province). Whatever these groups thought about one another during the Cold War, and however unpleasant life for them may have been, it is clear that their domination by the Soviet Union created a form of order from which they have now been released.

Summary

A major sociological perspective proposes that conflict is a source of both social order and social change. When individuals or groups contend for the resources that are valued within a culture, the result is either a balance of forces (neither side can gain a further advantage) or domination of one side by the other. In either case the result forms the basis for social stability. Conflicts take place at the individual, group, social class, and societal levels. When shifts occur in the balance of power between contending forces or in the domination of one side by another, social change occurs. Rather than seeing conflict as only a disruptive force, conflict theorists argue that conflict is central to the creation of social order.

The Illustration for this concept summarizes recent research by Murray Straus, a specialist in the study of family violence. It deals with the kind of conflict that we often do not think of as conflict at all: parents who physically punish their children for wrongdoing. His findings are interesting because they run counter to some commonly held American beliefs about the effects of physically disciplining children.

The Application for this concept is a content analysis of violence on television. Rather than relying on a subjective impression of how much interpersonal conflict there is on television, this application allows you to count more accurately the frequency of various levels of violence, including (1) murder and attempted murder, (2) physical violence clearly not intended to kill, and (3) nonphysical violence.

Illustration

Murray A. Straus, "Discipline and Deviance: Physical Punishment of Children and Violence and Other Crime in Adulthood," Social Problems *38 (1991): 133–54*

In the discussion of social conflict I tried to make it clear that, depending on the circumstances, conflict can contribute to, or detract from, social order. Sociologists have had a lively debate for many years over what these circumstances might be. For example, consider what happens when adults spank their four-year-old son for playing with matches in the attic. Clearly, they intend both to make their home safer, by preventing such dangerous play, and to make their child obey the rules of the home and the larger society. They may also hope that punishing the child will make him a better citizen when he grows up. That is what we mean when we quote the biblical saying "Spare the rod and spoil the child." But does their punishment accomplish these ends? In both the short and long terms, do parents increase or do they decrease social order by punishing their child? This article by Murray Straus, a specialist in the study of family violence, summarizes recent research on the subject. His findings may surprise you, because they run counter to a commonly held American belief about the effects of physical discipline on children.

Straus cites data from the 1975 and 1985 National Family Violence Surveys to show that over 90 percent of American parents physically punish their children to correct misbehavior. Figure 6-1 shows the reported rates of physical punishment of children for a 1985 national random sample of 3,229 American parents. Notice that almost all American families are reported to use physical punishment on their children who are between the ages of two and six. The near universality of this form of discipline in America is reflected in the fact that parents are permitted by law in all fifty states to punish their children physically. As of 1989 the right to punish children physically even extended to school employees in

FIGURE 6-1

Percent of American Families Using Physical Punishment to Discipline Their Children by Age of Punished Child

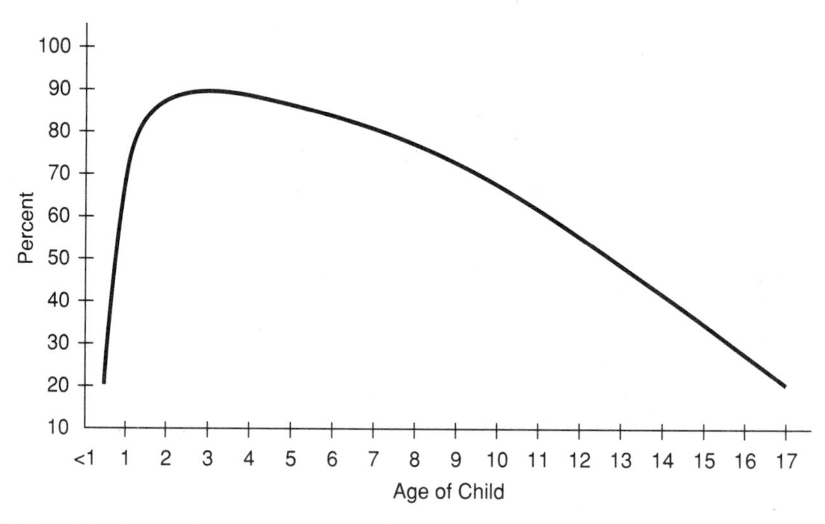

all but eleven of the states. According to Straus, the appropriateness of spanking, slapping, or shoving a child who has done wrong is justified by the belief that it "disciplines." That is, in sociological terms, punishment is applied to teach the child to obey, to be a better family member immediately, and, later, to be a better citizen.

However, Straus argues that the physical punishment of children has quite different consequences than Americans generally believe. He contends that punishing a child may produce obedience in the short term, but that in the long run it increases the probability that the individual will engage in violent crime in adolescence and adulthood. To support this belief Straus develops a theory he calls "Cultural Spillover Theory," and then he cites evidence, from a range of surveys, that is consistent with the theory.

Cultural Spillover Theory

Most Americans justify physically disciplining children on the grounds that it discourages wrongdoing, teaches them the difference between right and wrong and produces more law-abiding adults. However, violence, like other forms of behavior, is a liquid asset: once learned it can be used in many ways. According to Straus (1991: 137), "[V]iolence in one sphere of life tends to engender violence in other spheres, and . . . this carry-over process transcends the bounds between legitimate and criminal use of force." That is, parents who punish their children may use violence with admirable intentions, such as teaching them that their behavior has endangered others or that respect for authority is necessary for getting along in society. However, once parents have taught their children the lesson that violence has its uses, they no longer control how the lesson is applied. It is more than obvious to any American that violence is also used in a wide range of crime and deviance.

Straus is not alone in suggesting that deviant behavior can be learned just as we learn any other, more legitimate pattern of social behavior. Many students of deviance have found that children who grow up in environments in which violence is approved and rewarded come to see such behavior as normal and desirable. Early social learning theories of deviance focused on deviant subgroups. In the 1920s, for example, Edwin Sutherland found that people who had more contact with deviant subcultures were more likely to adopt their norms than those of the larger culture. If the norms of the group include crime and violence, then that is what its members learn. Straus's Cultural Spillover Theory extends the principles of social learning theories beyond subgroups. He suggests that American society teaches *all* its citizens that problems can be solved by violent means. The problem is that the lesson "spills over" into behaviors of which the society generally does not approve.

Straus's Evidence for Cultural Spillover

Straus was interested in demonstrating that when violence is approved of for achieving legitimate ends, it also is used in illegitimate behavior. To support this theory of cultural spillover Straus cites data from a number of studies. After I summarize some of the data here, I also want to convey how careful Straus is to acknowledge the important weaknesses in the research methods from which he draws his conclusions. Here are six of the specific research findings he used to support his theory.

1. In the late 1980s Straus, along with colleagues Larry Baron and David Jaffee, tried to account for the large differences in the rates of rape and murder in various areas of the United States. For their study they developed the Legitimate Violence Index, to measure the extent to which people in a given state approved of using violence to achieve socially legitimate goals ranging from disciplining children to punishing crime. They discovered a strong relationship between scores on the Legitimate Violence Index and rates of violent crime such as rape. That is, the more the people in a state approved of the use of violence for legitimate goals, the greater the rate of illegitimate violence in the state.
2. In 1987, using data from the New Hampshire Child Abuse Survey, Straus and David Moore published a study of the relationship between the legal and illegal uses of violence. They found that parents in the sample who approved of physically disciplining children for wrongdoing had a rate of child abuse four times higher than parents who did not approve of physically disciplining children.
3. Straus examined data from a 1972 study in which college students were asked to recall events from their high school years, including the extent to which they had been physically punished by their parents and the amount of delinquency in which they had taken part. Again, the link between the legitimate uses of violence (punishment by parents) and the extension of those lessons to illegitimate purposes (delinquency among the children) showed up in the data. The high school seniors who were physically punished reported having engaged in significantly more violent crime and property crime than those who were not physically punished by their parents.
4. Data from the 1985 National Family Violence Survey also showed the operation of cultural spillover. The more the respondents had been physically punished as teenagers, the greater the likelihood of their committing violent acts outside their families when they became adults.
5. Working with John Schwed in 1979, Straus found that the legitimation of violence can extend beyond the family. The researchers examined the rates of physical abuse by parents serving in various units of the

military. They reasoned that the use of violence would be seen as more legitimate in combat than in noncombat units. As predicted by Cultural Spillover Theory, they found that the reported rate of parental abuse was significantly higher among those serving in combat units than among those serving in noncombat units. And if you are thinking that this is merely the likely consequence of parents' spending so much time practicing for combat with varying sorts of weapons, Schwed and Straus also report that most members of combat units serve in noncombat jobs, such as mechanics, truck drivers, and cooks.

6. Straus cites a 1979 study conducted by Irwin Hyman in which he examined the rates of assaults taking place in schools. The data, taken from the 1978 National Safe School Study, again showed the relationship between the legitimate and the illegitimate uses of violence. Schools in which physical punishment of students was authorized for a large percentage of the school employees and in a wide range of circumstances had a significantly higher rate of assault by students than schools with much more restricted authorization of punishment of students by school employees.

A Note on the Development of Social Theories: Some Methodological Warnings

As you may have been able to figure out from the data Murray Straus used in his research review, he has spent his career producing and evaluating a great deal of data on the subject of violence. To that end, he has worked with a wide range of co-authors, each of whom has helped contribute another piece to the growing body of data used to construct Cultural Spillover Theory. Social theories, like theories in other disciplines, are attempts to find underlying forces that influence a large range of events. For example, the germ theory of disease argued that behind each of the many illnesses from which we suffer is a common force, the microbe. Another example, as discussed in Concept 21 of this book, is the social theory proposed by Emile Durkheim that the several types of suicide can be accounted for by the underlying fact of disrupted social belonging. Straus's search for an underlying theory of violence is in this same tradition. After years of thought and examination, he has come to believe that we can attribute a great deal of the violence in our culture to the fact that teaching violence as a strategy for achieving legitimate ends inadvertently spills over into the pursuit of illegitimate ends as well.

Like any good scientist, Straus understands that simply asserting the theory is not enough; he must also find data to support it. However, he understands, and clearly acknowledges, that the data he has produced and cited have serious shortcomings. The shortcoming on which Straus focuses is the lack of causality.

Examine all of the six findings I selected from Straus's research review and you will find a pattern. In every case it is impossible to determine which came first, the legitimation of violence or the illegitimate use of violence. For example, consider the finding that parents who approved of physically disciplining children for wrongdoing had a rate of child abuse four times higher than parents who did not so approve: isn't it likely that such parents also say that punishment of children is a legitimate, even good, thing to do? It is a shortcoming of much survey research that although the relationships between two variables can be established, one variable can rarely be shown to be a *cause* of the other. In Straus's words (1991: 133): "[S]uch studies cannot show that it is the physical punishment which causes the problems rather than the deviant behavior of the child which causes the problems." The solution to the shortcoming lies in the methods used to collect the data. Straus calls for more longitudinal surveys, in which specific subjects are studied over a long period of time. This approach could confirm or disprove the possibility that the approval of violence as legitimate can be established in given families, states, or cultures, and could demonstrate the eventual consequences of that approval for the rates of illegitimate violence.

Straus also suggests one other, somewhat less expensive way to collect the data needed to test his theory. Experimental studies (discussed in Concept 3) are particularly good at establishing causal relationships. However, they require that subjects be intentionally exposed, or not exposed, to the variables to be tested, and this often raises ethical problems. For example, Straus acknowledges that it certainly would be unethical, if not impossible, to assign some families to an experimental group in which they are asked to punish their children physically for wrongdoing, and other families to a condition in which they are asked not to administer such punishment. But Straus suggests that given the high percentage of American families that already physically punish their children, it might be possible to convince a number of them to stop doing so voluntarily in order to take part in a study of his theory. It certainly would be interesting to see, a generation later, whether the children who were not taught that violence has legitimate uses turned out as adults to be less prone to use violence illegitimately.

Application

According to national opinion polls, the form of conflict that worries Americans most is personal violence. The newspapers and talk shows are filled with concern about street muggings and murder rates. We are more and more worried that we are a violent society, and as the sociological perspective shows, perception can shape reality.

We have many "windows" through which we can view our culture. One of the most compelling, partly because it is found in virtually every home, is television. What does it show about violence in the United States? You have probably heard about congressional studies of violence on television and about groups like Action for Children's Television (ACT) that monitor and criticize television violence. But what have you seen in your own viewing?

In this application you can test for yourself the rate of violence on television, but not in the way most people do. Most people, if they have thought about it at all, have purely subjective impressions about the rate of violence on television. They may have seen only a narrow range of programs that are shown at certain hours. In addition, they do not take the time to distinguish various types of violence, opening the door to the tendency to lump all types under one heading. Lastly, an impression of violence may be greatly influenced by the individual's sensitivity to one type of violence in comparison with another. To deal with the problems raised by subjective evaluation of television content, this application uses the research technique of content analysis.

How Content Analysis Works

Content analysis is the systematic, quantitative analysis of textual (written) or pictorial data. Our normal method for evaluation of this type of information is to read a book or view a television program and then report on what we have learned. There is no way to test the accuracy of such a subjective, individual evaluation. We have to take the individual's word for what she or he has learned. By contrast, content analysis establishes categories of meaning that can be accurately described *before* the evaluation of text or pictures, allowing any trained coder to accurately count instances of each category of meaning. This is not as complicated as it may sound. An example should help.

Let's say a concerned citizen complains that the books in the teenage section of the local library are filled with "smut." This individual reader may have a special standard for what is "smutty," which may not fit the standards of the wider community. In addition, how many books, and which ones, has this person read? A more objective study can be done by content analysis. First, we develop a list of words that members of the community can agree are, in fact, "smutty." (This process can present some problems, because there will be some disagreement on some words in any community, but general agreement on a core of words can be arrived at.) Then, a representative sample of books from the teenage section of the library is selected. Finally, the books are closely examined, and the frequency with which each of the words on the list appears is coded. The end result is a quantitative summary of the ratio of "smutty" to "nonsmutty" words contained in the books in question. The community may even decide to have the rate for books in the adult section of the library calculated. It is up to members of the community to decide whether the rate of such "smutty" words offends them. But whatever they do, what they have to work with in making their decision is objectively collected data, not subjective impressions.

Conducting a Content Analysis of Televised Violence

We have been systematically studying television violence since at least the mid-1950s, and many methods of counting acts of violence have been devised. The most comprehensive and persistent data have come from the yearly studies conducted by George Gerbner and his associates since 1967. The earliest of these studies calculated the percentage of programs that contained *any* episode of violence. It reported that in the 1967–68 season 80 percent of the sampled programs contained violence. In a 1986 update of these studies Gerbner reported that the level of violence for 1984–85 was the fourth highest on record. (Gerbner et al., 1986). Though the Violence Index that Gerbner and his colleagues have used is too complex and time-consuming to use in this application, you will be able to measure televised violence by identifying three levels of violence, as follows:

1. **Murder and Attempted Murder.** This form of violence includes any acts by an individual or group of individuals resulting in or intended to result in the death of another person or persons, whether or not the intent to kill is clear. Examples include shootings, stabbings, attempts to run over another with a car or push someone over a cliff, or the like. Also included are shootings or other assaults that do not kill the target but that make the intent to kill clear. You will have to use judgment about intent.
2. **Physical Violence Clearly Not Intended to Kill That Does Not Result in Death.** This form of violence includes fistfights, wrestling, and other forms of beatings (including those with objects) in which the intent of the participants is clearly to injure or incapacitate the other but not to kill. It also includes physical violence done by agents other than persons, such as car accidents, falling objects, and so on, whether they cause death or not.
3. **Nonphysical Violence.** This is primarily verbal violence, or threat, including clearly menacing gestures and actual threats of harm by one person against another. A report of a threat would *not* be included.

These are the three categories of violence you will be looking for and counting in your viewing. They will be referred to on your coding sheet as (1) "Murder–death," (2) "Physical harm," and (3) "Threat." After a bit of practice it should become easy to code a given behavior into one of the three categories. You simply put a check mark in the appropriate box on the coding sheet. But no matter how familiar you become with the coding categories, there will be confusing cases. For example, what if a person is accidentally shot and is hospitalized in a coma but does not die during the program? Which category of violence is that, Murder–death or Physical harm? I would ultimately choose Physical harm, but it is a judgment call that I hope will be relatively rare. If you wish to test how clear the coding categories are, have someone else who is doing this application view and code the same television programs that you do. This test of "intercoder reliability" checks the rate of agreement about coding decisions. If the two of you are making the same coding decisions most of the time, things are going well. To illustrate how the acts of violence might be coded, I have filled out a sample coding sheet for three television programs I watched. This is how you should fill in your coding sheet for this application.

Sample Content Coding Sheet

Program number	Type of program	Duration in mins.	Instances of violence: Murder–death	Physical harm	Threat	Total
1	Crime	60	4	8	7	19
2	Action	30	2	12	9	23
3	Soap	30	0	1	2	3
Total for each category of violence =			6	21	18	

Grand total for acts of violence = 45

Sampling the Television Programs to Code

There are simply too many programs on television for all of them to be evaluated. You will have to sample a few. To keep things relatively simple, let's limit this content analysis to prime-time programs. It is not necessary to draw a representative sample of television programs. Merely choose your programs by when it is convenient for you to watch. Try not to watch all the programs on the same evening or two or in the same time slot, so that your sample is not too narrow. If you want to draw a representative sample, you can turn back to the Application in Concept 3 and use the table of random numbers to draw a list of twenty programs from a weekly television listing. (This application works best if you view and code at least ten programs, but twenty is better, so I have made space for entering that many programs on the coding sheet.)

Data Collection and Analysis

Using the Content Coding Sheet provided, count and record the instances of violence for each of the three categories of violence. Once you have recorded the information about the programs you watched and the instances of violence in them, add up the frequencies of violence in two ways. First, count the number of violent acts in each program by adding across all three categories of violence and putting the figure in the right-hand column. Do this for each program. Then, add down the columns for each type of violence, and put the totals in the spaces marked "Total for each category of violence." Lastly, add up the acts of violence of all types for all programs (it should be the same sum whether added across the bottom row or down the right-hand column) and put the total in the space at the right-bottom of the coding sheet marked "Grand total for acts of violence." Then you can calculate the total minutes you watched for this application (sum of the column for "Duration in minutes") and divide it by the figure for "Grand total for acts of violence." This gives you a figure for the average number of all violent acts per minute of television program viewed.

Interpreting the Results

Depending on the type of programs you watched, you may have found high or low rates of violence. For example, if you watched prime-time action programs like police or detective programs, you may have been impressed by the number of violent acts you found. By contrast, programs on public television and on commercial stations in the early evening have considerably fewer acts of violence. Also, soap operas, even the ones that are shown at night ("Beverly Hills 90210" is an example) do not have lots of shooting and fistfights. But they do have a good deal of suffering in hospitals and personal, emotional anguish. These are not examples of violence as defined here, but you can make your own case. What about films? Some, like the ones starring Sylvester Stallone, Arnold Schwarzenegger, Clint Eastwood, and Charles Bronson, can increase your murder–death violence count very quickly. I tried coding one of Stallone's and found that I was unable to keep up.

As a general criticism of using this technique, it is far from clear that television reflects an accurate picture of who we are. Because most programming is intended to make a profit, what we see on television is directed narrowly. We know, for example, that programming is aimed at the single largest segment of the population available during any viewing hour, assuming that it has money to spend on the products advertised.

Thus, television is a window on the culture, but one that gives a somewhat distorted look at only some of us.

Extending the Analysis

There are a number of ways of extending the analysis of the data you collect. For example, you can divide the programs you viewed in terms of the time of evening they came on. Were there more acts of violence later in the evening than there were earlier? Some activist groups have pressed programmers to move shows with violence to later in the evening. If you are particularly interested in some special aspect of this subject, you can add it to the coding sheet and analyze those data. For example, you can record the gender of the perpetrator and victim of each act of violence. (It takes some practice to become skilled at this, but not much.) Is it the case, as some evidence has shown in studies of violent films, that women are disproportionately the victims of violence on television? How frequently are women the perpetrators?

Lastly, you might choose to conduct a content analysis of violence in cartoons. These weekend-morning programs are directed at children and seem to have the potential to teach lessons about how problems can be solved, the nature of acceptable and unacceptable social behavior, and a range of other beliefs. From watching a few cartoons recently, I get the impression that they are less violent than they were when I was young, but this is one of those subjective impressions that can be checked only by a more objective content analysis. Note that if you choose to analyze cartoons, you will have to ignore the three categories of violence and count all instances of violence the same, because cartoon characters don't die or go to the hospital—they just pop back up the next frame and go on as if they had not been run over by a car or zapped by a ray gun.

Content Coding Sheet

Program number	Type of program	Duration, in mins.	Instances of violence: Murder–death	Physical harm	Threat	Total
1						
2						
3						
4						
5						
6						
7						
8						
9						
10						
11						
12						
13						
14						
15						
16						
17						
18						
19						
20						

Total for each category of violence = ___ ___ ___

Grand total for acts of violence = ___

PART TWO

Acquiring the Meanings of Social Membership

Animals other than humans are born into their social groupings already carrying with them many genetically transmitted instructions for dealing with one another and their environment. But humans must largely learn how to think and act through social interaction. Part Two deals with the various rules that groups of people develop for social interaction and the process by which the rules are passed from one generation to the next.

Concept 7 focuses on culture, the way of life that is learned and shared by people and taught to succeeding generations. Included in the discussion are the specific content of cultures, the qualities that all cultures have in common, and the way cultures develop to satisfy the needs of society's members.

Concept 8 examines norms, the rules for social behavior that express the beliefs shared within a culture. Two levels of norms are discussed: folkways, which provide guidelines for the style of everyday interaction, and mores, which are the rules for behavior that a group considers vital for its welfare and survival.

Concept 9 deals with socialization, the process by which individuals learn the rules of their society and develop an adult character. Through socialization the character of the society is maintained from one generation to the next. A number of theories about how socialization works are treated.

CONCEPT 7

Culture

Definition

Culture A way of life that is learned and shared by human beings and is taught by one generation to the next.

In the 1950s the English anthropologist Colin Turnbull lived with a group of pygmies in the Ituri Forest of Africa and, from his observations, wrote a wonderful study of their way of life, *The Forest People* (1962). He recounts not only daily life among the pygmies but also the special ceremonies in which they celebrate the forest as the protective, loving "father" of the tribe. In all, Turnbull presents an extremely affectionate and revealing picture of the pygmies and their life in the forest.

When he concluded his study, Turnbull left the Ituri Forest, but he took with him one of the pygmies, a particularly adventurous and sophisticated man named Kenge. They drove out of the forest, eventually pausing on the edge of a bluff that overlooked a broad valley. Turnbull pointed out a herd of water buffalo grazing in the distance. Kenge only laughed at the obvious joke Turnbull was trying to play, and he asked what kind of insects they were. After all, the pygmy knew water buffalo to be huge animals, and the animals Turnbull pointed at were tiny.

How could an intelligent man like Kenge have made such an error? The answer is that it was not an error at all. It was a judgment shaped by the experiences of his culture. As you will discover in this chapter, cultures serve to adapt people to the environments in which they live. In the forest, the pygmies could never get a view of objects from so far away that the objects appeared to decrease in size. The trees always got in the way first. Thus, when the water buffalo were seen at a distance, they were interpreted to be a different animal from the buffalo the pygmies were used to seeing full size. Culture helps us make sense of the environment in which we live, even to the extent of shaping our visual perceptions. It is therefore a serious error to judge other cultures in terms of our own experiences.

Yet cultures often *have* been condemned as inferior by people using their own experiences as the standard for judgment. For example, take the common distinction between what some people call "high" and "low" culture. In the United States, spray-paintings on the walls of buildings and subways are seen as low culture, and paintings in museums are considered high culture. Country music, rap, and rock are seen as low culture, whereas the works of Bach, Beethoven, and Verdi are viewed as high culture. In the evaluation of art and music, fans of high culture commonly condemn low culture as inferior, and control most of the positions from which these evaluations are publicized. These include, for example, newspaper and television reviewers, the boards of museums, art gallery directors and owners, and the councils that award grants for support of the arts.

The term *culture* obviously is part of our everyday vocabulary, but it is very often used inaccurately (from a sociological point of view). It is the aim of this concept to provide a sociological definition of this very abstract term—a definition in direct contrast with its everyday, colloquial meaning.

To begin with, the notion of high and low culture probably developed out of the work of nineteenth-century anthropologists. By midcentury, Charles Darwin had published his theory of evolution, and a number of unfortunate uses were made of the theory by people hitching a ride on the groundswell of its popularity. Anthropologists, especially in England, used it to support their belief that cultures evolve just as plants and animals do. According to the theory of cultural evolution, human societies can be arranged in order of advancement, with savages at the bottom, barbarians next, and, at the top of the hierarchy, civilized, or cultured, peoples. (Of course, the European and American anthropologists placed themselves in this last category.) So culture came to be equated with "advanced" forms of social life. Because the most apparent evidence of any people's way of life was the visible products and behaviors of daily living, culture also became equivalent to art, architecture, music, food, clothing, dance, and so on.

Culture from the Sociological Perspective

The sociological definition of culture differs from these other uses of the term in a very important way. No group of people is "uncultured," or without culture. In addition, there is no reasonable way to decide that one culture is better than another or representative of some higher way of life suggested by the term *high culture.* Just why this is true will become apparent from the following discussion.

Sociologists generally agree that humans differ from other social animals in the way they interact with their environment and in the origins of those interactions. Although other animals (such as ants) often live in complex social arrangements, with division of labor and other subtle relationships, these patterns of interaction are inborn, or instinctual. Most sociologists argue that humans are born with reflexes and drives (such as hunger) but are not born with complex instructions for interacting with one another or with the environment. These must be taught by one generation to the next and used every day in deciding how to act. (However, there is a school of thought that argues that the social behavior of humans, like that of other animals, is inherited. The **sociobiologists** contend that genetic instructions that promote the preservation and reproduction of the self influence directly our actions toward one another. A fuller discussion of sociobiology can be found in Concept 9 on socialization and the development on adult character.)

The way people in a group come to live their lives becomes a matter of general agreement among them that "this is what we believe in and the way we wish to live." Style of living can vary greatly from one group of people to another. Two groups may have entirely different ways of accomplishing the same goal (for example, bowing among Japanese versus hand-shaking among Americans), and the exact same behavior may mean entirely different things in two groups. For example, whistling at a performance expresses appreciation in the United States, but it shows strong displeasure in most of Europe. The rich variety of cultural behavior suggests that we humans invent ways of dealing with one another and the environment. Clyde Kluckhohn (1967:22) said it this way: "The concept of culture is made necessary by the observed fact of the plasticity of human beings. Newborn members of different groups are taught to carry out 'the same' acts in an almost infinite variety of different ways."

So the ideas and behaviors of a culture can vary greatly because they have no meaning independent of the meaning that is agreed on by the members of the group. There is, for example, no intrinsic, inevitable meaning to shaking hands, kissing, whistling, or wearing an earring. They mean what they mean in a given culture *only* because the members have come to agreement about what they will mean. Culture is symbolic. And one of the main things that having agreed-on symbols does for us is to make things seem orderly. When we all agree that it is a good thing to kiss, or shake hands, or go to work five days a week, we make living more predictable and stable. Even if we agree that we should insult one another whenever we come in contact, at least it is an agreement on which we can depend. Culture, whatever its character, creates order and makes sense of the world.

So far I have been trying to give you a general idea of what culture is like and what it does to create social order. But culture can be more closely defined if we look at its actual *content.* Once that is done, it is possible to see what *qualities* all cultures have in common and how they develop.

The Content of Cultures

Textbook discussions include a huge number of elements in their definitions of culture—knowledge, beliefs, art, mores, morals, laws, customs, symbols, values, norms, sanctions, folkways, artifacts, and a host of others. Some of these are merely synonyms for one another, and others are subcategories. To simplify things, I use Talcott Parsons's list of the elements of culture (Parsons et al., 1961:961–1204). All the elements that other writers list fit somewhere in Parsons's categories. He identified the components of culture as (1) knowledge (both empirical and existential), (2) values, and (3) forms of symbolic expression.

Knowledge (Empirical and Existential)

In American culture, our **empirical knowledge**—the accumulated information we share about how the world is constructed and operates—is generally a result of scientific inquiry. Other sources exist as well, including religious, folk (commonsense), and literary sources. In all cultures empirical knowledge is learned from generation to generation, sometimes taught primarily in schools, sometimes in the home or another place. The more detailed the body of empirical knowledge and the more rapidly it changes and increases, the more formal the structure for its transmission becomes, and the more specialized the members of a culture must be if that knowledge is to survive.

Questions about why we are here and where we came from are primarily the province of philosophy and religion in American culture. Such **existential knowledge** in other cultures differs greatly from our own, with many tracing the origins of human life to the earth, the oceans, or the transformation of animals to human forms. Once again, the important thing for members of the culture is that they agree that the story is true, whatever the story itself may be.

Given the importance people attach to their existential beliefs, it should come as no surprise that one culture might find the existential beliefs of another to be bizarre, contemptible, or even threatening. In our own history this tendency is reflected in the response of Europeans to the existential beliefs of the natives of the continent. Recent films such as *Dances with Wolves* dramatize the contempt with which the beliefs of these cultures were met, not only by Europeans but also by one tribe towards another.

Lastly, in America there is a significant battle being waged between two communities of authority for the right to address existential questions about the origins of life. The deeply held belief that life evolved by natural processes bumps directly against the equally deeply held belief in the biblical story of creation. In numerous court cases the differing sources of "knowledge" compete for space in school curricula. But questions about the meaning of life, the "why's" of our existence, are still overwhelmingly reserved for the realms of religion and philosophy.

Values

Values express the highest moral goods espoused by a culture. The sociologist Robin Williams (1970) listed some of the important American values: freedom, equality, democracy, individual success, progress, work, material comfort, efficiency, morality, science, patriotism, and the superiority of some groups of people to others. Because these values are so broadly and intensely shared in American culture and because they have endured so long, Williams termed them *dominant* values.

Different cultures can hold different values, such as values for belonging and loyalty rather than for independence and individualism; for consistency and conservation rather than for achievement and change through progress; for traditional ways rather than for innovation; or for religious faith rather than for rationality. Even within a single culture, the values are not all consistent with one another. For example, the American values of individual success and social equality often conflict. If, in order to get ahead, I seek special treatment at the expense of others, the value for equal treatment must be ignored in the interest of personal ambition.

American values are evident whenever statements of national purpose are expressed. They appear, for example, in the Bill of Rights of the U.S. Constitution. They are affirmed in Fourth of July speeches, high school civics classes, and Sunday sermons. Because values are the broadest, most abstract statements of social purpose in a culture, they are not normally expressed directly in everyday interaction. But from them flow a variety of beliefs, rules, and guidelines that regulate our daily lives. Many of the elements of culture that are listed in textbook definitions are examples of such beliefs, rules, and guidelines. **Norms,** for example, are the expression of values at the everyday, behavioral level. They are rules about how we are expected to behave in given situations. (For a fuller treatment of the subject of norms see Concept 8.) The value for achievement is expressed in the norm to pursue education and the norm for hard work. Because any statement of how we are supposed to behave is a moral statement (that is, it proposes what "ought to be"), one of the elements of norms is their moral content. **Mores** and **folkways** (Sumner, 1906), are two levels of norms. Folkways are everyday rules for behavior (or customs), such as fashion and manners. Mores are rules for behavior that must be obeyed to maintain the stability of the group. They are so important that they are usually codified in law, in which case the violation of mores is formally punished, for example, by jail terms or fines. Mores include the prohibitions against theft, murder, and threats to others and the responsibility to provide for dependents, fulfill contracts, and tell the truth. The rewards and punishments that enforce the norms of a culture (whether folkways or mores) are called **sanctions.** The terms that have been emphasized in this paragraph (norms, morals, mores, folkways, customs, and sanctions), can all be included in one category (values), because they are all outgrowths of the values of a culture.

Forms of Symbolic Expression

Certain objects are created to represent the feelings, experiences, and tastes of the people who create them. These are forms of **symbolic expression.** In American culture, the music we listen to (including rock) and the paintings we appreciate (even if they are on walls or subways) are meaningful expressions of our culture or, as you will see soon, a subculture. The objects we make, the songs we sing, the stories we tell "decorate" our lives while revealing much about what matters to us. Consider the way people decorate the places they live. Take, for example, the sculpture of the eighteenth century people of the African plain. The beings they carved are mixtures of human forms and the animals they ate, worshipped, feared: a square-bodied man has the head and face of an antelope; a crocodile has near-human arms and legs. In a real sense, those humans and animals were interdependent, and the art forms reflected this. In modern America we are so different from one another that there are countless forms of symbolic expression in our culture. But it is also clear that our decorations often reflect a central feature of our culture, commercial efficiency. The strip malls and fast-food restaurants that can be found across America express the spirit of our time just as the highly ornamented, massive architectural style of Victorian England reflected the moral severity and middle-class conservatism of that culture.

The Qualities of Culture

All cultures have the same qualities, even if they differ in the forms of their knowledge, values, or symbolic expressions. All cultures are (1) learned from previous generations, (2) shared to some degree by the members, (3) adaptive to the conditions in which the people live, and (4) focused on meanings and ideas rather than on the material production of the culture.

Culture Is Learned from Previous Generations

Culture is not inherited genetically; it is learned. In the United States, most of this learning takes place in the family, but some occurs in schools, religious settings, workplaces, and everyday interactions with friends. In America, existential knowledge is learned primarily from family and religion. A child who asks about death probably gets the answers from parents or a minister, priest, or rabbi. The transfer of empirical knowledge is accomplished by some mix of learning at home, from friends, and in school. As with existential knowledge, values fall largely within the domains of family and religion, though there has been an increasingly strident debate about the extent to which schools should teach values to children as part of their curricula. And lastly, forms of symbolic expression are learned everywhere. Parents teach children what they think is good taste in music, clothing, and so on. Then children get to be teenagers and often learn from one another and from the commercial interests (largely via television and movies) that market music and other tastes to them. And schools try to teach aesthetic sensibilities in art, music, English, and history classes. In short, we learn the culture from all our experiences with others.

Culture Is Broadly Shared

Because culture exists at the level of symbols, a behavior, an idea, or an object can have meaning only to the extent that a number of people *agree* about that meaning. When agreement exists about the meaning of a specific behavior, we can be confident that we know when it is acceptable to behave that way toward others. In small, preindustrial cultures, the culture can be uniformly shared. Deviation from the beliefs, values, and expressive behaviors of the culture can be closely monitored and punished. But as cultures get larger and more complex and change more rapidly, room for variations in the culture increases. Although the dominant culture persists, **subcultures**—groups within the larger culture whose beliefs, values, and styles of life are different in some respects from those of the dominant culture—spring up. The teenage subculture differs from the larger culture primarily in its folkways (dress, language, music, and so on), its emphasis on nonconformity with the larger culture (the adult world), and its value for loyalty within the subcultural world. The prison world can be seen as a subculture that emphasizes toughness, loyalty, and survival.

There is also a special category of subculture called the **counterculture.** It consists of groups within the larger culture whose beliefs, values, and styles of life are not only different in some respects from those of the larger culture but also in conflict with them. An example of a counterculture is the hippie movement of the 1960s. Hippies rejected what they saw as the militarism and materialism of the dominant culture, stressing, instead, pacifism and a spiritual style of life.

Culture Is Adaptive

Cultural characteristics, such as the values for individualism, progress, and styles of everyday interaction, do not develop at random. Even the characteristics of subcultures develop for good reasons. Each characteristic of a culture or subculture can be seen as helpful (or adaptive) for the survival of the people in the circumstances in which they live. Prisoners, for example, live in dangerous, unpleasant, stressful conditions. Emphasis on toughness, loyalty, and survival (as opposed to progress) helps prisoners adapt to the conditions of their incarceration. The dominant American culture, with its emphasis on progress, achievement, and independent action, developed out of the struggle to settle, dominate, and exploit the resources of a huge, wild, and already occupied continent. In Japan the values for respect, orderliness, group membership, and self-control are quite adaptive in a society with a high population density in a tiny habitable area. Even folkways such as tastes in food can be explained as adaptive. For example, could you eat a plate of whale blubber? No? Well, if that were all your environment could provide, you would be raised in a culture that taught you how good blubber is, and you would become a willing blubber-eater. In the United States, changes in the availability of certain foods have been quickly reflected in the changed images of such foods in the popular culture. For example, during a beef-price scare in the mid-1970s, prices for steak rose rapidly. Within weeks, reports were appearing on television and in print that horse meat could actually taste good and that soybean extenders diminished the quality and taste of hamburger very little. These were cultural adaptations to changed living circumstances. So culture changes; it is not static.

Culture Is Focused on Meanings, Not Objects

When we try to characterize a culture, we often look at the most obvious and fascinating evidence of its character. These are the products of symbolic expression by its people—their art, music, dance, drama, architecture, and so on. But the objects or events are merely representations of underlying feelings and values. Culture is not the objects themselves but the meanings that underlie them—their use, their value to the members of the group, the ideas they represent. The character of a culture is not always evident in its artifacts. Their use, value, and meaning to the people who made them must be known. A plain gold band might mean marriage in one culture (worn on a certain finger), manhood in another (worn through the nose), or aesthetic perfection in another (framed and hung on a wall).

Ethnocentrism and Cultural Relativism

Cultures help us make sense of the world and organize our interactions by making them predictable. We have agreed in our culture about what is true (our body of empirical and existential knowledge), what is good (our values), and how to express our feelings and experiences symbolically. To the extent that these agreements make life understandable and comfortable for us, we come to depend on them and to think they are pretty good things. In fact, there is a powerful tendency for cultural content to be considered not just *a* good way to live but *the only* way. This is especially true when members of a culture have limited, or no, contact with other cultures. The belief that one's own way of life is superior to that of any other culture is called **ethnocentrism.** The theory mentioned at the beginning of this concept—that cultures could be arranged in order of evolution, from savages at the bottom to civilized peoples at the top—was an example of ethnocentrism.

By contrast, I have tried here to express the view of **cultural relativism.** It argues that no culture is superior to any other, because each is a set of adaptations to the conditions in which its members live. If this is true, each cultural belief, value, or expression is equivalent to one of some other culture that serves the same purpose, no matter how bizarre it may appear to an outside observer.

Summary

Culture is a way of life that is learned and shared by groups of human beings and is taught by one generation to the next. Sociologists and anthropologists generally agree that, unlike other animals, humans interact at the level of symbols. That is, we attach meanings to our experiences that we can express to one another. Because two different peoples can attach very different symbolic meanings to the same experiences, cultures can (and do) vary greatly. Not only can one culture differ from another, but also one culture can change over time. Some cultures even tolerate the existence within their boundaries of subcultures, groups of people whose values, beliefs, and styles of life are different in some respects from those of the larger culture.

Culture consists of (1) knowledge—the accumulated information that people share about how the world is constructed, the principles by which it operates, and the meaning of existence; (2) values—the people's shared ideas about the most abstract goals they believe are worth achieving; and (3) forms of symbolic expression—the people's creative activities that result in the production of art, literature, dance, and so on.

Culture is adaptive, because it helps its members survive within their environment. But because cultures are symbolic (and therefore variable), two different cultures can be equally successful in aiding the survival of their people. This fact gives rise to the concept of cultural relativism, the belief that no culture is superior to any other in providing its people with the tools for environmental adaptation.

The Illustration for this concept summarizes a recent study of the extent to which the American South exhibits some of the characteristics of a subculture. Specifically, Christopher Ellison uses data from a large body of survey data (called the General Social Survey) to study the "subculture of violence thesis," which argues that people raised in the American South are taught norms that approve of violence more than is the case in the rest of America.

The Application for this concept focuses on cultural relativism. Ideally, the best way to understand that the same issues of survival can be dealt with in unique, but equally successful, ways in different cultures is to travel and live outside one's own culture. However, because this is not possible as a course assignment, this application takes advantage of the fact that summaries of such experiences are readily available in ethnographies in every library.

Illustration

Christopher G. Ellison, "An Eye for an Eye? A Note on the Southern Subculture of Violence Thesis," Social Forces *69 (1991): 1223–39*

On a recent visit to some college professor friends in Tennessee I inadvertently started a lively debate about whether the South was different in some way from the rest of the country. Since I was the only northerner in the group, and didn't know much about the subject, I tried to stay out of the discussion, but I admit that I was the one who got things going in the first place. All I did was to ask about the stock car racing that I had seen scheduled in town for the weekend. I wondered why it seemed to be so popular in the South.

Some claimed the South, especially southern cities, was just like everywhere else in America, even though it hadn't always been that way: the same television shows, the same fast-food places, the same toys that kids pester their parents to buy. Others claimed there were real regional differences, though the group was equally divided about whether or not this was a good thing. They talked about southern language, food, family, hunting (and the importance of dogs to it), religion, the pace of life, and too many other beliefs and styles to recall.

The heat of the discussion suggested to me that the issue of whether the South is a subculture matters a great deal to at least some southerners. Also, while all this was going on I kept remembering how I'd been involved in the very same debate when I was a graduate student in Boston in the late 1960s: We were discussing the ideas of culture and subculture, and someone suggested that the American South was a good example of a subculture. After all, wasn't it a place with different beliefs, values, and styles of living than the rest of the country? The fact that none of us was from the South kept disagreement to a minimum, but the contention that the South is a subculture has remained as controversial among sociologists as among any other segment of the population. In the reading for this concept, I present a recent study in which the debate is revisited.

Christopher Ellison began his 1991 study by citing the well-established facts that the American South has higher per capita rates of homicide and gun ownership than the rest of the country. In an effort to discover why this is the case, he examined national survey data (called the General Social Survey) collected in 1983. He focused on what has been called the "southern-subculture-of-violence thesis," which argues that people raised in the American South are taught norms that approve of violence more than is the case in the rest of America. But can the existence of such norms be demonstrated? And if they can, why do they exist in the South more than in the North?

According to Ellison, who at the time of the study's publication was a faculty member at Duke University in Durham, North Carolina (thus, not an inexperienced northerner characterizing life in the South from a distance), southerners tend to approve of a specific form of interpersonal violence, namely, violence that is defensive or in retaliation for some wrong. So the violence that is normative in the South is not random; rather, it is violence "for cause." Ellison speculated that such norms can be traced to two main sources: (1) southern ideas of chivalry and honor, and (2) the nature of southern religious images of God.

Some analysts, such as Eugene Genovese, have argued that the Plantation South was, in many ways, modeled after European aristocracies. Among the characteristics southerners adopted was a code of honor and loyalty among gentlemen. According to norms of the time, insults to the family name or to one's honor permitted, even required, violent acts such as dueling to set matters straight. In his study Ellison cited a number of analysts who have claimed that these norms for chivalry and revenge survive in modern times in the American South. That is, as a modern extension of the early, aristocratic codes of honor, southern norms for violence express the belief that if one is done some sort of wrong, it is permissible to resort to violence in response. Ellison therefore hypothesized that among southerners he would find greater support for acts of violence initiated in retaliation, or in defense, than he would find among northerners.

Second, Ellison hypothesized that something about the conservative Protestantism of the South contributes to the norms that approve of violence "for cause." He pointed to the belief, common in southern theology, in a personal God who specifically punishes people for their sins. Ellison speculated that southerners may extend these religious principles to everyday interactions. If this is so, then he expected to find that those who were more religious and saw God as a punishing "judge" would also tend to approve more of acts of violence "for cause" than those who were less religious and did not see God in this way.

The Measures Used in the Study

Ellison used questions from the 1983 General Social Survey to measure the variables needed to test his hypotheses. To determine who was to be classified as a southerner and who as a northerner, he did *not* use the residence of the respondent at the time of the survey. This was because he recognized that it was possible for a respondent to have been raised in the North and then to have moved to a southern state later. Such a person would be unlikely to have learned southern subcultural norms. Instead, Ellison counted as a southerner anyone who had lived in a southern state at the age of sixteen, an age by which the individual is likely to have strongly adopted their region's beliefs.

To measure belief in the use of violence "for cause," he created a scale by combining each person's responses to the following three situations:

"Are there situations that you can imagine in which you would approve of a man punching an adult male stranger?"

1. If the adult male stranger had broken into the man's house
2. If the adult male stranger had hit the man's child after the child accidentally damaged the stranger's car
3. If the adult male stranger was beating up a woman and the man saw it

Lastly, Ellison used two measures of religion. The first, which could be called *religiousness,* was a scale indicating the frequency with which a respondent claimed to attend religious services each week. The second was a measure of the respondent's belief about the nature of God. The question in the original survey was "When you think about God, how likely are each of these images to come to your mind?" Ellison's measure combined responses to four possible images of God: father, king, master, and judge. The idea was to determine which people saw God in terms of a personal God who—as judge, father, king, or master—is likely to punish people for their sins.

The Results of the Study

Ellison found strong support for the core idea of the study, that there is a southern subculture of violence, but of a special sort. Southern respondents in the sample (that is, those who were socialized in the South) expressed much stronger approval of violence "for cause" than did northern respondents. When presented with a situation in which the individual would be called on to defend her- or himself, or someone else, who was being wrongly attacked by "an adult male stranger," southerners, especially southern males, generally were significantly more likely than their northern counterparts to approve of a violent response. Ellison found that southern females approved of violence less than did southern males. He attributed this finding to the fact that the females in the study reported they had experienced less violence in their childhoods than the males reported, and there was a general tendency in the data for those who had experienced more violence when young to approve more of violence as an adult.

As to the hypothesis that "the violent attitudes of native southerners are shaped by a regional religious culture," Ellison found some relationships that will, no doubt, create even more spirited discussions of the sort I described at the beginning of this illustration. He found that in the population in general (southerners and northerners combined), there was a negative relationship between church attendance and support of violence "for cause." The relationship was not very strong statistically; nevertheless, the higher the reported rate of church attendance, the lower the approval of even defensive violence. This seems to make sense. The finding equates religiousness with a sort of pacifism. However, when Ellison examined the relationship between church attendance and approval of violence for the sample of native southerners, he found *exactly the reverse* of the previous finding: Among the southerners, the higher their reported church attendance, the *greater* their approval of defensive violence. This, Ellison claimed, was clear evidence for the thesis that the southern subculture of violence is supported by a religious culture.

He added evidence for the thesis in another finding of the study: The data showed that approval of defensive and retaliatory violence was significantly greater among southerners whose image of God emphasized power and authority (father, king, master, and judge) than among southerners whose image of God did not emphasize power and authority. This relationship did not show up among northerners.

In sum, Ellison claimed support in this study for the thesis that there is something special about the culture of the South, something that contributes to its high rate of approval for a specific sort of violence. It was not suggested that southerners approved of random, senseless violence. Rather, Ellison is careful to root the southern approval of violence in a tradition of honor, responsibility, and accountability for membership in the community. If honor is offended, rules of civility violated, or the defenseless attacked, then violent retribution is justified in the southern culture. It should be no surprise that this belief in the consequences of wrongdoing are reflected in the religious beliefs of the South. Not only did Ellison find more support for violence among southerners than among northerners, he also found that support to be related to a singularly southern notion of the consequences of human wrongdoing. I suppose I'll have to wait to see how my southern friends "discuss" these findings when I next visit them and toss the data on the table. That should be interesting.

The last finding I would like to summarize (I have discussed fewer than half the findings of the study) deals with the likelihood that the southern subculture of violence will weaken in time. There are two findings in the study that support this prediction. First, Ellison discovered that older southern respondents were significantly more supportive of defensive and retaliatory violence than were younger southerners. From this Ellison concluded that as younger cohorts of southerners age, they will bring the norms they learned in childhood with them into later life. So the southern subculture of violence will diminish. In addition, Ellison found that people who had moved to southern states did *not* adopt the norms for violence prevalent in their new communities. So continuing patterns of northerner retirement to the Sun Belt should also dilute the southern subculture of violence.

A Note on the Use of Available Data in Ellison's Study: Secondary Analysis

Although Ellison's study was published in 1991, the data he analyzed were collected in 1983. Social researchers usually want to study current data because they know that much can change in society in a short time, and old data can be unrelated to current conditions. So why did Ellison use data that were almost nine years old? The answer reveals something about the difficulties of conducting what is called secondary data analysis.

Secondary analysis uses data collected by someone other than the researcher. So if you were to analyze the personnel records of your college, census data collected by the U.S. government, or health records kept by your town, you would be conducting secondary analysis. The data used in this study were collected by the National Opinion Research Center (NORC), located at the University of Chicago. Each year since 1972 NORC has collected survey data on a wide range of topics from large, representative samples of Americans. The data, called the General Social Survey (GSS), is made available to researchers such as Ellison, who try to find questions from the surveys that they can use to test their hypotheses. Sometimes, however, the GSS questions do not measure the researcher's variables in exactly the way the researcher would have liked. This is one price paid for using existing data: the researcher saves a great deal of time and money, but often is forced to make do with less-than-ideal measures.

Another important shortcoming of secondary analysis is that a single source of data may not contain all the questions the researcher needs to test the study's hypotheses. This was the problem Ellison faced in using the GSS data. As he clearly acknowledged in the journal article reporting his findings, more recent GSS data were available, but only the 1983 data set included questions on *all* the variables he needed to test his hypotheses about the southern-subculture-of-violence thesis.

Application

The Definition section of this concept ended with a brief comparison of ethnocentrism and cultural relativism. The ethnocentric evaluates other people in terms of his or her own cultural values, while the cultural relativist recognizes that each culture can adapt to its world in its own, equally worthwhile ways.

If you live your life bound within a given culture and never have contact with any culture different from your own, it is easy to fall into the trap of ethnocentrism. I wish it were possible to have all students of sociology visit and observe the ways of life in cultures different from ours. Luckily, however, much of this sort of research has been done and is widely available for study. I refer to "ethnographies," studies of cultures in which the researcher lived with a group of people over an extended period and recorded, as much as possible without bias, how they live. Though most of these studies have been conducted by anthropologists among primitive and peasant peoples, there are also a number of what might be called "domestic" ethnographies, in which the subjects are subcultures within the United States.

Using an Ethnography to Compare Cultures

This application is designed to take advantage of the availability of such ethnographies in libraries, so that without having to conduct participant observations, you can compare one of these cultures or subcultures with mainstream American culture. The idea is for you to get an *almost* firsthand appreciation of what cultural relativism means. Listed below are more than a dozen ethnographies to choose from, at least one of which is very likely to be in your library. Some depict cultures outside the United States, and some portray subcultures within the larger American culture. You will be reading one of these studies (skimming at times, if you wish) to answer the questions about cultural differences that are posed below.

In order to keep this application to a manageable size, you should avoid trying to compare all the elements of the culture you read about with those of American culture. As you already know from reading this concept, culture consists of two sorts of knowledge (empirical and existential) as well as values and forms of artistic expression. In reading any study, you will find a great deal of information about each of these, but this application asks you to focus on cultural values, and a specific form of values at that.

Value Orientations

In this concept I list a number of specific values that are part of American culture, such as freedom, progress, individual achievement, and so on. Beyond these specific values we can also identify the broader "value orientations" of a culture (Kluckhohn & Strodtbeck, 1961). In this application you will be asked to focus on four of the value orientations identified by Kluckhohn and Strodtbeck. Value orientations reveal how the people in a culture deal with the important questions about their existence. Listed below are the major existential questions and the kinds of value orientations that can be adopted for each. (The value orientation that predominates in American culture is identified for each issue.)

1. **Relationship to Time.** Three orientations to time are possible: past, present, and future.
   a. A culture can be concerned primarily with the past, in which case the keeping of histories, stories, and traditions would be critical.
   b. A culture can be concerned primarily with the present, possibly emphasizing sensory experience and the sacrifice of some future benefit for a reward "today."
   c. A culture can be concerned primarily with the future. Evidence would include planning ahead in a number of ways and sacrificing present gratifications for future benefits. (This has been the dominant American value orientation. Think of our insurance policies, retirement plans, and savings accounts.)
2. **Relationship to the Physical World.** Three orientations to the physical world are possible: mastery, harmony, and subjugation.
   a. A culture can define itself as the master of the natural world, trying to shape the environment according to the needs of the people. (This is the dominant American value orientation. Think of all the floodplain control, air conditioning, and paving we do to try to "tame" nature and control its effects on us.)
   b. A culture can define itself as in harmony with nature, perhaps defining humans as equal with all other things in the natural world.
   c. A culture can define itself as subject to the whims of the natural world, accepting good weather or disastrous storms with a fatalistic acceptance of nature's domination over humanity.
3. **Relationship to Activity.** Three orientations to activity are possible: being, becoming, and doing.

a. Members of a culture can emphasize the process of being, valuing the expressions of self in the present.
b. Member of a culture can emphasize the process of becoming, focusing on perfecting oneself to some culturally accepted ideal.
c. Members of a culture can emphasize the process of doing, in which what one is like (being) or will be like (becoming) is not nearly so important as what one achieves. (This is the dominant American value orientation. Think of the way Americans stress that people should be judged only by their accomplishments.)

4. **Relationship of People to One Another.** Three orientations of relationships within a culture are possible: hierarchy, individualism, and collectivity.
   a. The relationships among the members of hierarchical cultures exist in terms of higher and lower degrees of status and power, and obedience to superiors is expected.
   b. The members of individualistic cultures are seen as responsible largely to themselves. (This is the dominant American value orientation. An American may belong to a hierarchical organization, but he or she is still expected to be independent.)
   c. The members of collectively oriented cultures are expected to relate to others as equals in achieving the goals of the culture. One is neither independent of nor inferior to others.

As you were reading about these value orientations, you may have noticed that I oversimplified somewhat in identifying the dominant American value orientations. I claimed that American culture is basically (1) future oriented, (2) committed to the mastery of nature, (3) concerned with doing rather than with being or becoming, and (4) individualistic. Clearly, large numbers of Americans are more concerned with the present than the future, just as some Americans believe that we should live in harmony with nature rather than attempt to master it. The fact is that the larger and more diverse a culture is, the greater its tolerance for cultural variation among its members. So if we are to be able to identify the characteristics of a broad American culture (and some people say we cannot, precisely because of our cultural differences), we must tolerate some generalizations. When you read your ethnography, you will almost certainly find less cultural variation described in it than you will find in American cultures.

Data Collection and Analysis

Listed here are nineteen ethnographies, the last seven of which describe American subcultures. You can choose from among these or find your own. While reading, keep in mind the four existential issues on which value orientations are based (relations to time, the physical world, activity, and other people). Try to determine the specific value orientations of the people you read about, and record them on the form provided. For example, what is their orientation to time? For each of the four areas of value orientation for the people you read about, circle it, and in the space provided list some specific evidence for your decision. For example, did the reactions of the people in your ethnography to natural events reveal their relationship to nature? If so, how?

Ethnographies About Other Cultures

Chagnon, Napoleon. 1984. *Yanomamo: The Fierce People.* New York: Holt, Rinehart & Winston.

Evans-Pritchard, E. 1940. *Nuer.* New York: Oxford University Press.

Fernea, Elizabeth Warnock. 1969. *Guests of the Sheik.* New York: Doubleday.

Freuchen, Peter. 1972. *Book of the Eskimos.* New York: Fawcett.

Hoebel, E. A. 1978. *Cheyennes.* New York: Holt, Rinehart & Winston.

Lobo, Susan. 1982. *A House of My Own.* Tucson: University of Arizona Press.

Redfield, Robert, and A. V. Rojas. 1962. *Chan Kom: A Maya Village.* Chicago: University of Chicago Press.

Seidel, Ruth. 1974. *Families of Fengsheng.* New York: Penguin Books.

Shostak, Marjorie. 1982. *Nisa: The Life and Words of a Kung Woman.* New York: Random House.

Tonkinson, Robert. 1978. *The Mardudjara.* New York: Holt, Rinehart & Winston.

Turnbull, Colin. 1968. *Forest People.* New York: Touchstone Press.

Turnbull, Colin. 1972. *The Mountain People.* New York: Touchstone Press.

Domestic Ethnographies

Erikson, Kai T. 1978. *Everything in Its Path.* New York: Simon & Schuster.

Gans, Herbert. 1982. *Urban Villagers.* New York: Free Press.

Howell, Joseph T. 1973. *Hard Living on Clay Street.* New York: Doubleday.

Liebow, Elliot. 1967. *Tally's Corner.* Boston: Little, Brown.

Rubin, Lillian. 1977. *Worlds of Pain.* New York: Basic Books.

Stack, Carol B. 1975. *All Our Kin.* New York: Harper & Row.

Susser, Ida. 1982. *Norman Street.* New York: Oxford University Press.

Recording Form

Name and author of ethnography ______________________________

Existential issue and orientation to issue in the culture	Evidence
Relationship to time: *past, present, and future.* (Circle one and cite evidence to support your choice in the spaces provided.)	
Relationship to the physical world: *mastery, harmony, and subjugation.* (Circle one and cite evidence to support your choice in the spaces provided.)	

Relationship to activity: *being*, *becoming*, and *doing*. (Circle one and cite evidence to support your choice in the spaces provided.)

Relationship of people to one another: *hierarchy*, *individualism*, and *collectivity*. (Circle one and cite evidence to support your choice in the spaces provided.)

CONCEPT 8

Norms: Folkways and Mores

Definition

Norm A widely accepted rule for social behavior that specifies how to act in given situations.

One year, while I was at a convention in Minneapolis, I noticed the following interesting social behavior: I was waiting for the hotel elevator along with a group of five other delegates to the convention. They were laughing, talking, and generally having a good time about something they had just heard. On the wall a light went on and a bell rang, softly announcing the arrival of an elevator. Suddenly, all five members of the group shut up as if their power supplies had been cut off. Soberly, they stepped through the elevator doors and carefully distributed themselves evenly among the four people already inside. Then, with the smiles entirely gone from their faces, they all turned to the front of the elevator and looked up at the lighted floor numbers to follow their progress to the lobby. The rest of the passengers did the same. As they all stepped out and dispersed, the group of delegates formed again and began to talk and laugh.

Sound familiar? I can't think how many times I have followed the rules of elevator behavior along with everyone else. I know that the blinking floor numbers are not as absorbing as the attention we pay to them would suggest. It just is not that important whether we are on the sixth or fourth floor at any given moment. So why do we all seem to follow these rules, and how do we learn them?

Norms as Social Rules

The uniformity of behavior in elevators makes it seem as though we had all read an elevator behavior manual: (1) distribute yourselves evenly throughout the car no matter how many people enter; (2) face front; (3) no touching; (4) no talking or laughing; and (5) read the floor numbers as you ascend or descend. But there is no such written set of rules. You probably can't find a written rule about how close to a stranger you should stand when talking, or how long it is permissible to look into someone's eyes before the other person is made uncomfortable, or what exactly constitutes "too personal" a question. All these are examples of the thousands of social rules for behavior called **norms.** We feel very deeply a need for socially approved guidelines for behavior. We feel secure when we know that what we are about to do will be approved of by others, or at the very least will not draw undue attention. The very basis of our social order is the ability of people to satisfy one another's expectations about how social interaction will proceed. In order for such expectations and predictions to be fulfilled a high percentage of the time, there must be agreement among groups of people about what constitutes acceptable behavior. It is not necessary that there be total agreement (or consensus), especially when the norms focus on less important issues such as everyday styles of interaction. But the more agreement there is about a norm, the more force it will have.

Norms as Social Morals

Norms reflect the underlying **values** of a culture, those deeply held beliefs about what is good or bad, right or wrong. In American society we believe in values such as individual responsibility, progress, and competition. Our norms express how these values are to be carried out in social interaction. For example, our strong tendency to keep careful score at athletic contests is a norm that expresses the underlying value for competition. Our values are very abstract statements of our convictions. But norms are concrete statements by a group of people about how its members ought to act in specific situations. Since any statement about what "ought to be" is a form of moral statement, norms are also moral statements. It is useful to distinguish two kinds of moral statements, social and theological (or philosophical) morals. **Social morals** are statements that apply to us because members of the society agree that something ought to be done, said, felt, or thought. Depending on the level of group agreement about the rule and its importance to the survival of the group, the social moral can have greater or lesser influence on our behavior. **Theological or philosophical morals** are those that have force because they are thought to stem from eternal or absolute truths. Sometimes it is difficult to separate social and theological-philosophical norms in our minds, especially because so many norms belong to all three worlds. The prohibition against murder, for example, is a rule common to religions, philosophies, and societies. But to understand the nature of social norms, it is helpful to keep the distinctions in mind. This is especially true because social norms can change over time and differ greatly among cultures.

In American society, for example, we generally shake hands in greeting, whereas in Japan the norm calls for bowing. Each behavior accomplishes the same purpose well. It is not as important *how* we agree to act toward one another as it is that we come to some agreement and act accordingly.

Folkways and Mores

Folkways

Norms contribute in varying degrees to the stability of a social order. In everyday life we have norms for specific kinds of social interactions; they tell us what kind of language to use, what kind of clothes to wear, or what eating utensils to use at a meal. The American sociologist William Graham Sumner (1906) called norms at this level **folkways.**

Fashion is a good example of norms at the level of the folkway. With experience we come to understand what kind of clothing is appropriate to each social situation in which we find ourselves. Dressing for the occasion allows us to fit in with others. When we arrive in some social setting wearing the appropriate clothing, we signal to others that we know the rules of the game and agree to play by them. This signal goes beyond just the rules of dress. In business, for example, a lawyer or banker who wears the conservative clothing of the profession is signaling to customers and colleagues that he or she probably also shares the other views and practices common to the occupation. By contrast, a banker who wears nonconformist styles of clothing (perhaps jeans and a T-shirt) suggests strongly that all assumptions about what bankers think might be wrong in this case. Generally, people dress to belong, even in college classrooms.

Folkways can change over time, as clothing fashions illustrate. The fashion industry, in fact, depends on seasonal changes, shaped (some would say dictated) by designers and manufacturers. Like fashion, argot (street language) is changed intentionally by those who invent it (such as jazz musicians), because its unavailability to the rest of the population distinguishes them (for jazz musicians, it identifies them to one another, and expresses an important value of theirs, that they be unique, experimental, at the leading edge of their form of cultural expression). By the time sociology teachers or *Time* magazine knows the latest language, it is automatically no longer "hip," "bad," or "cool." Other folkways, such as table manners and business customs, change much more slowly, but they do change.

When everyone follows the requirements of folkways, we feel comfortable because our social interactions are predictable. But it is not terrible if some of these rules are broken, as long as the general pattern of everyday interaction remains stable. Violation of folkways is usually punished, but only by informal means, such as ridicule, laughter, or embarrassment. Erving Goffman (1971), perhaps the best known expert on the sociology of everyday life, suggested that an effective way to highlight how folkways work is to intentionally violate some of them and to observe what happens to the character of the social situation as a result. (The Application in this concept takes advantage of this technique.) In the classroom, for example, the normative amount of time for eye contact is easy to establish. As I lecture in a large class, I have to move my glance continually from one side of the room to the other, "sweeping," so to speak, without dwelling on any one student for more than about two seconds. I know this to be the case, because, if I break the rule by staring into the eyes of a student for as long as six to ten seconds, both the student and I get very uncomfortable. Everyday interactions are shaped by many such rules for accepted styles of behavior.

Mores

Stable folkways make everyday social interactions more predictable and more comfortable. **Mores** do more than make interaction satisfying, they make it possible in the first place. When a society comes to associate certain norms with the welfare of the group, it takes measures to ensure that those rules will be obeyed by its members. Violators of mores are strongly disapproved of by society because they are seen as threats to the social order. Many mores are formalized in law, such as prohibitions against assault, theft, rape, murder, and incest. These are **proscriptions**—mores that focus on *disapproved* behavior. **Prescriptive** mores—those that focus on *approved* behavior—include laws demanding that contracts be fulfilled, that parents provide support for their children, and that employers pay a minimum wage. (There are also proscriptive folkways, such as the unwritten rule against picking your teeth at a dinner party, and prescriptive folkways, such as the custom of thanking the host for a dinner, even if you hated it.) Because of the strong belief that violations of mores threaten the welfare of the group, mores constitute much of our written, formal law.

Whether they are prescriptive or proscriptive, mores must be enforced, or social interaction itself would be impossible. If, for example, you had no reason to believe that the terms of a contract would be fulfilled, why would you enter into any contract? Business would stop. Or if there were no laws against murder or theft, we would operate by the rule of all against all. No groups could form for any purpose, because no one could afford to trust anyone else. Mores are those rules that members of a society believe *must* be maintained if social interaction is to take place.

We must feel secure that our lives, property, and other rights will be respected. Only then can social interaction proceed.

Because mores are so important to the survival of society, their violation is considered more serious than the violation of folkways. We maintain systems of police, courts, and prisons (and to some extent mental institutions) to ensure the punishment or isolation of those who break mores. The penalties range in severity from fines and brief jail terms to life sentences in prison and execution. We also maintain legislatures and courts to interpret and make changes in our formal system of mores. Although mores change more slowly than folkways, they do change. Sometimes we increase or decrease our emphasis on already established mores, and sometimes new ones come into existence.

Norms exist along a line, or continuum, ranging from everyday customs (folkways) at one end to our most strictly enforced laws (mores) at the other. Between the two extremes are norms that can be difficult to categorize as folkway or more. For example, how would you classify traffic regulations? Although they are formally established in law and are formally punishable, they clearly do not reflect our most deeply held values or severely threaten the stability of the community. They are mores to the extent that they must be followed if we are to have a safe and trustworthy traffic control system. After all, you can get killed out there. But they are also folkways to the extent that there are rules of courtesy on the roads that may be formally established somewhere but in practice are left up to drivers to enforce. For example, at an intersection with a four-way stop, each car has a stop sign to obey, and the norms for who proceeds next actually operate at the level of folkways, even though there are official rules. In the suburbs near where I live we sort of take turns, with the earliest to have arrived at the intersection going across next. The same sort of rules apply to merging traffic, unless it is in the city or during rush hour, in which case individual aggressiveness takes over and drivers seem not to acknowledge one another's existence very much. (In Boston, I have noticed, the same traffic folkways do not seem to apply as in Chicago, for example.)

There is, then, no extremely clear dividing line between folkways and mores. In fact, norms can, as a result of the efforts of various interested groups, move from the status of folkways to that of mores and back again. Rules about the use of alcohol and marijuana are good examples. Between 1920 and 1933, alcohol manufacture, distribution, sale, and use were illegal in the United States. What had, in many circles, been considered a distasteful practice was made a formally illegal one. Drinking passed from a proscriptive folkway to a proscriptive more and, with the repeal of Prohibition, back to a folkway. The same happened with marijuana. The Marijuana Tax Act, passed in 1937, made this plant a controlled substance and formalized punishment for its possession or use by unauthorized people (Becker, 1963). From time to time there are signs that, like alcohol, marijuana may one day pass back into the world of proscriptive folkways.

Summary

Norms are widely accepted rules for social behavior that specify how to act in given situations. They reflect the underlying values of a culture, those deeply held beliefs about what is good and bad, right and wrong. The cultural value for competition, for example, is expressed in the norm for keeping score during a game. There are two levels of norms. Folkways are the rules for everyday behavior, such as the manners, dress, and language appropriate to a specific social setting. Violation of folkways makes people feel uneasy and is usually informally punished, for example, by ridicule or laughter. In contrast, mores are thought by the members of a group to be vital to its welfare and survival. The violation of mores is, therefore, formally and severely punished. Examples of mores include the prohibitions against assault and murder and the demand that contracts be honored. Violation of such mores incur penalties such as fines or jail terms. Because mores are part of the deeply rooted values of a culture, they tend to change very slowly. By contrast, folkways (such as fashions) can change quite quickly.

The Illustration for this concept focuses on the operation of folkways. It summarizes some of the work of the sociologist Erving Goffman, who identified many of the less obvious rules for social interaction. Most people are aware of their rules for language, dress, or manners in social interaction, but most do not realize that they draw a variety of boundaries around themselves. These form what Goffman called the *territories of the self.*

The Application for this concept is divided into two parts. In the first, you will highlight folkways by selectively violating some everyday rules for interaction and will then carefully observe the reactions of those whose expectations have been upset. In the second part of the Application, you will make some careful measurements of a few territories of the self.

Illustration

Erving Goffman, "Territories of the Self," in Relations in Public *(New York: Basic Books, 1971)*

By the time we are teenagers, we have become very sophisticated about the rules that regulate everyday social interaction, although we don't always realize that we have. I am amazed at how well we learn the complexities of language while we are still tiny. Once we become aware of the complexities of our social rules, our early acquisition of them also seems surprising. The key to both sets of skills is, of course, the constant training we get, from the time we are infants. Parents repeat the words that children need for normal living and drill their children in social rules as well. By imitation and repetition we learn things that even our parents didn't realize they were teaching.

We are all aware of so many folkways: table manners, the clothing appropriate to specific settings, "thank-yous" and "pleases," handshakes. But the rules for interaction are much more complex than these, and sociology tries to see beyond the obvious. Goffman managed to describe some of the most frequently used folkways for social interaction that, in spite of their pervasiveness, are all but unknown to us.

When you have just been introduced to someone, how far away do you stand while you are talking? Are your faces six inches apart, or twelve, or twenty-four? Do you recall a time when you spoke with someone who tended to stand too close to you? It made you a little uncomfortable, didn't it? Perhaps you can think of a time when a person with whom you were talking seemed to be leaning backward, trying to get farther from you than you thought appropriate. Goffman explains such behavior in terms of folkways by thinking of people as constantly protecting or making claim to "territories." The kinds of territories involved vary, as you will see from the following examples, but once you start to think about them, you will realize that you know what they are. We respect one another's claims and have subtle rules for avoiding conflict over territorial claims.

Personal Space

Personal space is the immediate area around a person. Its boundaries form an irregular ellipse, reaching close behind the person, extending a little farther away around the sides, and then stretching out in front for a distance that depends on where the person is or what he or she is doing. The best way to determine your own boundaries in any situation is by realizing that, whenever another person breaks the boundaries physically, it makes you feel a little uneasy, as if someone were trespassing on your territory. In a classroom, for example, you have personal space that surrounds your seat closely on both sides (about halfway to the nearest occupied seat on either side) and stretches in front to just behind the next row of seats. Collectively, the personal space of a class of students stops just ahead of the first row of seats. I know that when I am teaching, I have a sort of "prowl area" within which it is safe for me to walk without seeming threatening to the students. It reaches from the front chalkboard to a few feet from the front row of seats. To illustrate the reality of personal space, all I have to do is to walk among the rows of students. It invariably makes both them and me a little uneasy. I have invaded their personal space and ignored my own.

Personal space changes size and shape depending on the situation. In normal conversation with one other person, your personal space is much smaller than it is in the classroom. In American society we tend to stand closer together than Japanese but farther apart than Italians, for example. The point is that, although there are some differences among the personal spaces that individuals demand, there is a folkway that is common within any culture.

The Stall

The stall is territory to which an individual must make an all-or-none claim. This is mainly because the boundaries of the stall are very clearly defined (unlike those of personal space, which are usually invisible). Examples of stalls include chairs, telephone booths, beds, picnic blankets (when you planned on being alone), and your side of a tennis court (assuming you are not playing doubles). In none of these examples is it reasonable for a stranger to ask you to share the stall. Because the boundaries of such territories are so clear, any request by another person to share the space is an extremely direct social encroachment and not likely to be accepted.

Use Space

Use space is the territory to which you lay claim because you may soon have to use it. As you are walking, you make constant calculations about where you soon will be. On an empty field there are no problems. But between classes in a crowded hallway you will see people weaving from one side to another, speeding up and slowing down, and generally enacting a sort of subconscious ballet to avoid one another's use space. Seen from above, a single individual's use space would form a constantly expanding and contracting ellipse. When miscalculations occur and territories overlap, you can see the discomfort on people's faces. Remember a time when you thought a person coming toward you was going to pass on your right, but she or he intended otherwise? Suddenly you were face-to-face, and then you did that little dance of embarrassed misdirection (first to one side and then to the other) until you had to stop completely, acknowledge each other, and (usually smiling some) pass.

The most elaborate mixtures of use space take place in games such as basketball. People stop, start, speed up, slow down, change direction, and jump, not only without warning but specifically to surprise and confuse others. Extremely close spacing is allowed, but not real contact. It is one of the few places in which the violation of use space is actually penalized (by foul shots).

The Sheath

The sheath is the skin covering your body and the clothes next to it—what literally sheathes you. Because this is the last boundary protecting you, encroachments against it by a stranger will, to say the least, be taken very personally. That is why, in crowded elevators, though extremely close proximity is allowed temporarily, touching is avoided. People seem to be able to make themselves thinner to avoid actual contact in such situations.

The Turn

The turn is the order in which an individual gets to do what he or she has been waiting to do. We stand in line to establish formally when our turn is. The invasion of this sort of territory has a special name, *cutting in*. When you step in front of others in line, you simultaneously violate the turns of everyone behind you. You can get into a great deal of trouble, depending on where you are. In my experience you should not do this in New York City at any time.

Possessional Territory

Possessional territory is very closely related to the sheath, because it consists of objects normally carried close to the body, such as a hat, gloves, and a wallet or handbag. Besides being close to the body, these objects are usually valuable to the person (that's why they are carried closely). In addition, handbags and wallets almost always carry information that is very private, which brings us to another territory of the self.

The Informational Preserve

The informational preserve consists of information about an individual that she or he considers private, such as age, income, religious or political beliefs, immediate plans, address, and telephone number. Because such information is considered personal, interviewers usually save questions on these topics for last. The idea is that, by the time they are asked to reveal personal information, respondents may have developed some sort of rapport with the interviewer and feel less threatened by revealing their age or income.

The Conversational Preserve

The conversational preserve is bounded by a person's willingness to enter into conversation with someone else. I'm sure there have been times when someone you really did not want to talk to kept pressing a conversation on you. You may have finally invented some excuse that you had to go or even resorted to being downright rude (saying, "Leave me alone"). The conversational preserve is, of course, one of the first barriers to meeting strangers. If you can't get a person to talk to you, that's about it for the relationship.

Once you get the feeling of some of these less-than-obvious folkways, you can start to discover others. For example, in the Definition section I gave the example of eye contact with my students. Eye-contact norms in our culture enable us to make subtle calculations of a person's intentions from this information alone. Consider a student who is sitting in the college library reading. She looks up from the book and sweeps the room from side to side, not really stopping to look at anyone closely. She notices that one person, a man, was looking at her the instant she looked at him. Was he "sweeping" the room too? A quick glance back to check reveals that he is still looking at her. Looking back into the book, she decides to wait a while before looking up again, just to make it impossible for a third coincidence to happen. If he's looking at her when she finally looks up again, there can be no doubt. Sure enough, he still is. Now she has to make a decision. Does she want to ignore him by never looking up again or leaving, or is she interested in finding out what is going on? He probably thinks he knows her, or wants to, and the casual-strangers relationship is threatened.

Eye contact is a territory of the self that regulates our everyday social interactions. We can discover other territories merely by observing closely how our interactions proceed. Advertisers have been clever enough to exploit the eye-contact preserve. That's why there are advertisements on buses and subways up near the roof line. That is where we look to avoid meeting the glances of all those strangers around us. We become a captive audience for the advertisers. In the Application for this concept some of these territories will become more apparent, and you may discover others that have not been mentioned.

Application

This application consists of two separate studies of folkways. The first is qualitative (no numbers are attached to the data you will collect). The second attempts to make a systematic and objective measurement of folkways.

A Qualitative Experiment with Folkways

Folkways are not usually written down the way formalized laws are. We learn how to deal with other people from our parents ("It isn't nice to wipe your nose on your sleeve") and from others ("I didn't realize you were supposed to wear a tie to the dean's party"). By the time we are adults, we have become so used to the accepted way of doing things in social situations that we usually take folkways for granted. They become "second nature" to us.

To highlight how such folkways operate, it is sometimes useful to break the rules intentionally. Once a folkway is violated, its function in the regulation of everyday interaction can become clearer. For this part of the application, select a few folkways, and violate them in interaction on the campus. (Usually it is better to do this among students, because you can later explain that it was "just an assignment." This does not always get you out of trouble in the "real world.") I suggest that you focus on just one or two norms to break. For example, you might try striking up a conversation with a stranger in an elevator, where the rules for behavior prescribe that you stare ahead or up at the floor indicator lights and keep quiet. Or you might try standing just a few inches too close to a person with whom you are talking. Another possibility is to choose a seat right next to someone in a room with a large number of empty seats, any one of which would have increased the distance between the two of you. Perhaps you could wear inappropriate clothes or respond to the normally automatic question "How are you?" with a long explanation of how badly you have been feeling. The possibilities are endless. If you need some additional ideas, look back through the Illustration from Goffman's "Territories of the Self" to find some.

When you break these folkways, observe as carefully as you can the reactions of the people with whom you are dealing. Are there pauses while they try to figure out what is going on? Do they seem to ignore the violation in a way that suggests they are trying to believe that it never occurred, hoping that it might go away? If you persist, does the person show some displeasure? People feel uncomfortable when the rules they operate by are suddenly upset. Is there a difference between the reactions of men and women, people you know well and relative strangers, or retiring and assertive individuals?

Record your observations of each violation in as much detail as possible. You should find that, after a few trials, you become more observant and the consequences of rule breaking become clearer.

One note of warning: please do not use this exercise as a license simply to be rude. It can yield legitimate results if done with discretion and concern for others' feelings.

A Quantitative Measurement of Folkways

No matter how experienced you become at observing the effects of violating folkways, your understanding of the consequences will necessarily be subjective. The short descriptions of your observations may be insightful, but they will lack precision and will be difficult to compare with the observations of others. So here is a technique for measuring the consequences of norm breaking more rigorously.

First, ask a sample of ten students (it is fine if they are friends of yours) to answer a few questions about two norms. Explain that we normally have rules for how we are expected to interact. Then ask the following two questions:

1. In American society how far away from another person are you supposed to keep your face when you are talking to him or her (assuming you know the person fairly well)? ——— inches.
2. For how many seconds is it normally permissible to maintain eye contact with a total stranger? That is, how long would you expect eye contact to be tolerated before the person being looked at looks away? ——— seconds.

Once you have obtained the responses of ten people, calculate the average estimate of distance in inches for the ten responses and the average time in seconds. Now you are ready to test their guesses against reality.

In conversation with someone you know fairly well, begin the conversation at what feels like a normal distance. Then slowly move closer and closer, until you notice that the person is backing off. Do this with about five people, and, in each trial, estimate as closely as you can the number of inches between your faces when the backing off begins. Calculate the average distance, and compare it with the estimate of your sample.

To do the same sort of test with eye contact, select a seat in a library reading area or a dining room, and make eye contact with a total stranger, counting the number of seconds before she or he looks away. Do this with a few other strangers. Then calculate the average time in seconds, and compare it with the estimate of your sample. Two important warnings are needed here: First, you may be surprised at how quickly people will break eye contact. It may be hard to measure. Try counting at about three counts per second to start. Second, it makes a big difference in our culture whether the person making the eye contact and the target person is a man or a woman. Men making eye contact with women will have much different results from women making eye contact with men. Take this into account in your evaluations, and use some restraint in conducting your study.

Comparing the estimates made by your sample with your actual test results for conversational distance and duration of acceptable eye contact, do you find that the estimates were very accurate? Some norms for interaction are so second nature to us that we are unaware they even exist, much less know precisely how they operate. Did you find that the estimates of women differed from those of men? There are a number of double-standard folkways—rules that are different for men and women.

How did you feel about breaking the rules of everyday interaction? Some people seem to be much more uncomfortable about such violations than others. Some students invariably love the opportunity to step outside the conventions for a while. Many parties are expressions of this feeling, legitimate excuses for people to ignore some of the rules, at least temporarily. Most people feel free to "let loose" and "act themselves" at a party, but not in class or at work. From the sociological point of view, people are "acting themselves" in both types of settings; it's just that the norms differ in the two situations.

CONCEPT 9

Socialization and the Development of Adult Character

Definition

Socialization The process by which an individual internalizes the rules of the social order in which he or she is raised and by which the character of the society is maintained from one generation to the next.

A few years ago I was introduced to a woman who was so beautiful that I couldn't help telling her I thought so, even though it was quite an awkward thing to say right away. But I figured that she was used to hearing it. Her response shocked me. A look of real hurt and dismay came over her face, and she asked me not to make fun of her. When I got to know her better, I learned that she had been reared by a mother who was quite beautiful herself. But the mother had always criticized the daughter, calling her awkward and ugly. She had dressed her daughter in homely clothes and generally tried to hide her from the view of friends and strangers. My guess was that the mother had been terrified of losing her looks with age and resented the attention paid to her increasingly attractive daughter. To eliminate the competition and maintain her self-image, the mother fed her daughter the lie that she was ugly. And it stuck! Nothing I or others could say basically changed her self-image as an ugly person.

Perhaps you know of a similar instance, maybe of the reverse variety: a person who is unattractive but, having received nothing but praise from doting parents, sees himself or herself as a wonderful gift to humanity.

This concept focuses on **socialization,** the process by which we come to develop our adult characters through interaction with the social environment. Physically it may be true that "you are what you eat," but in terms of the development of your social self "you are whom you meet." Understanding this process helps explain how we come to see ourselves, even in the more unusual cases just mentioned, and how social order is developed and maintained.

The Origins of Human Character

In the behavioral sciences, the "nature-nurture debate" has focused on whether human social behavior is inborn (nature) or learned from the environment (nurture). Most sociologists agree that humans differ from other social animals in the way they interact with their environment and in the origins of those interactions. Animals such as ants and musk-oxen live in very complex social groupings, but their basic patterns of interaction with one another and the environment are inborn **instincts.** Their instructions for how to behave are passed on genetically and are not free to vary. A stimulus (such as a change in the weather) triggers an inborn, automatic, unvarying response (such as migration). By comparison, sociologists generally argue that patterns of human social behavior are learned rather than inborn and that we learn at such an abstract level that we normally engage in **meaningful behavior;** that is, we evaluate the meaning of a stimulus before reacting to it. Asked to "sit," we may ask "Where?" or "Why?" or even refuse to do so because we resent the implied command in the tone of voice. We evaluate stimuli according to the rules our parents teach us for social behavior.

The position people take in the nature-nurture controversy directly influences their beliefs about how adult character develops and about the origins of social order itself. Imagine the nature-nurture controversy on this continuum of positions:

	Heredity theory (sociobiology)		Freud		Mead	
Nature	1	2	3	4	5	Nurture

At the left are those (such as advocates of sociobiology) who argue that the social patterns of adult behavior are inborn in them (Wilson, 1975). Occupying the middle positions are those, such as Sigmund Freud ([1909] 1957, 1923, 1930), who argue that some of the factors establishing human social order and adult character are inborn, but that they are shaped by a person's experiences with the environment. The positions on the extreme right of the continuum are taken by those, such as G. H. Mead (1934), who believe that people are born with no instincts for reaction to the social environment and that everything about human adult character and social order is learned from previous generations.

To illustrate how proponents of each position in the nature-nurture debate explain the acquisition of human social character, I will discuss each of the three positions, beginning with the sociobiological view.

Sociobiology

Sociobiology is a relatively recent theory that argues that human social behavior is inborn. It contends that human social behavior has evolved, in the same way as behavior patterns of other animals, as adaptations that promote survival in the environment (Wilson, 1975). Sociobiologists like E. O. Wilson have called the biological mechanism for human survival the "selfish gene." That is, people's behavior toward one another is believed to be the result of genetic instructions the purpose of which is the perpetuation of self. Genes, then, direct human behavior so as to increase the likelihood of their own reproduction. This principle is well established in the behavior of animals; for example, animals migrate to follow plentiful food resources or compete for territory and mates. These ideas become controversial, however, when they are applied to human social life.

For example, the incest taboo (the prohibition of sexual intercourse between closely related persons) is present in all cultures. Sociobiologists argue that incest is universally taboo because such inbreeding produces offspring who are likely to be genetically inferior to their parents and, therefore, less capable of passing on genetic information efficiently. The incest taboo, then, is seen as a set of genetic instructions for human behavior that promotes survival of the "selfish gene."

Much criticism has been directed against sociobiology on the grounds that it inappropriately applies theories of nonhuman social behavior to humans and that it fails to account for the rich variety of human social behavior. The dominant view in sociology is that human social behaviors are learned, not inborn. Sociologists generally argue that humans are uniquely capable of attaching meaning to their own behavior and teaching those meanings to succeeding generations. For example, the prohibition of incest is a rule with different meanings in different cultures. In one culture the incest taboo is maintained to protect the sanctity of the family. In another, incest is thought to cause a particularly unpleasant disease, whereas in some closely related tribes the incest taboo helps define marriage categories clearly and so clarifies tribal boundaries (Lévi-Strauss, 1963). Thus, the sociological view of the incest taboo is that it is a learned set of cultural expectations, with different meanings in different cultures.

If, as most sociologists believe, human social behavior is learned rather than inborn, then there must be a process by which cultural meanings are passed from one generation to the next. It is the process called socialization.

Freud's View of Socialization

Freud (1856–1939) was an Austrian physician and researcher who is now famous as the father of psychoanalysis. In his practice of treating patients' complaints of emotional problems, he came to develop a theory that their problems could be traced to the struggle between the child's desire for immediate gratification of its desires and society's demands that every individual accommodate the wishes of others. Freud proposed that all humans are born with a reservoir of sexual and aggressive urges that he named the **id.** Society, represented by the parents, cannot allow the child to seek pleasure endlessly and without regard to the well-being of others, so it regulates the behavior of the child. Of course, at first the infant can eat when it likes and eliminate when it likes. But, as soon as possible, parents begin to schedule meals for the convenience of the family (which means that the child's id is somewhat frustrated in seeking gratification of hunger) and toilet-train the child, also for the convenience of the rest of the family, now tired of changing diapers.

No longer able to seek pleasure without restrictions, the child learns how to satisfy those inborn urges in ways that are acceptable to others. Through application of the "reality principle" (Freud's term for the ability of an individual to defer physical gratification for social approval), the individual balances the demands of the id against the restrictions of the environment (primarily represented by parents). The component of the personality that develops out of this process Freud called the **ego.**

Eventually, the child takes on the rules for behavior that the parents have been enforcing from their positions of power. The internalized set of ideas about what constitutes good and bad behavior Freud called the **superego.** It is what people generally think of as the conscience. It is as if the society were now whispering in the child's ear advice about how to act.

According to this theory of how people are socialized, the demands of society are endlessly in conflict with the inborn desire of individuals to seek pleasure for themselves. Only by socialization and the maintenance of external, societal pressures are the egocentric demands of individuals held in check. But it is an uneasy balance between the id's continuing demand for gratification and the superego's unattainable demand for complete compliance with society's rules. The ego mediates between these extremes, but it can never fully resolve the basic conflict. Freud predicted continual discontent for us.

Mead's Theory of Socialization

Mead (1863–1931) developed a theory of socialization that differs from Freud's in two important ways. First, it assumes no inborn social characteristics. Second, because no inborn drives are hypothesized, the demands of society are not seen as inevitably conflicting with any individual demands. For Mead we are born like blank pages on which society writes its rules for behavior and its beliefs, using the process of socialization. That process has several stages.

As infants we are incapable of experiencing the environment in any organized way. Images and sensations rush in on us at random and out of control, in somewhat the way adults experience the world just before waking. But infants can learn. What eventually begins to organize and make sense of things is that people, especially family members, act toward us in consistent ways. They respond to our cries, talk to us, feed and hold us, and generally treat us as valued objects. Through interaction with others we learn to associate certain gestures (such as smiling), behaviors (such as crying), and sounds (including both words and baby noises) with specific outcomes. We cry and we get attention. We smile and adults make a big fuss. By a series of such interactions with others, we learn how to communicate. Without such interactions, neither sounds nor behaviors would have consistent meanings attached to them. As children we develop a vocabulary for the names and uses of the objects in our environment, including the people.

The next important step, once we have learned to become aware of the objects in the environment, is to realize that we are also objects in the environment. The most common word we hear as babies is usually our own name, but that is rarely the first word we say. I suspect this is because, until a certain point, we are unaware that we exist. We can see Mom, sure, or dolly. But from our baby point of view, where is this Martha they keep talking about? We can't see who that is until we discover that we also exist, the way other things do. And that discovery will never happen if no one ever acts toward us.

Awareness of this step in the process tells a great deal about how the adult self develops. Charles Horton Cooley (1864–1929) concluded that we really see ourselves not as some isolated entity but as a reflection of how we think others see us. He called this the **looking-glass self** (Cooley, [1902] 1964). If everyone acted toward me as if I were the most intelligent and handsome man in the world, that information would form all of my looking-glass self—I would have no other raw material with which to build a sense of who I am. Think back to the examples at the beginning of this section, and you can see what I meant by the expression "you are whom you meet."

While learning our basic vocabulary we may only imitate the sounds and behaviors of others without knowing what they mean. Later we develop the ability to attach meanings to them. Children practice meaningful adult behavior in the stage of development that Mead called **play.** Watch children between the ages of two and four playing alone. Often they seem to be taking on the role of a family member or other important person in their lives (such people are often called significant others). They may be just imitating, or they may be accurately fulfilling the expectations that others have of a person in that role. A child taking on the role of father may be delivering punishment: "You should not have done that!" That is one thing fathers are expected to do, in the experience of the child. Also, the child is seeing herself from the point of view of another person in a specific situation. In another situation she might portray the role of father as a teacher to his daughter, not as a disciplinarian.

During the play stage, what Mead called **role taking** involves the assumption of the expectations of only one person at a time (in Mead's term, **particular others**). During the next stage, the **game,** children learn to take into account the expectations of a group of roles simultaneously (what Mead called the **generalized other**). Games with rules place the impersonal authority of the group above that of any individual. If we act selfishly (cheat or ignore the rules, for example), we are chastised for spoiling the game. If things are to proceed smoothly, the rules must be followed. It is important to realize that this is not just a matter of placing ourselves at the command of the will of the group. Rather, it is a matter of having actually internalized that group will. What others once expected of us we now expect of ourselves. The social expectations (such as cleanliness) are now inside us. We once may have washed because we were told to by Mother, but now we do because we prefer to feel clean. We have internalized that societal expectation along with countless others, ranging from everyday manners to prohibitions against murder. In Mead's terms we have developed a "**Me,**" the internalized set of societal expectations that produces conformity to agreed on rules.

The other facet of the self, which Mead called the "**I,**" is the spontaneous, creative, impulsive, individualistic component. Mead's "Me" is roughly comparable to Freud's superego, and Mead's "I" is roughly comparable to Freud's id. But Mead did not believe that the "I" was inborn in humans the way Freud believed the id was biologically innate. Mead felt the "I" was social in origin, a product of interactions with the environment.

Agencies of Socialization: Family and Peer Groups

It is clear in both Freud's and Mead's models that the family is an extremely important agent of society in the process of socialization. In our earliest years, our contacts are overwhelmingly with family members, and these people interpret the rules of society to us. Even after we are grown and move away from the homes in which we were reared, the personal authority of our parents very often persists. They were, after all, the powerful models through whom we came to understand how individuals are supposed to behave. The authority of the group is more impersonal, as Mead's model illustrates, and is generally experienced outside the home in playing games with peers. **Peer groups,** composed of children of roughly the same ages, provide us with our first taste of nonpersonal authority. Our parents accept us even if we sometimes break the rules, because they love us. But acceptance in the peer group is much more dependent on our performance. So both families and peer groups are important agencies of socialization, but they are responsible for different stages of the development of adult character.

Adult Socialization

Everything we know about the process of socialization suggests that what happens to us while we are young is disproportionately influential in the shaping of our adult personalities. Both models of socialization described in this concept, Freud's and Mead's, focus on early childhood. But it is also clear that in a complex, urban society like ours things change too fast for us to stay the same throughout our adult years. We must adjust to changing demands in our environment, changes not only in knowledge but also in the everyday practices and beliefs that make life orderly. Consequently, we must **resocialize** to stay in touch with the new demands made on us. As you will see in the Illustration for this concept, the kinds of personalities that a society produces tend to fit the rate of resocialization demanded by the style of life prevalent in it.

One version of the kind of resocialization common in the United States is called **anticipatory socialization.** As we go through life, we can anticipate that we will change roles from time to time. You may know beforehand that you will move away to college, that you are going to get married, that you are going to become a parent, that you are going to go to work, that your children are going to move away, and so on. Of course, not everyone does all these things, but they are typical role changes in our society. In each case you can anticipate that you will be expected to act differently as you go from one role to another. Married people, for example, are expected to plan socially around each other's activities, rather than around those of old boyfriends or girlfriends.

The process of anticipatory socialization makes the transition to the new role smoother. We get to know before actual occupancy of the new role how we are supposed to act, so that we avoid the embarrassment of acting inappropriately. Problems arise when the new role is so unpleasant that we do not wish to anticipate it. We may then enter the role with no particular expectations in place and feel adrift. This is what often happens when we find that we have become old (in the eyes of others) or widowed. As a consequence, it is all too common for older Americans to enter old age without adequate preparation for the new role demands. In the absence of internalized role expectations, they may experience what has been called the "roleless role" of old age (Burgess, 1960), in which the role of the older American is defined largely by what she or he is expected *not* to do. "Don't go to work, don't worry, don't run the family, don't try to be a leader," and so on. Who would wish to anticipate such a role change?

One other form of adult socialization can be called **remedial resocialization.** Unlike anticipatory socialization, remedial resocialization is not a process we undertake ourselves; it is brought to bear from outside us. The clearest example is what sometimes happens in prisons. Convicted criminals in American society are often believed to have committed their crimes because of some flaw in their beliefs or values. Perhaps they do not respect the property rights of others adequately or have failed to internalize the work ethic fully. In theory, then, the American penal system tries to "correct" the flaws in the individual by *re*socializing him or her to the approved beliefs and values of the society. Society's ability to rehabilitate individuals in this way has been *extremely* limited. Prisoners who have undergone the experience report either that they simply don't believe that the values and beliefs taught in prison counseling will work to their benefit on the outside or that it is impossible to learn decent values in an absolutely indecent environment.

Summary

Socialization is the process by which an individual internalizes the rules of the social order in which he or she is reared and by which the character of the society is maintained from one generation to the next. There are a number of theories in the behavioral sciences about how this process occurs. They differ in the extent to which they hold that human behavior is inborn or learned through experiences with the environment. Extreme hereditarians believe that humans, like other animals, are born with instructions for behavior that are passed on genetically. By contrast, Sigmund Freud proposed that humans are born with some traits that influence, but do not wholly determine, social behavior. Freud termed the inborn reservoir of sexual and aggressive desires the *id* and the part of the self that acknowledges the demands of the adult social environment the *superego.* (The superego often is referred to as the *conscience*). The part of the self that mediates between the id's demands for immediate gratification and the limitations placed on the individual by the superego is termed the *ego.*

Most sociologists, however, believe that, unlike other animals, humans are born with no instincts or drives that influence their social behavior. This position is probably best represented by the work of the symbolic interactionist G. H. Mead. According to Mead, humans learn all the rules for social behavior through their experiences with other people. Tracing the process of socialization through the reactions of the child to the adult world, Mead argued that the rules for social behavior are represented by the "generalized other," a learned set of expectations people in a society have about one another's behavior.

Socialization is accomplished in a variety of settings. Although the most profound influences take place in childhood, it is a lifelong process. The primary childhood agencies of socialization are the family and the peer group (people who share important characteristics such as age or work). Later in life, adult socialization is made necessary by predictable changes in status (such as getting married or growing old) or by the fact that society changes and its members must take on the new social expectations of their environment. When such changes can be predicted, new expectations for behavior can be developed in advance by the process of anticipatory socialization.

The Illustration for this concept examines the way early schooling socializes children to some of the demands of society. According to the observations of Harry Gracey, kindergarten can be as demanding as a boot camp in teaching the lessons of regimentation and obedience to authority.

The Application for this concept is a survey designed to determine whether younger and older people differ in their attitudes toward sex roles. The debate about sex roles has focused on the behavior expected of American women. What has happened to the sex-role beliefs of people who were socialized before this debate gained public prominence? What are the beliefs of younger people? Are they being socialized to the newer role definitions?

Illustration

Harry L. Gracey, "Learning the Student Role: Kindergarten as Academic Boot Camp," in Readings in Introductory Sociology, *ed. Dennis H. Wrong and Harry L. Gracey (New York: Macmillan, 1968)*

We usually think of school as a place where the knowledge of a society is passed on from one generation to the next—how to add, how to write, how to dissect a worm. But much more is passed along in school. It is also a place where the values of the culture are taught to the young—what to believe in, what is good or bad to do, think, or feel. Whereas primary institutions, such as the economy and the political structure, create the values and beliefs of the culture, secondary institutions, such as the schools, help pass them on by socializing the young.

In observing the activities in a kindergarten class, Gracey chose a level of schooling at which the socializing process was most evident. Kindergarten is acknowledged by educators to be a year devoted to the preparation of five- and six-year-olds for the social and emotional demands of later, more subject-specific years of school. Kindergarten is almost solely aimed at teaching children that, in order to get along with the teacher and with one another, they must follow rules. It is the year of learning social meanings constructed by adults, and, as Gracey found, the method of teaching is much like that of the military boot camp. Kindergarten is one of society's basic training grounds, a little model of adult society.

In just one afternoon of observing a kindergarten class, Gracey recognized that the key to all the interaction was *regimentation*. The day was highly organized in three senses: (1) in physical space—*where* things occurred, (2) in time—*when* and *for how long* events occurred, and (3) in forms of behavior—*what* was allowed to occur.

Physical Regimentation in Kindergarten

On entering the classroom, Gracey immediately noticed that areas had been set apart for specific uses. There was a book corner; a teacher's area (with desk, files, and piano); a cleanup area around the sink; an open space in the center of the room for more active group play; a play store (featuring shelves, cash register, and things to buy and sell); a play kitchen with appropriate domestic items; a series of shelves on which toys, art supplies, and various other materials were neatly stored; a sandbox; and rows of hooks on which coats, hats, and sweaters were to be hung. There seemed to be a place for every kind of object and for every kind of behavior. Activities took place in the assigned areas, and objects were returned to specified places after use. The lesson underlying this physical arrangement was orderliness, not only in the "nuts and bolts" management of the day's activities but also in the way things generally ought to work.

Time Regimentation in Kindergarten

The day was divided into carefully organized periods, with a school bell marking the beginning and the end of the session. First, attendance was checked. The teacher then oriented the children in "real world" as opposed to class time. She quizzed them on the month, date, and day of the week for that afternoon. Then, job assignments were made, such as sweeping or wiping chalkboards. Next was 'serious time," which included a reading from the Bible, the pledge to the flag, and the singing of a song (which presumably also had to be serious), in this case, "America." "Show and tell" followed, during which children took turns talking about the objects they had brought to school. These included a toy helicopter, a bird's nest, and a toy gun. Next came a question-and-answer session run by the teacher, who asked individual children about a trip they had taken to the zoo. A session of directed play followed, during which children were called on to imitate the animals being sung about on a record the teacher selected. A five-minute exercise session then exhausted the teacher and delighted the children. The next period, approximately thirty minutes of "work time," included everything from painting pictures to cleaning the chalkboards. Cleanup from work time was accompanied, apparently automatically, by the singing of a song that went, "Clean up, clean up. Everybody clean up." (They must have heard it before somewhere.) Then came my old favorite, milk-and-cookies time, followed, of course, by rest time. Just before going home there was a final cleanup (with just two students assigned) and then a sing-along until the final bell rang and everyone lined up to leave.

The class had a time for every activity and every activity in its time. The emphasis on the orderliness of time was clear in the persistent use of the word *time* to name events. "Work time" and "serious time" not only describe the activities but also suggest that they are supposed to take place at specified times. By extension, this principle seemed to apply to the entire kindergarten day. The ringing bells and time-marking whistles that persist through high school (and in some cases into the workplace) repeat the lesson of time regimentation.

Behavior Regimentation in Kindergarten

Beyond the places and times for activities, the kinds of behavior allowed in the class were highly regulated. The interesting thing was that the most important behavior the teacher sought from the children was not concern for friends or interest in learning or respect for property, although all these did occur during the day, but rather *obedience.* The most approved behavior was the behavior the teacher required, even if it was totally lacking in meaning to the children. It was as if the teacher's guiding principle was "Do things where I say, when and for as long as I say, and the way I want them done."

Gracey reported that in several instances children asked why they were doing something, but the teacher consistently ignored such questions. The teacher also ignored comments by the children that were of no apparent interest to her or were too imaginative for her to deal with (such as the claim by one little girl that a bird nest had been made by a "rain bird"). At one point the teacher told a boy to "treat the flag carefully" and quizzed him about why the flag deserved such treatment. "Because it's our flag," he replied. "That's right," the teacher responded. It is easy to imagine the previous class lessons on respect for the flag, which had drilled into his brain the automatic (and certainly meaningless) response he gave.

In the kindergarten, a great deal of "what," "when," and "where," were emphasized at the almost total expense of "why." The overall lesson being taught, then, was obedience to authority. Gracey speculated that, from the point of view of adult culture, it is most important that children be socialized to submit willingly to the demands of authority and to often meaningless routine. After all, in large corporations and on manufacturing assembly lines, the ability to follow orders is critical to smooth operation of the business. The same ability is needed to complete tax forms, satisfy licensing procedures, and conform to all the other organizational demands of our large, complex culture. Certainly any of these behaviors requires the learning of specific skills, such as reading and writing, and of bodies of knowledge such as accounting or engineering. But the first lesson, at least the one we learn first in school, is *willingness to follow* the routines we are given to perform.

Application

Among the ideas we learn through socialization as children are ideas about how males and females are expected to behave. By the age of three, American children can recognize activities deemed "correct" for each sex: girls play with dolls, boys play with trucks; boys become doctors, girls become nurses; and so on. The influence of early socialization is very strong. But, as we saw in the Definition section, things can change. A debate about gender roles has been going strong for at least thirty years, and it promises to continue. What influence is it having? This application is designed to test the extent to which people have been influenced in their beliefs about gender roles. Specifically, you will conduct a survey to test whether older and younger people differ in their attitudes toward gender roles, reflecting new styles of socialization about this issue.

Measuring Ideas About Gender Roles

To test this question, you will need to measure people's ideas about American gender roles. Before the 1960s, the traditional role of the American woman was to be "passive, submissive, gentle, dependent, family oriented, emotional, sentimental, idealistic and intuitive" (Eitzen, 1980:154). I have chosen seven such female gender-role characteristics and supplied two opposite terms or descriptions for each characteristic, representing polar extremes along a continuum. Between these extremes I've left five spaces to indicate various levels of the characteristic. For example:

Independent ____:____:____:____:____ Dependent
1 2 3 4 5

A respondent can be asked to evaluate a number of objects, situations, or people using an item such as this (called a *semantic differential format*). For example, you could ask people to evaluate themselves on the scale. If they think of themselves as very independent, they would check space 1; if very dependent, they would check space 5. Space 2 represents more independent than dependent, space 3 represents a neutral position, and space 4 represents more dependent than independent.

A series of such semantic differential items can be used to measure respondents' agreement or disagreement with the traditional role of women. I have constructed a Survey Sheet of seven such items for you to use. Then, the responses of older and younger samples can be compared to see whether they differ. To make the comparison of samples, collect responses to the survey from ten students below the age of twenty-five and ten students older than forty. (If you can't find enough students over forty, you will have to interview staff members or other people in the community who are over forty. It is best to keep samples completely comparable—such as students with students—but it is not always possible.) If any of your respondents turns out to be between twenty-five and forty, ignore that survey sheet and find an additional respondent in the right age group. In selecting the samples, *men should interview only male respondents, and women should interview only female ones.* That way responses can be compared between both age groups and sex groups. In addition, you can avoid any hesitance respondents may have about discussing this issue with an opposite-sex interviewer.

The questionnaire is designed to be handed to the respondents while you read the instructions to them. This is because questions about their attitudes toward gender roles and about their age are somewhat sensitive to many people. If you allow them to place a check in a few spaces without being watched, fold their response sheet, and hand it to you, the likelihood that they will cooperate and be honest in their responses is increased.

Here are the introduction and instructions you should read to respondents (practice reading them beforehand until they sound fairly conversational):

Survey Sheet

In general, I think American women ought to:

	____ :	____ :	____ :	____ :	____	
1. Be aggressive	1	2	3	4	5	Be passive
2. Be independent	1	2	3	4	5	Be dependent
3. Rely on emotion and instinct for judgments	1	2	3	4	5	Rely on intellect and reason for judgments
4. Be willing to take risks or experiment	1	2	3	4	5	Not be willing to take risks or experiment

5. Mainly desire to be attractive	____	____	____	____	____	Mainly desire to be competent
	1	2	3	4	5	
6. Be tough	____	____	____	____	____	Be gentle
	1	2	3	4	5	
7. Be cooperative	____	____	____	____	____	Be competitive
	1	2	3	4	5	

8. I am:

________ less than 25 years of age
________ between 25 and 40 years of age
________ over 40 years of age

Introduction: I'm doing a survey of attitudes for a sociology class. It's about students' ideas about the role of women in the United States. Would you mind answering a few questions? We don't want anyone's name; it's completely anonymous and will take only a few minutes.

Instructions: I am going to give you a set of statements about the role of women in society. In the first seven questions there are spaces you can check to show how *you* generally expect American women to behave compared with the behavior of men. For example, in the first question do you generally expect American women to be very aggressive compared with the behavior you expect of men? If yes, then you should check the space numbered 1. If you expect them to be very passive, then check space 5. Space 2 indicates more aggressive than passive, space 4 indicates more passive than aggressive, and space 3 is right if you expect American women to be neither more aggressive nor more passive than American men. Just look at each of the first seven questions and, thinking about how you expect American women to behave in comparison with American men, check the space that best represents how you feel. OK. If that's clear, here are the statements. (Hand a copy of the survey sheet provided to each respondent. When respondents are done, have them fold the answer sheet and hand it to you. Put it in an envelope or folder among other papers, where it becomes anonymous.)

Analysis

To analyze the responses of the twenty people in your sample you will have to calculate a score for each respondent that we can call the *female role traditionalism* score. If you look at the first seven questions, you will notice that the most traditional characteristics for the female role were not all on the same side of the page. Some were put on the left (items 3, 5, and 7) and some on the right (items 1, 2, 4, and 6). Researchers usually mix the pattern of such items to keep a respondent from automatically checking all the spaces on one side of the page or being able to agree with every question without having to pay attention. So, to calculate a total scale score, the end of each item that represents traditional female role characteristics will have to be assigned the *higher* score value. Then the total scale will come out with higher scores indicating greater traditionalism about the female role. Items 1, 2, 4, and 6 can remain as they are, since the more traditional response category *already has* the higher score value in those items. But items 3, 5, and 7 need to be *reversed,* so that the most traditional response will be score-valued at 5, the next most traditional at 4, and so on. Therefore, on items 3, 5 and 7, replace the numbers under the spaces as follows:

_____ :	_____ :	_____ :	_____ :	_____
~~1~~	~~2~~	~~3~~	~~4~~	~~5~~
5	4	3	2	1

You then use the new numbers in figuring out the scale score. To calculate the female role traditionalism score for each respondent, add up the values of the spaces checked on the first seven questions. Because the lowest value for any item is 1 and the highest is 5, the lowest scale score possible is 7 (seven questions with a score of 1 on each), and the highest score possible is 35 (seven questions with a score of 5 on each). Figure 9-1 is an example of the scoring of one sample scale. Calculate the female role traditionalism score for each of your twenty respondents (the score in Figure 9-1 is 27) and transfer the scores to the appropriate spaces on the Data Analysis Sheet provided. (The score of 27 for the sample respondent would go in the column for older respondents, because this person checked "over 40 years of age.")

After you have recorded all twenty scores (ten from younger respondents, and ten from older ones), add up the scores in each column, and enter the total at the bottom of each column. Then divide each total by ten to get the average scale score for the younger sample and the average for the older sample.

Interpreting the Data

What did the comparison of the sample averages show? Were the older respondents more in agreement with the traditional female role expectations? If so, it is possible that they were expressing the kind of socialization that they received when young. Younger students are likely to have been subjected to a different kind of socialization on campus. There is a great deal more talk about female consciousness and equality of opportunity at colleges and universities than in the rest of the society. Do members of your class find that the scores of male students differ from those of female students?

Here is an interesting twist on this study that has shown up in my students' work. Older women who have returned to school after their children are grown are consistently more dissatisfied with the traditional female role than are younger female students of traditional college age. The older women seem to have more reason to express feminist opinions.

FIGURE 9-1

Sample survey sheet and scoring

In general, I think American women ought to:

1. Be aggressive	1	2	3	(X) 4	5	Be passive
2. Be independent	1	2	3	4	(X) 5	Be dependent
3. Rely on emotion and instinct for judgments	~~1~~ 5	~~2~~ 4	(X) ~~3~~ 3	~~4~~ 2	~~5~~ 1	Rely on intellect and reason for judgments
4. Be willing to take risks or experiment	1	2	3	(X) 4	5	Not be willing to take risks or experiment
5. Mainly desire to be attractive	(X) ~~1~~ 5	~~2~~ 4	~~3~~ 3	~~4~~ 2	~~5~~ 1	Mainly desire to be competent
6. Be tough	1	2	(X) 3	4	5	Be gentle
7. Be cooperative	~~1~~ 5	~~2~~ 4	(X) ~~3~~ 3	~~4~~ 2	~~5~~ 1	Be competitive

8. I am:

_____ less than 25 years of age
_____ between 25 and 40 years of age
__✓__ over 40 years of age

Score
4
5
3
4
5
3
3
Total=27

They have experienced some disappointments in trying to fulfill the traditional female role and want some relief from it. Now they are attempting to enter the world of work for pay. By contrast, younger women understand feminism in a more abstract way. Although they generally disapprove of the traditional female role, their sentiments are less strong and less willingly expressed. Older male students, on the other hand, tend to agree with the traditional female role more than their younger male counterparts. Both male groups, however, tend to agree with the traditional female role more than either group of females.

Data Analysis Sheet Survey of gender-role attitudes

Which did you interview?

________ All male respondents

________ All female respondents

Female Role Traditionalism Scores

Younger respondents	Scores	Older respondents	Scores
1	________	11	________
2	________	12	________
3	________	13	________
4	________	14	________
5	________	15	________
6	________	16	________
7	________	17	________
8	________	18	________
9	________	19	________
10	________	20	________
Total	________	Total	________
Average scale score for younger respondents (total divided by 10)	________	Average scale score for older respondents (total divided by 10)	________

PART THREE

Basic Social Forces

Sociologists are interested in discovering and understanding the underlying social forces that influence human behavior. Just as gravity is a force that cannot be observed directly (its existence and character must be inferred from the behavior of the objects on which it acts), so social forces must be inferred from their influence on the behavior of people. Part Three deals with some of these basic social forces.

Concept 10 looks at the ability of groups to stay together in spite of obstacles to their cohesion (or solidarity). This concept focuses on two basic sources of solidarity identified by the French sociologist Émile Durkheim: mechanical solidarity, which derives from the feelings of similarity among group members, and organic solidarity, which derives from the interdependence created by their differences.

Concept 11 turns to power, the ability to control or influence the behavior of others, even against their will. The benefits of social membership require people to relinquish some individual will to the group and acquiesce to the exercise of that power in the legitimated form called *authority*. This concept treats some of the types of authority and concludes with some theories about how power is distributed in American society.

Concept 12 concerns ideologies, the sets of ideas that attempt to explain how the world operates and that are used to justify a group's pursuit of its own interests. This discussion of ideology should be useful in making sense of terms such as *conservative, liberal, radical,* and *reactionary,* which are often used in everyday conversation but with vague meanings.

CONCEPT 10

Solidarity: Mechanical and Organic

Definition

Mechanical Solidarity Durkheim's term for the type of group cohesion that results from members' having similar values and ways of life.

Organic Solidarity Durkheim's term for the type of group cohesion that results from members' having complementary, or interdependent roles.

One winter I taught an eight o'clock class, and I remember one really rotten, snowy morning particularly well. Because I had to leave home before seven to get to class on time, I could not phone anyone at the college to find out if classes had been canceled for the day, so I drove in. As I walked to the classroom, I wondered whether many students had decided to attend. Of the fifty students enrolled, about half were there. Once the complaints about the weather had been expressed, we began to discuss why they had decided to come to class. Someone turned the question around and asked why I had bothered. After all, any of us could have stayed home with good reason, and with no penalty. The ideas we raised in the discussion had a great deal to do with the origins of group **cohesion,** or **solidarity.**

The Nature of Cohesion, or Solidarity

Any college class has some cohesion, or solidarity (I use these terms interchangeably). This is evident in the fact that the class meets at predictable times, meets in a predictable place, and has a stable pattern of interaction—lectures, exams, questions to the teacher, and so on. It is difficult to imagine a situation involving a number of people in which there is no cohesion at all. But interaction can occur largely by chance and then never occur again among people, such as those crossing at a traffic light. That temporary collection of individuals has almost total lack of cohesion. The people disband spontaneously. Though there is nothing preventing them from meeting again, neither is there anything causing them to do so. One measure of the cohesion of a group of people is the extent to which they are willing to overcome obstacles in order to continue to meet in their established pattern. For example, the fact that our class met, even though the weather was bad, showed it had some cohesion. A group that meets with 100 percent attendance in spite of a bomb threat has more cohesion yet. But measuring the amount of group solidarity this way tells nothing about where the solidarity comes from.

Two Sources of Solidarity

Just before the end of the nineteenth century, the French sociologist Émile Durkheim identified two primary sources of solidarity (Durkheim, [1893] 1960). The first, **mechanical solidarity,** he thought was characteristic of feudal, aristocratic societies. The second, **organic solidarity,** he applied to the forms of social order that developed out of the Industrial Revolution.

Mechanical solidarity results from the feelings of belonging that people experience when they see themselves as similar to everyone else around them. Durkheim recognized that in feudal society there was very little differentiation in the way jobs were done (little division of labor, to use his terms). In addition, there was little variation in the values held by members of a community or in the way they lived their daily lives. To get a feeling for how mechanical solidarity operates, just think of how comforting it can be to be among people who are just like you, all doing something together. When this is the permanent condition, as it was for members of feudal communities, it becomes the main source of identification and solidarity for the members.

By contrast, organic solidarity results from complementarity rather than similarity. Durkheim saw that societies were becoming increasingly complex, with highly specialized divisions of labor and members who emphasized individualism rather than similarity to others. He argued that the cohesion of such a social order would depend not on the similarity of the members but rather on their special differences from one another. When a task is divided into small, interdependent jobs, its completion requires that every person play his or her role reliably. Cohesion in this situation results from the fact that people *need* one another's specialized behaviors for the tasks of the society to be accomplished. As a model for how organic solidarity works, think of any assembly line, whose smooth operation depends on the completion of all the different jobs along its length. Or consider the college or university you attend: Its organization may include a board of trustees, a chief executive (president, chancellor, or provost), various deans and/or vice presidents, associate deans, and administrators at all levels. Then there is the faculty, with its various ranks, students at each level of study, personnel who maintain the physical institution, and so on. Without these subgroups of people, no complex institution could deliver the service on which its existence depends. They are interdependent.

Combining Mechanical and Organic Solidarity

The shift away from mechanical and to organic sources of solidarity in Western societies was quite dramatic. The technical and social revolutions that brought these changes about were quite dramatic themselves. Yet the change was not total. Prerevolutionary social order did have some division of labor and role complementarity, so some small amount of solidarity must have been organic in origin. And postrevolutionary societies still do have some shared values and styles of life that contribute to the feelings of belonging in the communities in which they occur. There is, then, some mechanical solidarity even in modern life. The fact that both mechanical and organic solidarity can exist in one situation is illustrated by the example of classroom solidarity with which this concept began.

Mechanical Sources of Class Cohesion What did the students feel in common with one another and with me as their teacher that made us all feel we ought to show up in class (even in the snow)? In other words, were there any mechanical sources of class cohesion? We agreed that we believed in the value of education and in the idea that once you start something, you ought to see it through. On further examination of these beliefs it was clear that we all meant the same thing by commitment to a project. Once a student had enrolled in the course, he or she wanted to finish it. Once I had contracted to teach a course, I meant to do the job. But we meant different things by the value of education. Some people meant that education makes a person smarter or improved in some way. (That's what I had in mind.) Others meant that education can increase the amount of money a person can earn or the quality of job she or he can get. But even though we understood the idea somewhat differently, our belief in education contributed to the decision by all of us to go to class. The result was mechanical solidarity in the class.

Organic Sources of Class Cohesion Was there any organic solidarity in the class? What complementary roles had to be performed if the task of a college class was to be accomplished? There are only two such roles, student and teacher. Although it is not a very complex division of labor, unless both sides participate, no class is held. I have a very strong sense of responsibility toward the class, because if I don't show up, quite a few people have wasted their time, and the job I'm being paid to do is not getting done. Students feel much the same way about attending, but because some students can fail to show up without causing the class to be canceled, their feeling of responsibility to attend is less strong.

Notice that in the class, as in other segments of society (such as work and home), systems of punishment compel people to fulfill their roles. Workers who don't show up are not paid or may even be fired, family members who don't do their chores around the house can be reprimanded, and students who don't come to class can be given lower grades. (Teachers who don't come to class can be fired also.) So the organic sources of class solidarity are as clear as the mechanical sources. We go to class because we need one another in order for the job to get done, and we have ways to punish those who fail to do their part of the job properly.

How Much Mechanical, How Much Organic?

It is a matter of speculation exactly how much of the solidarity of any group is due to mechanical sources and how much to organic sources. In the classroom, for example, I would guess that there is a roughly even split between them. It is interesting to make guesses like this for the many groups with which we are familiar. Is a given family one in which values, beliefs, and ways of living are strongly shared and similar among the members? Or is the performance of everyone's special family job heavily emphasized? How about in an office? Is there an emphasis on everyone being "in the same boat" or "shooting for the same goal"? Or is the emphasis on individual competence and individual responsibility? Over the years I have sensed great differences in the kinds of solidarity that operate on various college and university campuses. Those larger, city schools I have attended or visited seem to depend on organic sources of solidarity. People are working to provide and to get degrees, to earn livings and to generate new knowledge, rather the way workers in a large business interact to make products and earn livings. They share a belief in the value of what they are doing (mechanical solidarity), but the core of their cohesion seems to be organic. By contrast, I have also seen many colleges (usually the smaller and less urban ones) in which the shared (mechanical) sense of community is powerful. It can be seen clearly at times of collective experience, such as sports events and graduations. For these people the interdependence of campus roles is important, but the mechanical source of cohesion is much greater than in the larger, more urban schools.

A Note on Mechanical and Organic Solidarity

Sometimes people get the terms *mechanical* and *organic* reversed in discussing solidarity. I think this is because *mechanical* reminds people of machines, so it is mistakenly associated with postrevolutionary social order in which machine power dominates. At the same time, *organic* is sometimes mistakenly associated with prerevolutionary social orders because the term reminds people of animals (organisms), which supplied the power before the Industrial Revolution.

Here is how to keep them straight. Mechanical solidarity suggests the repetitive sameness of machines, and sameness is the key to mechanical solidarity. Organic solidarity suggests the interdependence of specialized parts seen in complex organisms, such as humans. The interdependence of heart, lungs, skeleton, brain, and so on create the overall, organic system. So, mechanical solidarity is prerevolutionary—the sameness of repetitive machinery. Organic solidarity is postrevolutionary—the interdependence of parts in complex higher organisms.

The terms *mechanical solidarity* and *organic solidarity* are part of a tradition of dichotomies, categories of opposites that represent the sides of an intellectual debate. A number of social theorists who analyzed the revolutionary changes in Western societies during the last century used dichotomies such as this one to clarify the differences between pre- and postrevolutionary social orders. They are discussed fully in Concept 19, on social change, and mechanical and organic solidarity are mentioned there, as well.

Summary

Solidarity (or cohesion) is the ability of a group to hold together in spite of obstacles. The French sociologist Émile Durkheim distinguished between two sources of solidarity, which he called *mechanical* and *organic.* Mechanical solidarity derives from the feelings of belonging that people experience when they see that they are like others in their group. The key to mechanical solidarity is similarity. It is the type of cohesion that Durkheim believed dominated in feudal, aristocratic Western societies before the Industrial Revolution and the social changes that accompanied it. By contrast, organic solidarity derives from the interdependence of group members who do different jobs or play different social roles, all of which must be accomplished if a shared group objective is to be achieved. The key to organic solidarity, then, is complementarity rather than similarity. Durkheim believed organic solidarity was to dominate in post–Industrial Revolution societies, with their high degree of division of labor and emphasis on individualism.

Although mechanical and organic solidarity are generally thought of as opposite sources of cohesion, groups can benefit from both at the same time. A group may, for example, share beliefs and have members who have similar characteristics (sources of mechanical solidarity) and at the same time use a complex division of labor to achieve the group's goals (a source of organic solidarity). In addition, independent of its source, the level of group solidarity can vary. Some groups may disband when faced with only slight obstacles, whereas others may go to great lengths to continue to exist.

The Illustration for this concept summarizes a classic work on solidarity in which Lewis Coser (working from the ideas of Georg Simmel) proposes that conflict can be an important source of group cohesion. Coser suggests that conflict may clarify the beliefs of a group, highlight the boundaries around it, act as a safety valve to release pressures within it, mobilize the energies of members to confront a common enemy, and even create new alliances to form new groups.

The Application for this concept is a relatively unstructured survey in which you will ask people what they think are the primary reasons for their memberships in various groups. You can then analyze their responses to discover the extent to which people attribute social cohesion to mechanical or organic sources.

Illustration

Lewis Coser, The Functions of Social Conflict *(New York: Free Press, 1956)*

Groups range in size and complexity from the dyad (with just two members) to the society (the largest group, with many subgroups within it). Discussions of the ability of groups to hold together—their stability or solidarity—tend to focus on the extent to which members share goals and rules for behavior. Thus, a study of American society might emphasize the values of progress, science, and individual achievement that Americans have in common. We are taught such values by our parents and other members of the society, and it makes sense that having values in common makes us feel a common bond. But there is another, less obvious source of solidarity that is often overlooked: conflict.

Frequently, conflict is treated as if it caused nothing but disruption to social order. In this sense, conflict means trouble—trouble between wife and husband, trouble between neighbors, trouble when it erupts in the streets. Viewed this way, conflict is something to be avoided or contained to control its harmful effects.

An alternative view of conflict was first outlined in detail near the beginning of the twentieth century by the German sociologist Georg Simmel (1858–1918). Some years later the American sociologist Lewis Coser provided a valuable service by pulling together and reformulating

Simmel's work in his book *The Functions of Social Conflict* (1956). (Note: Although most of the ideas discussed in this reading originated with Simmel, I have summarized them from Coser's book and thus refer to them as Coser's.)

To begin with, Coser defined social conflict as "a struggle over values and claims to scarce status, power and resources in which the aims of the opponents are to neutralize, injure or eliminate their rivals" (1956: 8). He then proposed that social conflict can actually increase group solidarity at many levels.

Conflict Within Groups

Conflict Clarifies Norms

When conflicts occur within a group, its members tend to refer to the norms on which their association is based. For example, if a wife and husband argue about how money is to be spent in the family, they are likely to support their different positions by asserting that "we never wanted to own too many things" or that "we have to keep the house up to neighborhood standards." Statements such as these make family norms subject to reexamination. If one party responds, "We never said such a thing," negotiation about the continuing validity of the norm is likely. In some cases, the norms in question are reaffirmed as still appropriate, but in other cases, conflict leads to the modification of existing norms or the creation of new ones. As a result of a conflict, a family might decide to become more or less concerned with money. Conflict contributes to the maintenance or enhancement of group solidarity by forcing groups to focus on the question "What do we actually believe?"

Conflict Acts as a Safety Valve

When there are tensions within a group, conflict brings the disagreements into the open, allowing the contending parties to resolve their differences. This is a "hydraulic" notion of conflict, a belief that tensions in a relationship build up or accumulate, like water pressure in a heated container. If some of the pressure is not released, the survival of the relationship is threatened. But Coser notes that conflict can act as a safety valve only if two conditions are met.

First, the conflict must concern only those "goals, values or interests that do not contradict the basic assumptions upon which the relationship is based." If conflicts about such "noncore" matters are allowed to be expressed in open conflict, issues at the core of the relationship are less likely to become matters of controversy within the group. An example of a noncore conflict is a disagreement between married people about who should do which household chores. A core conflict might concern the desirability of living together.

Second, conflicts must be "realistic"; that is, they must concern frustrations that result from the competing demands within the relationship. A disagreement about household chores is an example of a realistic conflict. Because realistic conflicts arise from the interactions between group members, they can be resolved by them. For group solidarity to benefit from internal conflicts, the conflicts must become resolved. Otherwise tensions in the relationship are not diminished and continue to accumulate. Certain kinds of conflicts, which Coser calls "nonrealistic," arise from the experiences of group members outside the group—for example, from frustrations they experienced when young or in the performance of roles outside the group. Nonrealistic conflicts can result from a frustrating day at work or from a personal intolerance learned during childhood. Because they are not generated within the group, they cannot be resolved within it. Nonrealistic conflicts cannot contribute to group solidarity.

Just because a group has no conflicts does not mean that it is stable. Some groups, especially those in which the members feel that the relationships are weak or unstable, will prevent disagreements from being expressed. They are afraid that conflict will destroy the group. Groups whose members feel secure about the stability of the relationships feel able to allow disagreements to be expressed. In fact, Coser suggests that the frequency of realistic conflicts may actually be an index of a group's stability. The greater the number of conflicts, the greater the stability of the group.

Conflict Between Groups

Conflict Mobilizes Energy and Increases Internal Solidarity

Conflict between groups mobilizes the energies of group members and leads to increased internal solidarity in two ways. First, it provides group members with an intensely shared experience, and this alone brings group members closer together. Second, conflict between groups reinforces group norms and bonds. Norms can become unclear, stagnant, or taken for granted if they are not expressed in behavior. Conflict puts norms and bonds into action, thus reinforcing their usefulness and clarifying them.

Think, for example, of how a fight with a landlord or a neighbor makes a family closer. Afterward, family members might even refer to the conflict with fondness: "Remember when we really took care of that rotten . . . !" Before the conflict, interaction in the family may have been just drifting along, the family losing some of its closeness. A conflict can clarify and reaffirm group solidarity.

Conflict Clarifies Differences

Conflict between groups establishes or clarifies the differences between them and increases the awareness by group members of the distinction between "us" and "them." This process not only increases the clarity of internal norms (by providing an example of out-group norms with which they can be compared) but also stabilizes the relationship between groups (by clearly marking boundaries and establishing the balance of forces between them).

Think of two countries locked in a border dispute. The conflict exaggerates the characteristic beliefs of each country, especially in the area of the dispute. It also establishes rules by which they can understand each other's interests. The enemy about whom we are totally uninformed is more threatening than one who follows rules, even if they are rules of conflict. Coser goes so far as to suggest that contending parties wish each other to be fairly unified. The point is that each group must be able to count on the other to follow some predictable rules for conflict. If either group becomes too unstable, it is subject to desperate, unpredictable attempts to gain advantage. That would disrupt the balance of forces between the contending groups. According to this view, because a balance (of sorts) has been maintained between the United States and the former Soviet Union for some time, instability in either country would present a great threat to that balance, because it might lead to extreme measures by one side to "catch up," including possible nuclear war. (The ultimate threat to group stability is, of course, annihilation.)

Conflict Creates New Alliances

Conflict between groups creates new alliances and associations and also leads to types of interaction between antagonists other than conflict. When groups conflict they often seek help in their struggle. Groups whose interests seem similar are contacted, and alliances of groups are formed. During recent years the world has seen the development of a European Community (formed to compete with other economic forces such as the United States, the former Soviet Union, and Japan) and the Organization of Petroleum Exporting Countries (OPEC). Cooperation among countries in associations like these increases stability at the societal level by binding together groups that might otherwise have more divergent interests.

When groups are drawn into persistent, stable conflict relationships, they often try to reduce the conflict level and stabilize their relationships further by engaging in nonconflict activities. Even during the most tense years of the Cold War, the United States and the Soviet Union, for example, took part in athletic contests and cultural exchanges designed to lessen feelings of hostility. Even if these are thought of as competitive activities, they nevertheless bind the opponents together in shared undertakings that must be planned and financed cooperatively.

In sum, as Coser points out, conflict can be viewed as more than just trouble. It may be the basis on which much group solidarity is built.

Application

From the work of Durkheim we know that social cohesion can stem from at least two sources. As you saw, however, the contrasting of mechanical solidarity before the Industrial Revolution and organic solidarity after it is something of an oversimplification. First of all, in prerevolutionary societies there was a significant amount of interdependence. It stands to reason, therefore, that there was a significant amount of organic solidarity. Similarly, in postindustrial societies, such as our own, people do share many values and ways of life, which suggests that some part of our solidarity is mechanical in nature. Second, it is possible for both varieties of cohesion to simultaneously contribute to the overall solidarity of a single group (I used the example of a college class).

This application is designed to demonstrate that the concepts of mechanical and organic solidarity are not limited to the analysis of historic events. They are useful today in our efforts to understand what binds people in groups. They may even reveal that members of the same group persist in their membership for entirely different reasons. However, we must be able to recognize each type of solidarity if we are to demonstrate that both still operate in everyday life.

One way to collect data on this subject would be to design a highly structured set of questions containing a comprehensive list of reasons people might give for belonging to a given group. To use such a list, you would have to ask the respondents to check off those reasons that applied to them. You would then calculate the percentage of the reasons each person checked off that were "mechanical" in nature and the percentage that were "organic." The problem is that I would never be able to come up with all the possible reasons for belonging to a group, and such a list would be awkwardly long anyway. Just imagine reading a list of thirty items to a respondent and asking him or her to check however many apply.

An Unstructured Survey of Reasons for Solidarity

Often such a highly structured form of survey is the most appropriate one for a given research aim. In fact, quite a few of the applications in this book are structured surveys. However, in this case it makes sense to use an unstructured interview technique. *Unstructured interviews* consist of open-ended questions and prompts for further information ("Could you tell me more?") that the researcher asks without limiting the kinds of answers the respondent may give. Unstructured interviews often are used by researchers specifically because they allow respondents to identify important variables without their being prompted. In this case, I am assuming that the people who belong to groups have their reasons for belonging to them, but that if you were to ask them whether those reasons are "mechanical" or "organic" in nature, they would not know what you were talking about. The open-ended technique will seem more natural to you *and* to the respondent. In addition, if you ask people to explain the main reasons why they belong to groups, you will not artificially limit the kind of information you can get. In this application, unlike many of the others in this book, you will not be calculating percentages or scaling scores. You will be learning how to code, or categorize, responses.

Data Collection

You should conduct brief interviews with ten people. They can be students on your campus, friends from home, or your parents and their friends—in fact, any adults who are likely to belong to groups. (This includes just about everyone.) You will be asking people to name two groups they belong to and the things that bind each group together. The second question is the basis of the interview, but some people may give you very sparse answers, and you may have to follow up with "prompting" questions or statements. I will explain more about this soon. First, here is what you should say to each respondent to introduce yourself and the study. (Practice reading it until it sounds fairly conversational.)

Introduction:	I'm doing a survey about the groups that people belong to. Would you mind answering a few questions? It will take only a few minutes.
Instructions and questions:	Everyone belongs to a number of groups. For example, there is the family, the work group, a group of friends, community or church groups, hobby clubs, volunteer groups, maybe fraternity, sorority, or social groups, and so on. I'd like you to choose one such group that you belong to and answer just two questions about it.

1. What is the group you are thinking of? ______________

2. Now, thinking about this group, what would you say are the most important reasons that people belong to it? That is, what keeps these people together as a group? (Record the reasons the respondent gives, up to a maximum of seven. If the respondent stops after one or two reasons, simply follow up by asking, "Can you think of any more?" If the respondent cannot, try asking, "What reasons did you have for joining this group, and do they still apply to your current membership?" or another prompting question that makes sense to you.)

Reason 1: ______________________________

Reason 2: ______________________________

Reason 3: ______________________________

Reason 4: ______________________________

Reason 5: ______________________________

Reason 6: ______________________________

Reason 7: ______________________________

Now, I'd like you to name another group you belong to and answer the same questions about it.

3. What is the group you are thinking of? ______________________________

4. Now, thinking about this group, what would you say are the most important reasons that people belong to it?

   Reason 1: ______________________________

   Reason 2: ______________________________

   Reason 3: ______________________________

   Reason 4: ______________________________

   Reason 5: ______________________________

   Reason 6: ______________________________

   Reason 7: ______________________________

Analysis

The purpose of this study is to determine whether people attribute their group memberships to both mechanical and organic reasons. To do this, you must be capable of categorizing each response as mechanical or organic. This process, known as **coding,** requires some practice. It will not always be clear how to code an answer. Some answers will seem to belong to neither category. Record your coding decisions on the Data Analysis Sheet provided. Here are a few tips on coding the responses you get.

The key to the nature of mechanical solidarity is *what people feel they share* or have in common. Mechanical solidarity responses should be statements referring to shared values and ways of doing things, such as "We all like the same things," "We think alike," "We all love dogs," and so on.

Organic solidarity is about *interdependence,* the extent to which people associate because they need one another to get things done. Look for responses such as "We need each other," "We can count on one another," or "I've got to make a living."

Some responses will not fit into either category—for example, "We have no choice," "No particular reason," or "My friend is in the group, so I joined."

What did you find out about the process of coding? Did it become easier after some practice? Were there a large number of responses that did not fit well into either the mechanical or the organic category? What about the frequency of the responses—were there more mechanical or more organic reasons given? Did the response seem to vary with the types of groups about which your respondents were talking? If so, how?

Data Analysis Sheet

Reasons for group cohesion	Your coding decision (check one)	
	Mechanical	Organic
1. ____________	______	______
2. ____________	______	______
3. ____________	______	______
4. ____________	______	______
5. ____________	______	______
6. ____________	______	______
7. ____________	______	______
8. ____________	______	______
9. ____________	______	______
10. ____________	______	______
11. ____________	______	______
12. ____________	______	______
13. ____________	______	______
14. ____________	______	______
15. ____________	______	______
16. ____________	______	______
17. ____________	______	______
18. ____________	______	______
19. ____________	______	______
20. ____________	______	______
21. ____________	______	______
22. ____________	______	______
23. ____________	______	______
24. ____________	______	______
25. ____________	______	______
26. ____________	______	______

27. ______________________________ __________ __________

28. ______________________________ __________ __________

29. ______________________________ __________ __________

30. ______________________________ __________ __________

CONCEPT 11

Power and Legitimacy

Definition

Power The ability to control or influence the behavior of others, even against their will.

Legitimacy The agreement among people that the exercise of power in a given situation is appropriate.

The typical pattern of events in the Newman family in my hometown used to go like this:

Father says to oldest son, "Trim the hedges before you go play baseball."

Son complains, "It's not my turn."

Father is unimpressed: "I don't care whose turn it is. Do it anyway."

"Why?" asks nervy oldest son.

With the ultimate fixing stare, father closes the conversation: "Because I say so."

Oldest son then finds younger brother and convinces him, through not-so-subtle threats ("Trim the hedges or I'll smash you!") to take over the odious task. Second son seeks out younger sister. Taking extreme advantage of her love for him and her still undeveloped debating skills, he convinces her that not only is it time for her to learn how to trim hedges but also she will enjoy the job and benefit greatly from doing it. By the time she takes up the clippers, she is deeply grateful for having been given the opportunity.

Right down the line, each person wanted the unpleasant job done by someone else; but how could each get another to do it against his or her will? In each case a different form of **power** was applied to overcome the will of the less powerful person. The father had the use of his socially approved authority to get what he wanted because of his position in the family. The oldest son applied raw physical threat to coerce his younger brother. And the younger brother used subtle persuasion to convince the younger sister that she should do the task, without her realizing that it was for his benefit rather than her own.

Power is the ability to get others to do what we want, even against their will. Among the assets that we value in American society (the others are money and the esteem of others), power is the most liquid; that is, it is most easily used to gain the other valued goods. And it is at the very base of social order.

Social Order and Legitimate Power

In theory, if people lived completely isolated from one another, they would be entirely free to do as they liked. But humans are social, and the groups that humans form, as Plato noted, are essentially agreements, or contracts, among the members. In forming groups, individuals give up to the group some of their free will. The group then has some control over the actions of its members and is intended to exercise that control for the benefit of the membership (although it doesn't always work out that way). Rules against theft, murder, and a variety of other actions are agreed on within the group, and all members are made subject to them. According to this contract view, social order results from removing a portion of the power that individuals have to do as they like and giving that portion to the group. To be expressed, the collective will of the group must be invested in individuals (a process that happens in a variety of ways, discussed next). When people agree that the exercise of power by an individual in a given situation is appropriate, that power is considered to have **legitimacy.** Legitimated power is called **authority.** In the case of the Newman family, the oldest son used raw, unjustified power to get his younger brother to trim the hedge. There was no attempt to get the victim to feel that it was right that he should take such orders. The father, by contrast, had the authority of his position in the family. He applied legitimated power. Power is both social in origin and positional, because people are put in social positions from which they can wield power.

Most of the legitimate authorities with which we are familiar, such as parents, political officeholders, and teachers, are publicly and intentionally placed in these positions. Processes such as elections and the interviewing and evaluation of applicants for advertised jobs are generally accepted means of establishing authority. And the ceremonies, such as the presidential inauguration, and even brief introductions around an office, serve officially and publicly to legitimate the power of a new officeholder. But power is not always intentionally and publicly entrusted to specific individuals. The people who are subject to authority do not always get to decide who will hold that authority or even to know how the person is chosen. Sometimes people are privately appointed to positions of power or inherit such positions from other powerful persons. This is especially true in private businesses, in which ownership and management of even large corporations often remain within families. The fact that people acquire power this way does not prevent that power from being accepted as legitimate. Once a person is in a position of power, no matter how he or she got there, the first task is to do what is necessary to legitimate that power. People in positions of power have special access to methods of communication, such as the mass media and the educational system. Through these, powerful people can promote themselves as effective and valuable officeholders, thus justifying and solidifying their power. The most common examples of this process are the political figures who attempt to solidify their hold on a country immediately after the successful overthrow of a previous leadership. Suddenly, their voices and faces are everywhere. But the process is not always so dramatic. It also works in a quieter, more familiar way in American business. Perhaps you have noticed advertisements on television and in magazines promoting the beneficial activities of the largest corporations in the United States, such as those dealing in oil, timber, and other natural resources. These companies know that the average television viewer is not likely to buy gypsum, bauxite, or any product directly from this advertiser. What they are doing, at least in part, is legitimating their power to manage and exploit natural resources and the environment. They want Americans to believe in their capacity to do their work well and in the public interest, since challenges to that legitimacy would threaten their business interests.

Types of Authority

In his classic study *The Theory of Social and Economic Organization* ([1925] 1964), Max Weber identified three types of authority:

1. **Charismatic Authority.** This is the type of authority in which an individual, because of personal appeal, is entrusted with the legitimated power of a group. Such people are capable of persuading others, by their forcefulness, attractiveness, or other special qualities, that they should be in control. John F. Kennedy and Winston Churchill were good examples of people with charismatic authority. On the more menacing side of the charisma ledger, so were Hitler and Mussolini.
2. **Traditional Authority.** According to Weber, charismatic authority is unstable, because no one can predict when a charismatic leader will appear or how long he or she will survive. Authority can be made

more stable by vesting power in a category of people who can pass control on within their group. For example, if only elders are given authority, the succession of power is stabilized. Weber called this process the *routinization of authority*. The term *traditional authority* indicates that a specific method for succession of power is explained or justified on the ground that it was traditional, the way things had been done for generations.

3. **Rational-Legal Authority.** Traditional authority has, in general, been superseded in our society by rational-legal authority. With the rise of the middle class and the decay of feudalism, national systems of law replaced the traditional authority of divine-right rulers. Simultaneously there developed a belief that a person's worth could be rationally evaluated in terms of the merit (productivity, really) of his or her actions and ideas. The form of authority that Weber termed *rational-legal* was an expression of these developments. Such authority is given to an individual who is rationally judged to be expert in some specific area and is the holder of a legally established office. The authority is given to the person not as an individual but as an officeholder. Because the authority is attached to the office rather than the individual, a judgment may be made that a more expert person should be put into the position of authority.

Weber's three types of authority are based on different forms of justification, but all can be considered legitimate by those who are subject to the will of the authority. Notice that a person can have more than one type of authority. John F. Kennedy, for example, was considered to have legitimate authority partly as a consequence of his charisma, partly because he had rational-legal legitimacy, and partly because traditionally white men over thirty-five years old have held the American presidency.

Two Models of the Distribution of Power

Who has the power in the United States? Certainly it is not held only by the president or even by members of government alone. Businesspeople must have some, because they determine who gets ahead financially; teachers must also have some; parents some; and others some. Power can be held in varying degrees. Over how many people does one person have control or influence? If only one, then little power is held. If one person can hire or fire thousands of workers, then he or she is quite powerful. Rather than trying to identify every individual who has some power, sociologists have developed two broad theories about the way power is distributed in society. One argues that power is held by a small, elite group of people, and the other contends that power is distributed across a wide variety of different (or pluralistic) groups.

The Elite Model

In his book *The Power Elite* (1956), C. Wright Mills claimed that a small group of leaders of corporations, government, and the military basically ran the United States. He saw these leaders as unified by common interests and tastes, fostered by similar institutional goals and shared social backgrounds. As a result of their agreements with one another, Mills argued, their power would continue to concentrate, and public political debate would have less and less influence on policy decisions in the country.

Mills is not the only person to propose that power in the United States is concentrated in relatively few hands. A number of people, some of whom have founded organizations based on their beliefs, maintain that the controlling power is in the hands of any one of a number of groups, such as an eastern liberal establishment; a coalition of right-wing fundamentalist conservatives; Jews; Communists; and probably a number of others with which I am unfamiliar.

Obviously, the problem in evaluating such theories is to find concrete evidence that a theory has grounds. Many imaginative conspiracy theories lack documentation because there is simply no foundation for them. But even a well-researched theory such as Mills's is difficult to document, because powerful people who might attempt to establish a concentration of control would be very secretive, and their actions would be difficult to reveal. The evidence produced in support of Mills's thesis points to the interlocking of executive positions in the military, corporate, and governmental sectors. Though government is charged with the responsibility of regulating business and military practices, it is common for people to move from one segment to the other and back again. In common language the resulting relationship is termed "one hand washing the other." It is evidence of results rather than of the actions that produced those results.

Another theory of elites expands the number of people who hold the power in the United States to about 5 percent of the population. William Domhoff (1967) identified a small "governing class" composed mainly of very wealthy business owners and operators who also influence the executive branch of government and, through it, the most important regulatory agencies. Once again, the evidence is primarily of results rather than of specific actions. The actions taken by powerful people do seem to benefit the relatively small group of people identified as elites; but it is possible that they benefit from some other system of decision making.

The Pluralistic Model

By contrast, the pluralistic model argues that power is distributed among a wide variety of interest groups that assert themselves either by voting or by lobbying efforts. In his statement of this model, David Riesman (1961) proposed that none of these interest groups can actually control the decision-making process by itself, but that each can wield a sort of veto when issues of special concern to its well-being are decided. In order for decisions to be made, coalitions of otherwise separate interest groups must be forged, and government must mediate between contending interest groups.

According to the pluralistic view, power shifts depending on the specific issue. The large mass of citizens is not dominated by a unified elite but is actually courted by contending interest groups searching for allies in the settlement of a particular issue. Accordingly, this model sees a progressive dispersion of power in the United States, not the progressive concentration proposed by the elite model.

Once again, the problem is the contradictory (or absent) evidence. On the one hand, it is clear that there are disagreements between competing segments of the society. Sometimes even large corporations lose decisions, and choices in the public interest are made. For example, laws for the protection of public lands and of air quality have been passed over industry protests. If there were a wholly dominant power elite, wouldn't it have had its way in every issue? On the other hand, as the elite model has shown quite clearly, such decisions are relatively rare. Although I am generally a proponent of the elite model, especially Domhoff's (could you tell?), I must admit that this is a confused and continuing debate. The best I can do is to outline the two positions and leave it to you to decide your own view.

Summary

Power is the ability to control or influence the behavior of others, even against their will. When people agree that the exercise of power is appropriate for a given situation, that power is said to be *legitimized.* Groups sometimes choose individuals to hold legitimized power (called *authority*) by election or some other method in which members can influence the assignment of power. However, people sometimes have no say about who will have power over them (for example, when an individual inherits the presidency of the family corporation).

The sociologist Max Weber identified three types of authority: (1) charismatic authority, in which an individual gains a position of power due to personal appeal; (2) traditional authority, in which power is routinely given only to members of a certain category of individuals, such as elders; and (3) rational-legal authority, in which power is given to an individual who is rationally judged to be expert in some specific area and who is the holder of a legally established office. This authority is attached not to the individual but to the office.

Two well-known models have been proposed to describe the distribution of power in the United States. The elite model, most often associated with the work of C. Wright Mills, asserts that a small group of leaders of corporations, the government, and the military hold the most important positions of power in the United States. By contrast, David Riesman's pluralistic model does not find power so concentrated in the country but rather distributed among a wide variety of interest groups. These groups assert themselves (depending on the specific issue) either by voting or by lobbying efforts. No group is seen as capable of controlling the decision-making process by itself, but each can block the efforts of others to have their way. Thus, for actions to be taken, coalitions among the groups must be forged.

The Illustration for this concept summarizes the controversial studies conducted by Stanley Milgram and recounted in *Obedience to Authority* (1974). In these startling experiments, Milgram showed that the willingness to follow the instructions of an authority figure (in this case, an experimenter with a clipboard and a lab coat) depends on the structure of the social situation rather than on the personality of the obedient individual. In fact, ordinary citizens who volunteered for the study were found willing to do great harm to total strangers, merely because they were instructed by an authority figure that the study required them to do so. The implications of these studies for understanding social (as opposed to personality) forces is dramatic.

The Application for this concept is a study of the distribution of power at your school. By carefully examining your college catalog, you can learn enough to assign specific values to the relative power of members of the faculty and the administration. Then, by dividing the personnel by sex, you can determine the relative power of men and women on your campus.

Illustration

Stanley Milgram, Obedience to Authority: An Experimental View *(New York: Harper & Row, 1974)*

It is estimated that during World War II a total of more than 6 million Eastern European Jews, Poles, Catholics, Gypsies, and homosexuals were murdered in Nazi concentration camps. The American soldiers who liberated death camps at Dachau, Sachsenhausen, Buchenwald, Ravensbruck (for women), and Auschwitz saw sights and heard tales horrifying beyond the scale of human cruelty even in war experiences. How could humans do such things to other humans? After the war, Nazis were tried at Nuremberg for "crimes against humanity." The defense most of them chose to present has become familiar to us all: "I was just following orders."

The world tried to make sense of the horrors of death camps by arguing that the Nazis had been monsters, madmen so evil that they actually preferred to destroy life and cause pain. This was a consoling belief, because it freed the rest of humanity to believe that the Third Reich was a special, unrepeatable case of human nature gone berserk. It could, therefore, never happen among the rest of us. *We* could never be that way. Yet, watching Adolph Eichmann (the architect of the extermination plan) sitting in the witness chair at his trial years later in Jerusalem, the world saw an extremely plain, sane-looking man. In her book *Eichmann in Jerusalem,* Hannah Arendt ([1963] 1977) described the evils of Nazi behavior as "banal," meaning they were, in a strange sense, unremarkable, even conventional. If the Nazis were not, as the prosecution contended, brutal, sadistic, and twisted, if their behavior was in some sense ordinary, then what conditions could have led to the Holocaust? One compelling answer lies in the work of Milgram, who, from 1960 to 1963, conducted a series of experiments in people's obedience to authority.

The Experiments

Milgram placed an advertisement in a New Haven, Connecticut, newspaper offering four dollars for volunteers to participate in a one-hour "study of memory." The hundreds of people who responded came from a broad range of occupations, from unskilled blue-collar workers to professionals and white-collar executives. Ages ranged from the twenties to the forties. (Both men and women took part in the studies, but for the purpose of clarity, I will discuss only the studies involving men.) In fact, the experiment was not really about memory but about the willingness of subjects to follow the orders of a person they had never met before.

In each experimental session there were three participants. (1) The experimenter, dressed in a long, white lab coat and holding a clipboard, presented a fairly imposing image of authority. (2) The subject (a person who had responded to the ad) was always given the role of "teacher." (3) A "learner," played by a compatriot of the experimenter, was well rehearsed in the lines he was supposed to say during the experimental sessions. So the real focus of the study was on the behavior of the teacher. Both the experimenter and the learner were in on the real purpose of the study.

In a single experimental session the following took place. The experimental subject was told that a coin flip had determined that he was to play the role of the teacher and that another volunteer had been assigned the role of learner. The experimenter explained that the purpose of the study was to discover how punishment influenced memory and learning. The teacher was given a list of vocabulary questions with which to quiz the learner. For every incorrect answer that the learner gave, the teacher was to administer a shock to the learner.

The teacher was then shown an impressive-looking electronic console, which featured a panel of thirty lever switches. The switches were labeled with increasing levels of voltage, ranging from 15 to 450 volts. In addition to the voltage markings, the switches were labeled from "Slight shock" at the lower end, through "Intense shock" at the 255-volt level, to "Danger: severe shock" at 375 volts. (The top two switches, at over 435 volts, were ominously marked "XXX.") In fact, the shock generator was a phony, with the exception of the 45-volt switch, which was used to give the teacher a slight "sample shock" to make the equipment absolutely convincing.

The experimenter brought the teacher and the learner to a booth that was out of sight of the shock generator panel. There the teacher saw the learner attached to the shock generator's business end. The teacher then returned with the experimenter to the panel of switches. The teacher was given a list of vocabulary questions with which to quiz the learner. For every incorrect answer that the learner gave, the teacher was told to punish the learner with a shock, proceeding from the lightest shock and continuing to the most severe shock with each wrong response. (Unknown to the teacher, the learner was going to intentionally give wrong answers so as to force the teacher to administer shocks.)

Remember that the teacher had never met the learner and did not know that the shock generator was a phony. No learner ever got a shock at all, but to make the deceit believable, the learner (using a microphone) responded to the supposed shocks according to the following script: At 75 volts the learner grunted. At 120 volts the learner started to yell about the pain and to groan. At 150 volts the learner yelled and demanded to be released from the experiment. At 270 volts the learner screamed in agony. At 300 volts the learner desperately refused to answer any more quiz questions. From 330 volts on, nothing at all was heard from the learner. He might, presumably, have passed out or even died.

The experimenter stood next to the teacher at the panel, and the vocabulary quiz began. If a teacher resisted administering shocks for wrong answers, the experimenter asserted his authority, beginning with mild statements and proceeding, when necessary, to stronger ones. The four levels of "prodding" were: (1) "Please continue," (2) "The experiment requires that you continue," (3) "It is absolutely essential that you continue," (4) "You have no choice, you *must* go on." These statements were made firmly, but without shouting or using a threatening tone. There was no direct statement that the experimenter was responsible for what happened to the learner or that the teacher was not responsible—only the four statements listed and the imposing appearance of the experimenter.

Results of the Experiments

If you were in the position of the teacher, do you think you would have obeyed the authority of the experimenter? Would you have administered shocks to a person you had never met, merely because he got some vocabulary questions wrong? Well, there is evidence that the subjects in Milgram's studies didn't think they would have either. Milgram described the experiment in detail to a group of people who were similar to the subjects in the experiment, and he asked them to predict how they would react. They were also asked to predict how the subjects in the experiment would react. Every one of them predicted they and the experimental subjects would defy the experimenter's demands that they continue with the experiment, long before the shocks became dangerous.

Yet, in fact, in the version of the experiment I have described, over 60 percent of the teachers continued shocking the learners *right to the very end of the scale,* the end marked with "XXX," at 450 volts! Just in case you think this result occurred because the teachers thought the learner was not actually harmed, you should know that Milgram ran the experiments another time with the learner screaming through his microphone (at 195 volts) that he had a heart condition and that it was hurting him. In spite of this added factor, over 60 percent of the subjects continued to obey the authority of the experimenter at all times, punishing the learner for incorrect responses right to the end of the voltage scale. Even when the learner was brought right into the room with the teacher, in fact, seated right next to and touching the teacher, 30 percent of the teachers administered maximum shocks.

Why did they do it? Assuming that these people believed they were causing great pain to other people (and all of Milgram's questions to the subjects confirmed that they did believe they were administering real shocks at the levels marked on the panel), how could their behavior be explained? If we were to use the approach of the prosecutors at Nuremberg, we would argue that these were psychologically, morally warped people. Evil, cruel people must have answered the ads. But remember that the ads had said nothing about administering shocks, only that a study on memory was being conducted. And the results were that in many conditions almost two-thirds of the subjects complied with the demands of the experimenter to the extreme. Could New Haven have had that high a percentage of madmen?

In light of these findings, Arendt's notion of the "banality of evil" seems much more likely. That is how Milgram interpreted his results. When circumstances exist in which authoritative figures make demands (even evil ones), ordinary people are all too willing to believe that they are not responsible for their actions. They "just follow orders." The same social situations that lead people to obey authority in everyday life (at work, at home, or in the community) also operate when authority makes objectively immoral demands. If Milgram is right, what happened in

Germany during World War II could happen here or anywhere that social circumstances permit. We do not have to have evil populations for evil deeds to occur.

Application

When people agree with one another that the exercise of power in a given situation is appropriate, they consider the power to be legitimate. In many cases the distribution of legitimate power in organizations is formally established. This is true of all the colleges and universities with which I am familiar. To uncover the distribution of power in a college, you can look at the college catalog, where everyone is listed by position. The president of the college has the most power, because he or she has the greatest influence in matters of budget, hiring, firing, promotions, and so on. Below the president, in decreasing order of power, are deans or vice presidents, department chairpersons, full professors, associate professors, assistant professors, and last, instructors.

It is interesting to see how many people are in each of these positions. Usually we expect more people in positions of less power than in positions of greater power. But another interesting fact about the distribution of power is that it is often distributed unevenly by categories of people. For example, women in American colleges and universities are found disproportionately in positions of lower power.

Studying Your College's Distribution of Power

In this exercise you will analyze the distribution of power in your college or university with respect to the sex of the people who work there. To do this you have to obtain a current college catalog. Usually at the back of the catalog (or introducing the sections on each academic department) are the names of the faculty members and their qualifications. The administrators are usually listed either at the front of the catalog or with the lists of faculty.

I have provided a model Data Recording Table to help you record how many men and women are in each of the positions at your school. Use the first name of the administrator or faculty member to identify the person's sex, and record the number of men and women who are at each level of power. Gender is not always clear from a person's first name. The name might be used for either sex (such as "Leslie"), or the person might be listed only by initials. For the purposes of this application you should keep a count of the number of names whose gender you cannot easily determine. The percentage of such cases should not normally be high enough to threaten the validity of the method. However, you should know that a more rigorous study using this method would require further research to discover the sex of each person the coder could not easily categorize. If you like, you can probably get this information from the personnel office at your school.

On the Data Recording Table fill in the number of people in each category (the columns labeled "Number"), and in the column at the right fill in the total number of people in each position. There will be only one president and a few deans (at some schools they are called vice presidents rather than deans), but there will be many professors at each level. (If your school is huge, you may have to limit your analysis to one division, such as liberal arts.)

Analysis

Have you noticed a pattern yet? If not, a pattern might emerge in the next step. Using the figures for the total number in each position, calculate what percentage is male and what percentage female. For example, to calculate the percentage of assistant professors who are female, divide the number of female assistant professors by the total number of assistant professors. Enter the percentages of males and females in each position in the columns labeled "Percentage" on the table.

If in your school there is a tendency for more powerful positions to be held by one sex more than the other, it should have become clear in your analysis. But to amplify the point, here is a further calculation that can be done. So far the listing of positions has suggested that full professors are more powerful than those below them on the list. But we do not know by how much. Each position has special, legitimated powers associated with it. College presidents control budgets and fates of entire departments. Department chairs and full professors are tenured faculty members who sit in judgment on the applications for promotion and tenure by junior faculty members. Assistant professors and instructors control only what happens in their classes. To reflect these differences in power, I have assigned a number to each position that serves as an index of its relative power, and I have listed these numbers in a column labeled "Power index for position" in a Data Analysis Table. (These are my own evaluations of relative power. They are based on my observations of the relative ability of each position in the college hierarchy to control the fate of those at lower levels.) From the Data Recording Table fill in the number of men and women in each position in the Data Analysis Table. To calculate the total power of men on campus, multiply the power index for each position by the number of males in that position, and enter the figure in the column on the right labeled "Male power." Do the same for the women. Then sum all the male power figures and female power figures, and compare the two sums. (To illustrate the procedure I have provided a Sample Table analyzing a small college in Massachusetts.)

SAMPLE TABLE

Distribution of power by sex in a Massachusetts college

Position	Power index for position	Number of males	Number of females	Male power	Female power
President	50	1	0	50 (50 × 1)	0 (50 × 0)
Deans or vice presidents	20	6	1	120 (20 × 6)	20 (20 × 1)
Department chairpersons	10	18	3	180 (10 × 18)	30 (10 × 3)
Full professors	8	62	18	496 (8 × 62)	144 (8 × 18)
Associate professors	6	44	19	264 (6 × 44)	114 (6 × 19)
Assistant professors	2	41	51	82 (2 × 41)	102 (2 × 51)
Instructors	1	9	18	9 (1 × 9)	18 (1 × 18)
Totals		181	102	1,201	428

Interpreting the Results

If your school is anything like the school in the sample analysis, you should see a dramatic difference between the power of men and that of women. (At the sample college, men are almost three times as powerful as women.) Most colleges and universities are aware of this disparity, and, in advertising for new adminstrators and faculty members, they make special mention of their desire to hire women and other minority-group members. In fact, most are required by law to demonstrate that they not only do not discriminate in hiring and promotion but also follow affirmative-action guidelines to increase the percentage of minority-group members on staff.

You can easily check the progress your school has made in this regard (at least in the hiring and promotion of women) by comparing the results you get in this application with data from catalogs of previous years. Most schools put out catalogs every year. A comparison of figures for male and female power (calculated as a percentage of total power in the school to account for changing numbers of employees) might be quite revealing. How much progress, if any, has been made at your school?

DATA RECORDING TABLE

Number of male and female administrators and faculty members at your college

Position	Males		Females		Total number of people in each position
	Number males	Percentage males	Number females	Percentage females	
President					1 (There should be only one president)
Deans and vice presidents					
Department chairpersons					
Full professors					
Associate professors					
Assistant professors					
Instructors					
Totals					

DATA ANALYSIS TABLE

Distribution of power by sex in your college

Position	Power index for position	Number of males	Number of females	Male power	Female power
President	50				
Deans or vice presidents	20				
Department chairpersons	10				
Full professors	8				
Associate professors	6				
Assistant professors	2				
Instructors	1				
Totals					

CONCEPT 12

Ideology

Definition

Ideology A set of ideas that explains how the world operates and is used to justify a group's actions in pursuing its own interests.

Young Rodney Ropes, a lowly but honest and aspiring clerk, has been wrongly accused of embezzling from the company of Otis Goodnow, and he has been summarily fired. After numerous adventures across the country, he is kidnapped and held in a cave for ransom. The captors decide to request $5,000 of a new, wealthy friend of Rodney's and demand that he write the ransom note (Alger, 1974:234):

> *Rodney knew his danger, but he looked resolutely into the eyes of the men who held his life in their hands. His voice did not waver, for he was a manly and courageous boy.*
>
> *"The boy's got grit!" said one of the men to the other.*
>
> *"Yes, but it won't save him. Boy, are you going to write what I told you?"*
>
> *"No."*

Just so you won't be kept in permanent suspense, I'll tell you that Rodney gets away, and the kidnappers are jailed. The real embezzlers confess out of a sense of guilt and shame (both apologize to Rodney) and are forgiven by Rodney and the employer. Rodney is offered his old job back but turns it down, because he is now rich—while held captive, he discovered gold in the cave, for which information he was paid $75,000 and given a part-ownership in the gold mine.

Pretty nice result for the once-poor, young Rodney. But he was not unique. The author of this exciting tale was Horatio Alger (1834–99), and he wrote 108 other stories just like this one during the last quarter of the nineteenth century. In every one, a young man overcomes his poverty by a combination of hard work, honesty, adventurousness, bravery, and a strong dose of good luck (often the leading character is discovered to be the long lost child of a millionaire). Lots of people must have wanted to read this sort of stuff, because Alger's books sold over 40 million copies. (Interestingly, Alger himself appears to have been nothing like the upstanding, frugal, and moralistic heroes of his novels. He is reported to have been fired from his job as minister of a congregation on Massachusetts's Cape Cod after allegations of improper relationships with some boys of the choir, and died penniless, having squandered his money.)

Alger's novels present one set of ideas about what causes poverty and about how it can be overcome. By strenuous individual effort, a willingness to take chances, and the right qualities of character, Alger contended, anyone could make it. During the last decades of the nineteenth century another, quite different theory about how poverty might be overcome was also widely read (Mao, 1963:197):

> *In a very short time, several hundred million peasants . . . will rise like a tornado or tempest—a force so extraordinarily swift and violent that no power, however great, will be able to suppress it. They will break through all the trammels that now bind them and push forward along the road to liberation. They will send all imperialists, warlords, corrupt officials, local bullies, and evil gentry to their graves.*

This quotation is from the writings of Mao Tse-tung, leader of the Chinese revolution and chairman of China's Communist government until his death in 1976. He was a student of the work of Karl Marx, who was writing in Europe at the same time that Alger was writing in the United States. But notice how different these ideas were. Where Alger focused on the qualities and activities of individuals, Marx and Mao focused on the need for collective action. For them, poverty could be overcome only by mass revolution. Individual initiative would only weaken the collective will of the working poor. Alger saw the rich of the business world as deserving their successes, and he preached that poverty could be individually overcome by entering the world of business. By comparison, Marx and Mao saw the rich of the business world as the oppressors of the poor. Far from providing opportunity for the poor, this segment of society would have to be destroyed to eliminate poverty. These two views (or **ideologies**) could not have been more opposite. One stressed individual initiative aimed at upward mobility within the capitalist system. The other stressed collective action aimed at the overthrow of the capitalist system. But what these, and all, ideologies have in common is that each works to justify the actions of a group of people.

Ideology and Justification

The term *ideology* has been used in many different ways. In fact, I once saw it defined simply as any set of ideas. In this definition the term is too broad to be useful. Ideology has a specific meaning in sociology that can help make sense of a number of issues.

Every group of people invents, collects, or otherwise borrows ideas to help its members explain how things operate in their world. Such ideas may deal with the physical events surrounding them, the social circumstances in which they find themselves, or a variety of other issues. To go back to the difference between the ideologies I described earlier, how do poor people explain their poverty? How do rich people explain their advantages? To explain how they got into the circumstance in which they find themselves, they have to explain how the entire system of income distribution operates.

Clearly, not all groups of people come up with the same set of ideas about how things work. For example, rich people are likely to come up with an explanation (or subscribe to an existing one) that justifies *continuation* of the circumstances that made them rich and allows them to stay advantaged. Their ideology is likely to argue that those who are rich deserve to be and that others can get rich too, if they follow the example of successful people. When Alger's stories came along, their popularity was a measure of the American willingness to promote and believe in this ideology.

Poor people, on the other hand, are likely to justify a *change* in the way things work. Their ideology will suggest that the way things work is neither fair nor inevitable and that those who are rich do not deserve to be—they actually deserve to be removed. Marx's work has been studied and adopted by millions who see this ideology as promoting their own interests.

It is important to understand that ideologies are not just simple statements of how things *ought* to be in society. They are much more subtle than that and, therefore, much more influential. Ideologies imply the way things ought to be only as the logical consequence of the way things are seen as working. For example, the ideology expressed by Alger proposed that poverty and wealth are distributed on the basis of individual merit. Income results from differences between individuals in industriousness, intelligence, adventurousness, and so on. The inevitable conclusion is that the system of rewards is fair, because it rewards merit, and ought to remain as it is. It is individuals who must try harder.

If, however, Marx's ideology is correct—that poverty and wealth are distributed by power—an individual's merit in terms of work, intelligence, and so on has no impact on his or her well-being. The inevitable conclusion of this ideology is that the system must be changed to make economic advancement possible. By expressing a group's self-interests as the consequence of the group's explanation for how things operate, ideologies act as a sort of mediator. Talcott Parsons (1902–79) saw how ideologies linked the empirical world (the way we believe things actually operate) with the world of values (our broadest ideas about how things ought to be) (Parsons et al., 1961:96).

There is a wide variety of ideologies, ranging from the small-scale ideology of a group of people who want to save a stream from destruction to the global ideology that Marxism became. One of the most destructive and pervasive ideologies in America is racism. It fits the definition of an ideology. When racism argues that black Americans are disproportionately poor because they are inferior, it is a set of ideas explaining how the world works. When it employs those ideas to deny black Americans equal opportunities, it is being used to justify a group's actions in pursuing its own interests. But whatever the specific ideology, all ideologies have one thing in common: Every ideology has consequences for the stability of, or changes in, the way things are. In *Ideology and Utopia* (1936), the German sociologist Karl Mannheim (1893–1947) distinguished ideologies (which serve to maintain the social order) from utopian ideas (which serve to bring about a new order). It is possible to expand this distinction by making special use of some terms from political science. This will also have the benefit of clarifying some confusing everyday terms.

Politics and Ideology

If we think of ideologies in terms of how they express the appropriateness of changing a given social structure, a logical set of categories for kinds of ideologies emerges. We can label them with the political terms *conservative, liberal, radical,* and *reactionary.*

1. *A conservative ideology* supports the conclusion that things work pretty well as they are and that the structure of society ought to remain unchanged (ought to be conserved).
2. *A liberal ideology* supports the conclusion that the structure of society is basically sound but that some things could be improved. A wide range of people call themselves liberals, probably because they vary in the degree of change they feel is necessary in the structure of society. Such changes range from minute tinkering with details of the social order to large-scale adjustments to the basic nature of institutions. Everyone who thinks change can be profitably made, but thinks that the basic structures of society should remain, qualifies as a liberal.
3. *A radical ideology* supports the conclusion that the structure of society is basically unsound and unfair and that it ought to be entirely changed (radical change, or change from the roots). It should be no surprise that people who express such ideologies make both conservatives and liberals (to a slightly smaller degree) nervous.

4. *A reactionary ideology* supports the conclusion that the structure of society is basically unsound and that it ought to be replaced by a form of social order that existed in the past.

Because the circumstances of life change from time to time, it is also true that groups of people change their ideas about what is in their best interests. Consequently, ideologies change to suit the times, and people can change the ideologies to which they subscribe. During the 1970s and early 1980s, for example, it was clear that economic hard times in the United States were causing a number of people to become either increasingly conservative (in protection of what they perceived as their relatively advantaged positions) or increasingly radical (in pursuit of the upward mobility forestalled by inflation and rising unemployment). With the election of a new, Democratic president in 1992, it remains to be seen whether a swing away from the political conservatism of the 1980s has begun.

Summary

An ideology is a set of ideas that attempts to explain how the world operates and is used to justify a group's actions in pursuing its own interests. Two well-known examples are the democratic-capitalist ideology (expressed in the stories written by Horatio Alger) and the communist-revolutionary ideology (expressed in the work of Karl Marx and Mao Tse-tung). The democratic-capitalist ideology argues that wealth is distributed by merit in a capitalist society and that those who work hardest and are the most talented and adventurous will inevitably be rewarded for their efforts. By contrast, the communist-revolutionary ideology argues that wealth is distributed by raw power in a capitalist society. The people who are the wealthiest are not necessarily the most talented or hard-working, merely those in positions of control. An ideology not only explains how the world operates but also serves to justify the actions of the group that holds it. Thus, the democratic-capitalist ideology tends to be held by those who would like the social system to continue as it is, in the belief that it will benefit them. Such people tend to be well off within the society. By contrast, poorer people are more likely to hold an ideology such as the communist-revolutionary ideology, because it argues that the system that creates such poverty must change.

The tendency of an ideology to either support existing social systems or advocate change in them is reflected in the political terms *conservative, liberal, radical,* and *reactionary.* A conservative ideology suggests that a social system ought to continue as it is. A liberal ideology suggests that the social system is basically sound but that it needs to be changed to some extent. (The extent of the change called for determines the degree of liberalism.) A radical ideology suggests that the social system is basically unsound and ought to be replaced with an entirely new one. And a reactionary ideology suggests that the social system is basically unsound and ought to be replaced with one that existed in the past.

The Illustration for this concept summarizes David Prindle's study of ideology in the Screen Actors Guild. His interviews with guild members reveal a nearly even split among the actors between conservatives and progressives. He suggests a number of factors to account for these differences.

The Application for this concept is a survey in which you ask people to respond to five statements, the first four of which are designed to measure the kind of ideology they have about the reasons why some people succeed in America. The fifth question is intended to measure the respondents' attitude about the extent to which opportunities for success should change. The idea is to see if ideology and support for social change are related.

Illustration

David F. Prindle, "Labor Union Ideology in the Screen Actors Guild," Social Science Quarterly *69 (1988): 675–86*

If the reaction of my nonsociologist friends is any indication, the term *ideology* shares with terms such as *faith* and *meaning* an ability to bring an otherwise-normal conversation to a halt. Ideas such as these seem so large and timeless that many people feel they can be used only when abstract and theoretical issues are at stake. And yet only recently an American president, Ronald Reagan, completed what many have described as the most ideological presidency of this century. During this period, ideology was anything but a remote and abstract concept, because virtually all Americans were materially affected by it.

Reagan was elected (and reelected) on the basis of his conservative ideological beliefs rather than on the basis of specific plans and programs. According to recent "insider" books about him, once in office he was guided exclusively by his abiding faith in the capacity of the individual, who must be allowed to operate in the free-market atmosphere of capitalism, unencumbered by interference (especially by government). This conservative ideology drove virtually all his decisions, and far from disapproving of Reagan's operation of a government by ideology, the voting public loved it. Had he been allowed to run for office again (assuming he had wanted to) public-opinion polls suggested that he would have won again, by a landslide.

At the time of his first election there was a good deal of speculation in the news media about Reagan's ideology. For one thing, he had once been a liberal, and for another, he had twice served as president of the Screen Actors Guild, from 1947 to 1952 and then again in 1959. How was it possible that an actors' union could produce this most influential conservative? After all, aren't unions, especially unions of artists, likely to be radical?

A Study of Union Ideology

To understand how ideologies can develop and operate within unions, consider the data collected by Prindle for his study of the Screen Actors Guild. He interviewed fifty-six members of the guild because he had reason to believe that it was an excellent "laboratory" for the examination of ideologies. That is, he thought the union would place few restraints on the ideological choices of its members. First, the guild is a rarity among American unions, one in which all officers and members of the board of directors are unpaid. Second, union higher-ups could have no expectation that union activity would be of special help in getting acting work from producers. On the contrary, it was perceived as potentially harmful. Both of these facts meant that the ideologies of the most active union members need never be tailored to any union career or professional career interests.

To measure the ideologies of the people he spoke to, Prindle used a definition of ideology much like the one in the Definition section of this concept. He saw that in the case of labor unions one's belief about the nature of the relationship between the workers and the management leads to beliefs about how the union should act in the interests of the workers. He distinguished two categories of ideology, as follows:

1. **Progressive Ideology.** "Some activists are relatively hostile to management and believe that only a vigilant and confrontational work force can hope to protect itself from exploitation. This group is usually ready to strike to force its will upon the bosses." Prindle calls this ideology "progressive," though the appropriate term from this concept would probably be "liberal."
2. **Conservative Ideology.** "At the other philosophical pole some workers see management and labor as partners cooperating for mutual benefit. In their view reason and good faith will allow negotiators to work out accommodations that are profitable to both sides. They abhor striking."

Prindle found equal numbers of progressives and conservatives (twenty-five each) in his sample of guild activists. The remaining six people in his sample were labeled "centrist." But what was there about these actors that led some to develop such opposing ideologies? The researcher asked a series of questions to try to find out, including questions about:

1. Personal background, including religion, social class, and amount of education

TABLE 12-1
Variables Prindle found to be associated with ideology

	Percentage of progressives	Percentage of conservatives
Personal background variables		
Religious background ($N = 50$)		
Protestants	27.8%	72.2%
Catholic	53.8%	46.2%
Jewish	81.8%	18.2%
Professional experience variables		
Medium of preference ($N = 42$)		
Film	23.8%	76.2%
Television	100.0%	0.0%
Stage	83.3%	16.7%
Medium of first job ($N = 48$)		
Film	18.2%	81.8%
Television	37.5%	62.5%
Stage	62.1%	37.9%
City where career began ($N = 50$)		
Los Angeles	12.5%	87.5%
New York	70.6%	29.4%
Other	64.7%	35.3%
Medium in which they worked most ($N = 31$)		
Film	10.0%	90.0%
Television	60.0%	40.0%
Stage	83.3%	16.7%
Did they attend acting school? ($N = 29$)		
Yes	70.0%	30.0%
No	33.3%	66.7%

2. Professional experience, including the medium (film, television, or stage) they preferred, the medium in which they got their first job, the city in which their career began, the medium in which they had worked most, and whether they had gone to acting school
3. The degree of their success, measured by income and whether they had played mostly leading or character roles

Prindle found that several factors were closely related to ideological choices by these actors (see Table 12-1).

Understanding and Interpreting the Data

The data presented in the table summarize only the significant variables from Prindle's study; that is, these factors are the only ones that he could be confident were actually associated with ideological position. Other variables, such as the respondent's income, father's political party, father's occupation, whether one came from a union family, and years of employment were not found to be related to union ideology. However, it is apparent from the table that quite a few variables were found to be significantly related to ideology. Here is a summary of the data and some of Prindle's speculations about why the responses turned out as they did.

First, actors who came from Protestant backgrounds were much more likely to be conservative (72.2 percent) than progressive (27.8 percent), and the exact opposite was true of those who were raised in Jewish homes (81.8 percent progressive versus 18.2 conservative). Here the researcher suggests that because the Protestant religions are highly individualistic, conservative assumptions of individual responsibility (which are, as Max Weber noted, at the root of Protestantism) would make collective actions such as strikes abhorrent. By contrast, the Jewish response to centuries of persecution has necessarily been collective. Jewish people's willingness to band together in the interest of survival, according to Prindle, makes their more progressive union ideology understandable.

Second, those with careers in film were consistently more likely to be conservative than progressive, whereas those with careers on the stage were consistently more likely to be progressive than conservative. This difference was evident in data for (a) medium of preference, (b) medium of first job, and (c) medium in which they had worked most. Prindle attributes this clear effect to the nature of the work in each medium. He notes that films are controlled largely by directors, producers, and the technicians that put them into final form. Actors work before a camera and crew, not an audience, and sometimes do not even see the other actors to whom they are supposed to be directing their lines. That is, film acting is an individual and compartmentalized endeavor, again more in keeping with the individual emphasis of conservatism.

By contrast, acting on the stage is intensely collective. The actor must be in intimate touch with both the audience and all the other actors, or a performance can fail. Additionally, because the audience changes every night, sensitivity to audience characteristics requires a special awareness of the "group enterprise." The nature of work on the stage, then, is much more attuned to the collective sensibilities of the progressive ideology. It seems likely that the association of city of first job with ideology is due to the fact that New York is dominated by work on the stage (70.6 percent progressives), whereas Los Angeles is dominated by film work (87.5 percent conservatives).

Lastly, Prindle suggests that the tendency of those who attended acting school to be more progressive (70 percent) than those who did not (66.7 percent conservative) results because those who attend acting school experience extensive socialization into the profession. That is, they learn not only how to act but also much about the "craft and culture of acting." This professional socialization is likely to contribute to an actor's sense that she or he is a member of a large, interdependent community, encouraging a progressive ideology.

There are some important limitations to these findings i.e., the sample is only of one union and consists of only fifty-six people. This does not mean that the data are useless. Prindle is careful not to generalize the findings to other types of unions, and the statistical tests he used take into account the small sample size. Associations between the variables had to be very strong for chance variation to be ruled out. But you must keep in mind that associations between variables are not causes. Just because the medium in which one's career is focused is closely correlated with union ideology does not prove that this factor is a cause of the respondent's ideological choice. Consider that one's ideology may come first and then lead to career choices; that is, conservatives may not become stage actors, and progressives may hate film (or Los Angeles).

Having noted these cautions, it is, however, interesting to notice that the data have a broader, underlying theme. Prindle notes that many occupations require collective and communal behaviors and that individual responsibility and isolation are the rule in many others. Thus, film acting is more individual, and stage acting more communal. Might further research find, then, more conservative ideologies among truck drivers than railroad workers, among home-care professionals than hospital employees, and among golfers than football players?

Application

During the presidential campaign of 1992 many political commentators again tried to determine who was a liberal and who a conservative, and whether Americans were becoming less conservative than was the case during the 1980s, when Ronald Reagan and George Bush were in office. Republican candidates (including George Bush) usually were quite willing to call themselves conservatives while accusing their opponents of being liberals. Meanwhile, Democrats, almost without exception, resisted the label of liberal. When terms like these are used in the political arena, they often are applied narrowly. For example, a person who wishes to restrict access to abortion is likely to be labeled conservative. A person who wishes to spend money to improve the quality of life in poor areas of cities is likely to be called a liberal.

The sociological meaning of the term *ideology* is broader than its meaning in politics. For sociologists *ideology* refers to the set of ideas that explains how the world operates and that is used to justify a group's actions in pursuing its own interests. I began the Definition section of this concept by distinguishing between two, clearly opposed, ideologies. One was based in the belief that the lives people live result from their merit as individuals. The other was based in the belief that the lives people live result from the opportunities made available to them by circumstances often beyond their direct control. It was further suggested that people who are more satisfied with their lives are likely to hold the first ideology and to oppose much change in the way society operates. Those dissatisfied with their lives would be more likely to hold the second type of ideology and to support change.

The language that expresses the extent to which people support or oppose social change is common in political jargon. In everyday conversation, people often label beliefs about change using terms such as *conservative, liberal,* and *radical.* So a person who thinks that the basic structure of society is sound and that things such as the distribution of rewards and the structure of the family ought not to change is likely to be identified as a conservative. A person who thinks the structure of society is basically sound but that it needs some adjustment or change is likely to be identified as a liberal. And a person who thinks the structure of society is basically unsound and that there ought to be fundamental change in it is likely to be identified as a radical.

A Survey of Ideology and Support for General Social Change

This application is designed to help you to see if you can find evidence of the relationship between people's beliefs about the origins of success and failure and their support for social change. In order to do this you will have to interview ten people, asking each to agree or disagree with five statements. Responses to the first four statements, which are adapted from a number of social research measures of political liberalism-conservatism, together form a measure of ideology. That is, the responses to the first four statements will be scored and these scores added together to produce a single measure of "Beliefs about the origins of success and failure in America." The fifth statement measures the respondent's support for general social change.

You can interview any ten people you like. It does not matter if they are on your campus or from your neighborhood or apartment building or dormitory, or some other setting. It will help in testing the hypothesis if the respondents differ in their ideologies. That is, you want to find some people who believe that success is due to individual effort and some who believe that forces beyond the individual influence chances for success. (The reason for this is explained later in this application in the section called "Some Potential Problems in Doing This Kind of Research.")

Next I present both an example of what you should say in introducing yourself and the research statements to present to each respondent. As with other such surveys in this book, it is best to become familiar with the wording of the introduction and the research items so you don't stumble over them and can sound as conversational as possible.

Introduction: I'm doing a study of what people believe about what it takes to succeed in America today. Would you be willing to answer questions about this? I don't need your name, and your answers will be completely anonymous.

Instructions: I'm going to read you five statements, and after each one please say whether you agree strongly with the statement, just agree with it, neither agree nor disagree, disagree, or disagree strongly with it. Okay?

Then read each statement to the respondent, and record the answers in the spaces provided on the data recording sheet. The numerical values to record for each response are as follows:

"Agree strongly" = 1
"Agree" = 2
"Neither agree nor disagree"= 3
"Disagree" = 4
"Disagree strongly" = 5

Here are the five statements you will be presenting. (Take a few minutes reading them aloud before conducting the actual interviews.)

1. Poor people usually are poor through no real fault of their own.
2. There are plenty of people in America who succeed mainly because of connections.
3. If people want to succeed badly enough, they usually can.
4. Those self-help books and television programs often work if a person takes them seriously.
5. It would be a good idea if we changed a lot of the rules in society so that more Americans could have a chance to succeed.

Recording the Responses

Record the responses of the ten people in your sample on the data recording sheet. You will need to summarize their responses before they can be analyzed. For each respondent, calculate a score (from items 1–4) that we can call the "Beliefs about the origins of success and failure in America" scale. If you look at the four statements that form the scale, you will see that statements 1 and 2 argue that success is influenced mainly by forces beyond individuals and that statements 3 and 4 argue that success is due largely to individual merit. The reason there are two of each is to keep a respondent from agreeing automatically with all the statements, a tendency called "agreement set" or "yea-saying." Thus, to express their ideology strongly, people who think success is due to social forces must agree strongly with statements 1 and 2 and disagree strongly with statements 3 and 4. Reversing the expression of half the questions is intended to counter the "yea-saying" tendency. And as you can see in the following sample response, you score these items by reversing the number values below items 3 and 4. This way, higher scores on the items indicate greater belief that success is due to the characteristics of individuals. Be sure to reverse the values of the responses to items 3 and 4 before recording them on the data recording sheet. Here is an example of the scoring for one response:

Scoring a sample response

1. Poor people usually are poor through no real fault of their own.

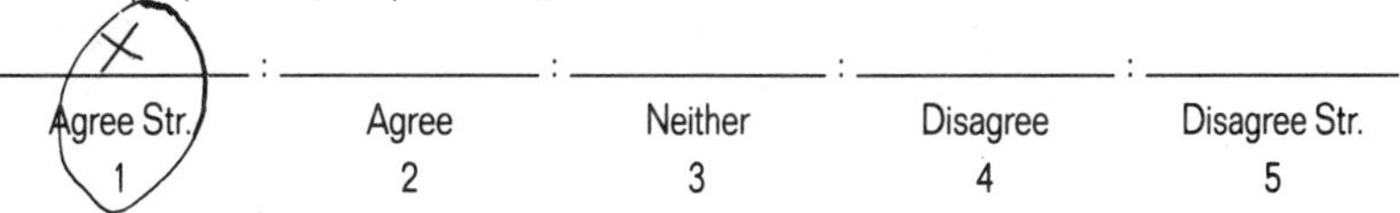

2. There are plenty of people in America who succeed mainly because of connections.

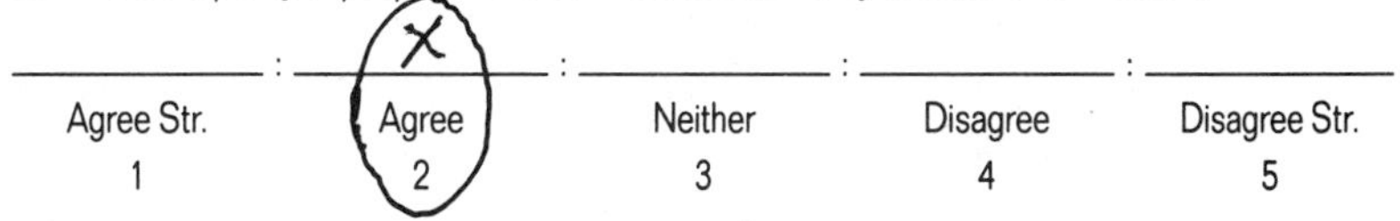

3. If people want to succeed badly enough, they usually can.

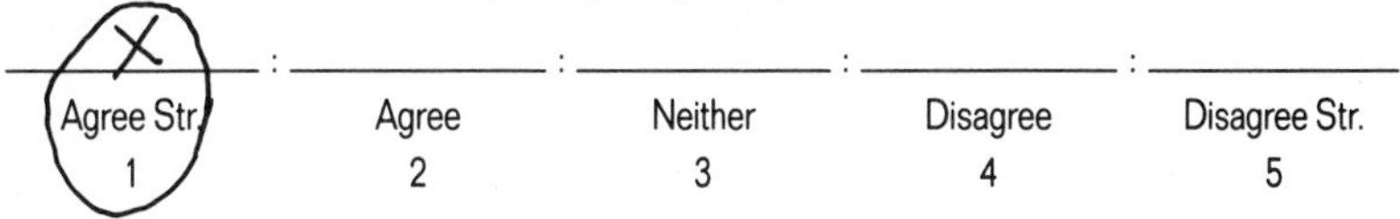

4. Those self-help books and television programs often work if a person takes them seriously.

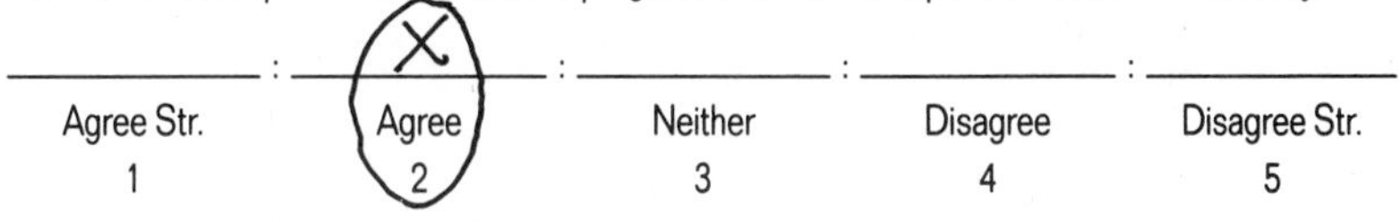

Here is the ideology score of this sample response, to illustrate how to score items 1–4.

Item No.	*Score*
1	1
2	2
3	1
4	2
Total =	6

(This is the scale score for the respondent's ideology.)

The responses to item 5 are not reversed. So if this same person had responded to item 5 by strongly disagreeing, you would score it with a "5," as follows:

5. It would be a good idea if we changed a lot of the rules in society so that more Americans could have a chance to succeed.

______ :	______ :	______ :	______ :	X
Agree Str.	Agree	Neither	Disagree	Disagree Str.
1	2	3	4	5

Data Analysis Instructions

Once you have recorded the responses of the ten people in your sample to all five questions and summed each respondent's adjusted scores for items 1–4, you are ready to analyze the data. The idea is to see if people who scored highest on the ideology measure have different attitudes toward social change than those who scored lowest on the ideology measure. To find this out, identify the five highest-scoring respondents on the measure of "Beliefs about the origins of success and failure in America." These are the people who believe that "success is due to individual effort." On the data analysis sheet record these people's scale scores (items 1–4) in the spaces on the left, and also record their scores for item 5. In the spaces on the right of the data analysis sheet, put the scores of the five respondents with the lowest scale scores, along with their scores for item 5. These are the people who believe that "success is due to forces beyond the individual." Now add up the item 5 scores for the people with the "individual effort" ideology, and put this sum in the space at the bottom of the column. Do the same for the people with the "social forces influence success" ideology. What do you find? Is there an apparent relationship between the ideology of the respondents and their belief about the advisability of change?

Some Potential Problems in Doing This Kind of Research

One problem that can occur in doing this application is not finding different ideologies represented in your sample. For example, if none of the ten respondents think that success is due to individual effort, then you cannot attribute differences in support for social change to differences in ideology. You might have to keep looking for people with differing ideologies to make the application work. Another problem is people's willingness to answer questions like these. For one thing, the questions sound artificial: this simply is not how such questions normally are raised or discussed. Also, some people don't like to give short, simple responses to complex statements. It is a shortcoming of research that uses this kind of question that certain people will not take part. It is a good idea to recognize the weaknesses in any type of research. Keep in mind that these are just brief illustrations of the research process. Full-scale research is much more exacting.

Data Recording Sheet Scores for items 1–4, with items 3 and 4 adjusted for "yea-saying"

Respondent No.	Item 1	Item 2	Item 3	Item 4	Sum of 1–4	Score for Item 5
1	______	______	______	______	______	______
2	______	______	______	______	______	______
3	______	______	______	______	______	______
4	______	______	______	______	______	______
5	______	______	______	______	______	______
6	______	______	______	______	______	______
7	______	______	______	______	______	______
8	______	______	______	______	______	______
9	______	______	______	______	______	______
10	______	______	______	______	______	______

Data Analysis Sheet

Five highest scores on the ideology measure (1–4)	The same respondents' scores on item 5 (Column A)	Five lowest scorers on the ideology measure (1–4)	The same respondents' scores on item 5 (Column B)
______	______	______	______
______	______	______	______
______	______	______	______
______	______	______	______
______	______	______	______
	Sum of column A =		Sum of column B =

Structural Components of Society

Any structured system, whether it is a small animal, a huge building, a family, or social order as a whole, has component elements. This part deals with the structural components of society, including he various ways they fit together to form substructures ranging in size and complexity from small groups to large institutions.

Concept 13 examines social status and social role and their relationship to social structure. A social status is a position within a social strucure, and a social role is the behavior expected of an individual in a particular status. Status and role form the basic building blocks of social structures, whatever form those structures take.

Concept 14 deals with groups, ranging in size from the dyad (a twoperson group) to the society. In everyday use, the word *group* can refer o almost any collection of people. But the sociological definition requires hat group members share interests, interact with one another, share ights and obligations of membership, and have a sense of belonging to he group.

Concept 15 begins with a discussion of formal organizations, groups organized to achieve some specific goal by coordinated, collective effort. t then focuses on bureaucracies, special types of formal organizations eaturing an extensive hierarchy of authority and expertise, a division of abor, and explicit rules of performance for its members.

Concept 16 focuses on institutions, which are sets of rules establishng how group members agree to accomplish universal issues of survival, uch as the procreation, care, education, and socialization of the young or the distribution of power. Several specific institutions, such as the amily, schools, government, and the economy, are discussed.

CONCEPT 13

Social Status and Social Role

Definition

Social Status A position or place within a social structure to which certain rights and obligations apply.

Social Role The behavior expected of an individual because of the social status he or she occupies in a given social structure.

During high school I worked part time as a file clerk for a large insurance company. On the wall of one room was an organizational chart of the company. It looked like a huge pyramid with some complex jigsaw pieces sticking out of the sides at a few spots. At the tip was the chairman of the board; below were all the board members and the presidents of various divisions; a number of department heads came under each president; supervisors were under these; and so on. At the bottom layer were the line workers, such as secretaries and clerks (including me). Anyone who wanted to could find his or her own position in the company by looking at the chart. Perhaps it was meant to give employees a sense of belonging in the company or a desire to advance within its structure. Some people like having very concrete goals, and the chart may have given them a clear picture of the future they could shoot for in the corporate structure. (It only made me feel insignificant to see where I was on the chart and how far it was to the top. Maybe that's why I didn't go into the insurance business.) From that chart an employee could tell exactly where she or he was in the organizational structure and who was in a position to give orders to whom (social status). Before long it became apparent just how each employee was expected to behave in his or her position (social role). I was expected to shut up and take orders—from everyone.

Social Structure, Social Status, and Social Role: An Overview

Social structure, status, and role are among the most basic and useful concepts in sociology. They form important building blocks for understanding the way social interaction takes place. I want to take a little time to define them and show how they relate to one another before discussing them in greater detail.

Social Structure

To begin with, relationships in society tend to be patterned and stable. To describe the overall pattern of such relationships, sociologists use the concept of **social structure.** A social structure is a pattern of social relationships that forms the stable framework within which social interaction takes place. Social structures range in size and complexity, from the friendship or marriage of two people, through small, voluntary organizations, such as clubs and fraternities, to larger, more structured organizations, such as colleges and businesses, to huge organizations, such as the largest corporations in America, the American military, and the executive branch of the federal government. The relationships in social structures can be depicted in the same way that the insurance company I worked for depicted its relationships in its organizational chart. The same kind of schematic could be drawn for other social structures, for example, friendships (Friend #1—Friend #2), families, colleges, and the executive branch of the federal government (*big* schematic).

Social Status

Each position in a social structure is a **social status.** The status gives a name to a spot within the structure (such as "file clerk," "lieutenant in the navy," or "fraternity pledge") and suggests the rights and obligations that apply to the position. For example, in the insurance company I knew not only where I fit in the structure of the company but also who was allowed to give me orders (anyone) and to whom I could give orders (no one). Social statuses, then, are consequences of the structures in which they are located. If the company had hired office messengers of even lower status than clerks, my status would have changed. In fact, just this year the college where I work hired two new deans. The organization of the college had been changed to create colleges of arts and sciences and of education and the responsibilities of the deans of these new colleges were spelled out in a series of documents that also specified to whom everybody would be responsible from that point. Statuses were changing all over the place, and with them, the behaviors of employees were expected to change.

Social Role

Social structure and *social status* are terms that are useful for describing *relationships* but do not focus on *behavior*. That is where the concept of **social role** comes in. Social roles are sets of expectations for the behavior of a person in a given social status. To use the example of my job in the insurance company, I was expected to do filing and be willing to take orders from anyone who wanted to ask work of me—not a very complicated role. As you will see later, some roles are composed of much more complicated sets of expected behavior. For now, it is enough to realize that *a social role is basically a social status in action*. Because of the status a person occupies in a social structure, certain kinds of behavior are expected of her or him by others (a role). That is how social interaction is generally organized.

The concepts of status and role are so basic to the sociological vocabulary that they deserve more extensive treatment than I have given them so far. Also, each has become part of the everyday vocabulary used in ways other than those used by sociologists. Now I want to give a more detailed and specific discussion of each.

Social Status as a Relational Concept

A status within a social structure makes sense only in relation to the other statuses in that structure. Ralph Linton, who did most of the work in defining these ideas, put it this way: "The position of the quarterback is meaningless except in relation to the other positions. From the point of view of the quarterback himself it is a distinct and important entity. It determines where he shall take his place in the lineup and what he shall do in the various plays" (Linton, 1936:114). A quarterback without the game of football is as meaningless as the president of a nonexistent country. (This always reminds me of Gary Trudeau's "Doonesbury" comic strip character B.D., who was quarterback of his college football team and who, ever since graduating, has never taken off his helmet. In real life, in fact, former governors and presidents often are addressed by their old titles, long after they have retired from office or been defeated. This may seem awkward; for instance, when Richard Nixon is now addressed as "Mister President," everyone knows he is no longer in the governmental social structure and no longer has the formal status of president. Of course, there may be some other social structure, such as "former holders of high office," in which the status has meaning, which is why the title is still applied.

The status a person has is defined by his or her relationship to others within the social structure, which is expressed in behaviors toward them. So the status of the quarterback is given meaning only in terms of the things he or she can and must do *because of his or her relationship to the other positions on the team* (the ideal or theoretical rights and obligations that apply to the position). As Linton recognized, because rights and obligations "can find expression only through the medium of individuals, it is extremely hard for us to maintain a distinction in our thinking between statuses, and the people who hold them and exercise the rights and duties which constitute them" (Linton, 1936:113).

Everyday Meanings of Status

Knowing the sociological meaning of the term *social status* can help us understand everyday uses of the term, in expressions such as *high status, low status, status job,* or *status seeker.* When we think of a person with status, we usually mean someone who is respected and has a good deal of influence. In this sense, status is related to a person's position not only in a structure but also in a *hierarchy* of power or esteem. Sometimes status is defined in terms of this sense of higher and lower positions. For example, one sociologist (Henslin, 1975:33) defines status as "the standing or position one occupies within his or her social class." High-status people, then, are those near the top of some social structure in which differences in wealth, power, or esteem are considered important. Low-status people are members of the lowest social class and those held in low regard by others in the hierarchy.

Generally, a person who is high in status within a system of relations has high status in all the elements or measures of that system. For example, a person who is very wealthy probably also commands both power and the esteem of others. But sometimes a person experiences what is called **status inconsistency,** high status in some measures of social standing and low status in other measures. A person may have high status in one regard (perhaps he or she earns a great deal of money) but low status in other areas of life (for example, because he or she earned all that money as a contract murderer or by some other reprehensible means). Some American self-made millionaires, who have little education but a great deal of drive, seem to feel this status inconsistency. Once they become wealthy, they get little thrill from making more money. Instead, they may seek to raise their esteem by purchasing aristocratic titles in Europe or by donating money to universities with the expectation of receiving an honorary degree or two in return. This also illustrates the term *status seeker,* someone who wishes to associate with high-status individuals, such as judges, surgeons, or quarterbacks.

Such everyday uses of the term *status* to denote a position in a hierarchy can, at least for the sociologist, be misleading. Almost all American social structures seem to have the quality of hierarchy to one degree or another, possibly because Americans are so competitive and concerned with performance. But hierarchy is much more important in some social structures than in others. In the military, for example, rank is formalized, and each status carries extremely specific rights and obligations, which are visibly symbolized by uniforms. Officers enjoy a wide range of advantages and bear correspondingly greater responsibilities than enlisted personnel. In families, however, a hierarchical definition of status is too narrow. Parents may have more power than children, but children are held in as high regard as parents within the family and may actually exercise more rights (while parents are heavy on obligations). In the family, son or daughter is not necessarily a *lower* status than mother or father. It is merely a *different* set of rights and obligations implied by the position. It makes little sense to say that a particular status in the family is more desirable or higher in esteem than another. Here, the sociological definition of status is much more useful than the everyday definition.

Ascribed Versus Achieved Status

How do people come to occupy a particular status? You are born a son or a daughter, a male or a female, a black or brown or white or other race. You have no say in the matter. Those are your **ascribed statuses.** In contrast, statuses that you have as a consequence of what you have *done,* either intentionally or not, are termed **achieved statuses.**

I achieved my status as a file clerk as a consequence of applying and being hired for the position. People act in order to get married and have children, so father or mother is an achieved status. Quarterbacks, contrary to what you may have heard from television commentators, are made—not born. They practice, exercise, and work to earn the opportunities to play the position. This distinction between achievement and ascription, for which Linton (1936) was also responsible, becomes very important when we realize that some people's opportunities are either limited or expanded, automatically, by the statuses into which they are born. In American society we believe that a person's well-being ought to be a consequence of her or his own efforts. We are justifiably angered when something other than actual merit determines our own opportunities. We ought to be just as angry when something other than merit (such as sex, race, age, or any other ascribed status) limits the opportunities of other people in society. (For a more complete discussion of this issue, see Concept 18, on discrimination and prejudice.)

Social Role as a Behavioral Concept

Statuses get their behavioral content in the form of role expectations. Just as the lines in a play "fill in" the character of the actor, so the social role attached to a status establishes expectations for the behavior of individuals, making their behavior predictable and orderly for the people with whom they interact.

Everyday Life as Role Playing

In Shakespeare's play *As You Like It* (1975:239) the character Jaques begins a famous speech with the following lines:

> *All the world's a stage,*
> *And all the men and women merely players.*
> *They have their exits and their entrances,*
> *And one man in his time plays many parts.*

Sociology has borrowed directly from theater in developing the concept of the social role. In films or plays the author sets the part in a dramatic situation. In everyday life we take on the role we are expected to play, depending on the situation we enter. We understand the demands of a particular social role only to the extent that we are familiar with the situation; once we are familiar, we play our part as if it were scripted.

We act differently from one moment to the next depending on both the situation and the people involved. At school, for example, you probably speak and act differently with your friends than with your teachers or with the people who work in the cafeteria or bookstore. In each role you play, people expect you to act toward them the way you do. In fact, throughout the day it is normal for you to change roles many times. You may begin the day by playing the part of roommate, friend, sister or brother, son or daughter, or parent. On the way to school you may be a driver or a commuter on public transportation. Later you are a student and, if you have a job, an employee, a coworker, or perhaps a boss. If you go out to dinner or shopping, you will play the part of a consumer or customer. At a party you could be either a guest or a host, and different types of behavior would be appropriate for different types of parties. Is it an evening tea gathering with the college dean and her husband or a bring-your-own-beer pizza party with your friends? You certainly would not be expected to act the same at both.

The stability of social life depends on the predictability of others' behavior. We are most comfortable when our guesses about what will happen are accurate and most upset when they are wrong. To avoid the embarrassment of acting "incorrectly" in a social situation, we look for cues beforehand. If you were invited to a dinner party, you would probably need to know what time it would start, who would be there, what kind of dinner it would be (a formal, six-course dinner or a cookout), and so on. Once you have pinned down what kind of event it will be, you can prepare to dress and behave correctly (or to develop an unfortunate illness that prevents your attendance).

From infancy, our training in social roles begins. "No, Charlie. Eat with your spoon at the table." Notice that Charlie, who is two, is allowed to eat with his fingers when he is sitting on the floor. Or "I know we told you to tell the truth all the time, Jennie, but you shouldn't have told Aunt Harriet she is fat, even if she is." It is a very subtle social skill for a nine-year-old to learn the situations in which the truth is permissible. By the time we are adults, we can deal efficiently with hundreds of different social situations, switching smoothly from one to another, as long as we know beforehand something of what the new situation will be like.

The Flexibility of Roles

Knowing the rules for behavior in a given situation makes it possible for us to fulfill one another's expectations. These rules are socially agreed on (normative), but they are not totally uniform and inflexible. There is a range of behaviors within which we can satisfy the requirements of a role. The flexibility of role requirements is evident in the behavior of teachers. I have had teachers who seated students alphabetically, took careful attendance, and structured every minute of every class. I have also had teachers who asked the class to meet in the cafeteria for very open discussions, loosely based on assigned readings. Both styles satisfied my expectations of teacher behavior, although they were from the opposite extremes of the range of role behaviors.

Social roles are even flexible enough to allow for the expression of our individuality by a process that the sociologist Erving Goffman (1961b) called **role distancing.** We do not all enter the same situation the same way. We are different from one another. This is one of the main reasons why teachers fulfill the role of teacher in such different ways. But sometimes we wish to deliberately separate ourselves from the role in which we are acting. We may dislike playing the role but find ourselves in it due to unavoidable circumstances. According to Goffman, we may express our individual characters within the role by behaviors that show who we are "underneath." For example, a young boy may distance himself from what he sees as the childishness of riding a tame merry-go-round by performing daredevil stunts, whereas an adult man might express the same sort of role distancing by feigning an exaggerated concern with the security of his seat belt. Still, our fulfillment of others' expectations, within the broad range of performance for the role, expresses the stability and predictability of human social behavior.

Role Set

In a given social situation the statuses of the individuals are spelled out in each person's rights and obligations. They imply the behaviors each is expected to exhibit. In a college, for example, one status I understand well is that of faculty member. As a faculty member I am expected to act in certain ways depending on the people with whom I am interacting. Students expect me to hold classes, lecture, answer questions, give exams, know something about the subject I am teaching, and give out grades. Other teachers *do not* expect me to lecture them, give them exams, or record their attendance; they expect me to discuss teaching or research and writing or to collaborate in professional projects. I am expected to act in yet other ways toward administrators, staff members at the college,

FIGURE 13-1
Role set of a faculty member

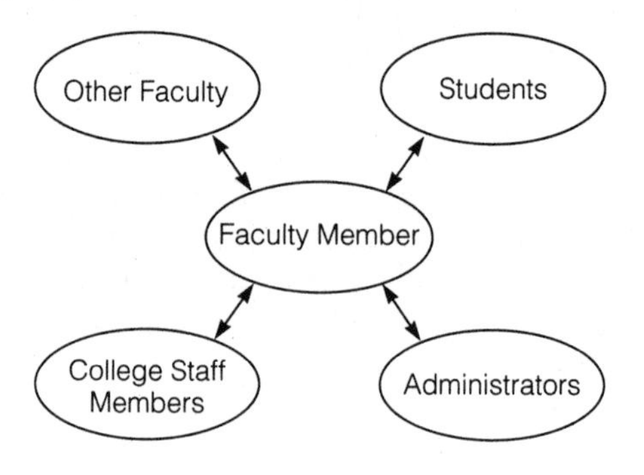

and so on. For every status, then, there is a set of role relations, a different one oriented toward each other role. This is called a **role set** (Merton, [1949] 1968).

Figure 13-1 is a picture of what the role set of a faculty member might look like. Notice that each role relationship is reciprocal (the arrows go in both directions). This indicates that, just as the faculty member is expected to act in specific ways toward the person occupying each other role, so that person is expected to fulfill the expectations held for him or her by the faculty member. In life, as in a play on the stage, when everyone knows his or her part, everything goes well. People get very uneasy when someone steps out of character, as if she or he were saying lines from the wrong play.

Role Confusion, Role Strain, and Role Conflict

Because we all have many statuses and each status has a number of roles related to it, it is inevitable that we will run into problems trying to fulfill all the expectations for our behavior in all situations. For example, at the college where I teach, most of the students have part-time jobs. Fairly often I meet a student of mine waiting on tables in a restaurant where I am eating or working as a sales clerk in a store where I shop. "What are *you* doing here?" is the students' most common reaction to my presence, as if they were amazed to learn that college teachers eat or find it necessary to buy underwear. But their surprise is understandable, since the social status situation is confusing. At the first moment we see each other, am I a teacher or a customer? Should the student behave as a student or a sales clerk? Our "regular" classroom roles are wrong for the situation, but these are the roles to which we have become accustomed. This is an example of **role confusion.** Because the situation is not very complex, it takes only a few moments to settle into the familiar roles of customer and sales clerk. We can figure out what roles we should have been playing.

Sometimes, however, the difficulty of knowing how to behave is built right into a role set. I once saw my eight-year-old niece trying to decide whether to tell her parents that her older brother was doing something dangerous. She simultaneously felt loyalty to her brother and a responsibility to keep him from harm. (She resolved the problem by telling her brother that if he didn't stop, she would tell on him.) She had experienced a **role strain,** conflicting expectations about her behavior within a single role set (Goode, 1960). As a family member she was expected by her parents to behave one way and by her brother to act another way. Another example of role strain occurs when a shop supervisor is expected both to implement management's policies for cutting costs and increasing productivity and to safeguard the workers for whom he or she is responsible.

Each member of society has many social statuses and therefore must fulfill the expectations of behavior for many role sets. Sometimes difficulties in role performance are brought about by conflicts between a person's different role sets rather than within them. Such a situation is called a **role conflict.** A very common (and very serious) example is conflict between an individual's roles as worker and as family member. The employer expects workers to put in long hours and to place the company's interests ahead of individual interests. The family expects its members to be available for both work at home and leisure time. How the conflict is resolved (if it is) depends on how the individual defines the legitimacy of the competing demands. Many marriages have ended when the demands of a job took precedence, and many careers have failed to advance when the needs of the family were made primary.

Lastly, it is all too easy these days to see all three versions of role difficulties—confusion, strain, and conflict—as operating within the life of one person. For example, a number of studies of the "superwoman" syndrome suggest that some women have piled new opportunities for careers and independence on top of old expectations of female role performance in an attempt to "do it all." Trying to fulfill all the role expectations of family and career can create (1) role conflict (Do you stay at work longer, or leave work to go see your child in the school play?); (2) role strain (Do you behave as the boss expects, or as the people who work for you expect?); (3) and role confusion (How do others know when to act toward you in your role as mother versus your role as career woman?)

Summary

Human relations do not occur at random. They are patterned and stable. Sociologists use the term *social structure* to refer to a pattern of social relationships that forms the stable framework within which social interaction takes place. Social structures are similar to the blueprints for a building. They depict the design for a stable structure, showing all the elements and how they fit together. Each position within a social structure is called a *social status.* As Ralph Linton pointed out, social statuses are relational; that is, they make sense only "in relation" to other statuses. Within social structures, each status has attached to it certain rights and obligations.

Social structure and *social status* are analytical terms. That is, they are used to describe relationships but do not focus on behavior. Sociologists use the term *social role* to refer to the actual behaviors expected of a person because of the social status he or she occupies in a social structure. A social role, then, is a social status in action.

Social role and social status are very basic to the sociological vocabulary. In a sense they are the core elements of social relations, just as bricks and beams are the core elements of physical buildings. Beyond their sociological meanings, each of these terms has everyday uses as well. For example, status has come to refer to the fact that people occupying positions near the top of social structures are able to command higher incomes and to wield power over others. They are often called *high-status* people. In addition, sociologists distinguish between ascribed status (a status into which one is born) and achieved status (a status one occupies as a result of what one has accomplished). Status inconsistency is a condition in which an individual has a high status in one measure of social standing but low status in another.

The term *social role* has some special uses as well. People can fulfill the expectations of a role with a wide variety of behaviors. Thus, roles have some flexibility. Sometimes people wish to express their individuality while still fulfilling the expectations attached to a role; Erving Goffman called this practice *role distancing.*

Because any social status establishes relationships with many other statuses in a social structure, many different kinds of behavior can be expected of the occupant of a status. Sociologists use the term *role set* to refer to all the roles associated with a particular status. Different kinds of behaviors are expected of a person in a given role. In addition, all individuals belong to a number of social structures and so must fulfill the expectations of a number of role sets. Problems are inevitable. *Role confusion* is the problem of knowing what role to play at a given time. *Role strain* is a conflict in the expected behaviors of an individual within one role set. *Role conflict* is a conflict in the expectations for a person's behavior when he or she occupies two or more role sets simultaneously.

The Illustration for this concept summarizes a debate about the statuses and roles of older Americans. On one side of the debate is disengagement theory, which argues that it is best for the well-being of the society and the elderly person to leave the statuses and roles of mid-life. On the other side of the debate is activity theory, which claims that it is best for the older person to maintain activity and role engagement into late life. The illustration also shows how such debates are sometimes argued in the sociological literature.

The Application for this concept is a survey of the relative status of a number of American occupations. Over the years, surveys of occupational prestige have shown that some occupations are consistently held in high regard by most Americans (physician and U.S. Supreme Court justice, for example), whereas others are evaluated as having low status (garbage collector and bartender). Still others may change in status ranking (the prestige of lawyers and politicians suffered greatly after the Watergate scandals). Some occupations have been selected from previous national surveys for reexamination in the Application. Using the survey materials provided, you will determine the relative rankings of sixteen occupations and examine whether these rankings have changed over the years.

Illustration

Elaine Cumming and W. E. Henry, Growing Old: The Process of Disengagement *(New York: Basic Books, 1961)*

Elaine Cumming, "Further Thoughts on the Theory of Disengagement," International Social Science Journal *15 (1963): 377–93*

B. W. Lemon, K. L. Bengtson, and J. A. Peterson, "An Exploration of the Activity Theory of Aging: Activity Types and Life Satisfaction Among In-Movers to a Retirement Community," Journal of Gerontology *27 (1972): 511–23*

C. F. Longino and C. Kart, "Explicating Activity Theory: A Formal Replication," Journal of Gerontology *36 (1982): 713–22*

What do you think you will be doing when you are seventy years old? Will you still be working, the head of your family, the head of a corporation, retired and doing whatever you like (so long as it has no deadlines or responsibilities attached), or simply inactive and bored? Perhaps you think it depends on the kind of person you are. Well, yes, it probably does, in part. But one of the main benefits of the sociological perspective is to point out that the circumstances of our lives are powerfully influenced by the social forces around us. In this case you might get a good hint about what is in store for you later in life by looking at the broad pattern of social relations between older Americans and the rest of society. What sort of status and role does an older person have in American society?

A social status is a position within a social structure to which certain rights and obligations apply, and a social role is the behavior expected of individuals because of their social status. In the Definition section of this chapter, status and role were discussed as held by individuals. However, social statuses, and the roles associated with them, are distributed in patterns in society. For example, positions of power within American corporations tend to be held by white males. The status and role of homemaker (caring for home and children) are overwhelmingly held by females. So, to get back to predicting your future, are there particular positions Americans are expected to occupy, and certain kinds of behaviors they are expected to exhibit, when they reach the age of sixty-five or seventy? Are older people generally expected to be active participants, or even powerful leaders, in society, or are they expected to withdraw to some degree from engagement in the economy, the government, and families?

This illustration summarizes one of the classic debates in sociology. Its participants (activity theorists versus disengagement theorists) represent opposite positions about the proper status and role for older Americans. This debate illustrates two things: (1) that the concepts of *status* and *role* have application beyond describing the situations of individuals; and (2) how debate is conducted in sociology's books and journals. This illustration is about more than just ideas and data; it also has consequences for the social policies we devise to deal with issues such as aging.

Disengagement Theory

In 1961 Elaine Cumming and William Henry published *Growing Old: The Process of Disengagement* (1961). They began their work by disagreeing with what they saw as the commonly held belief that the best thing for the individual to do in late life is to remain active. According to this "activity" theory, our social role engagements are seen as good for us in much the same way that physical exercise is. This idea is commonly expressed in the saying "Use it or lose it." But Cumming and Henry argued that the inevitability of illness and death in late life made disengagement from social involvements functional for both society and the disengaging individual. They defined *disengagement* as "an inevitable process in which many of the relationships between a person and the other members of the society are severed, and those remaining are altered in quality" (1961:211). Their book formed the basis for the perspective on aging called disengagement theory.

Their ideas were based on what they thought it takes for people to fulfill social roles. During most of our lives we are actively engaged with the rest of society through our social roles. These "knit" us into the social system because other people have expectations for our role performances and because so many roles are interdependent. But our social roles require energy and skills if we are to operate in them satisfactorily. Cumming and Henry cited the abundant data about physical and psychological declines in later life to support their belief that the older we get, the less able we are to fulfill the expectations others have of us in roles such as head of family, worker, and community leader. For example, people who are forgetful or ill or distracted by the difficulties of late life presumably are less able to perform social roles. Cumming and Henry suggested that it would be disruptive for both society and the aging individual if older people tried to maintain into later life the role involvements of their mid-lives. Society would be disrupted because its jobs would be performed less efficiently. For example, a critically important older worker who stays on the job might take more sick days, forget important information, or be unable to learn new skills when necessary. For the older individual, maintaining mid-life roles into late life would be dysfunctional in two important ways: First, Cumming and Henry reasoned, late life requires a person to face different sorts of tasks than do earlier stages of life. These can include straightening out the material affairs of one's life (such as arranging for life in retirement and putting an estate in order for heirs) and coming to terms with philosophical issues (such as weighing one's accomplishments or evaluating the meaning of the life one has lived). Second, they suggested, a person who stays engaged in critical role involvements too late in life risks feeling guilty for the disruptions caused to others by his or her inefficient role performances. So disengagement is seen partly as a sacrifice the aging individual is motivated (as a member of society) to make for the good of others and partly as an act of self-interest so she or he can face the special challenges of late life. To sum up, then, disengagement theory contends that disengagement is functional for society because it removes people from important roles before they become inefficient, and functional for aging individuals because it frees them (1) from feeling guilty for disrupting society, and (2) to face the challenges of late life.

Publication of this statement of disengagement theory immediately stimulated energetic criticism. For example, some critics pointed out that if disengagement was so functional for the disengaging individual, why did some people seem so reluctant to give up lifelong roles? And why did other people seem to have such a difficult time in retirement? Two years after publication of *Growing Old*, Elaine Cumming published a journal article entitled "Further Thoughts on the Theory of Disengagement" (1963), in which she made adjustments to the original theory in response to some of the criticisms. Primary among these changes was her acknowledgment that some people have a more difficult time with disengagement than others. Specifically, Cumming said, men have more difficulty with disengagement than women because men tend to play the "instrumental," or provider, roles in society, that is, the roles that focus on the accomplishment of tasks in the workplace. By contrast, women typically spend their lives performing the "socioemotional" roles in society, those focused on relationships in the home and community. When a man disengages, (especially when the holder of an instrumental role is forced to retire from it), he faces the tasks of later life that require socioemotional skills he often lacks. He is at a loss. For the woman, the loss of mid-life roles such as family or community involvement are not so disruptive. This is because she still faces tasks that require her socio-emotional skills, such as developing new relationships in old age and integrating her retired husband into the home and retirement.

Activity Theory

Disengagement theory began with a criticism of what Cumming and Henry saw in 1961 as a generally unstated assumption in American society (they called it an "implicit" theory) that remaining active would create the greatest satisfaction in late life. But activity theory was not stated explicitly in the journal literature until 1972, in an article entitled "An Exploration of the Activity Theory of Aging: Activity Types and Life Satisfaction Among In-Movers to a Retirement Community" (Lemon, Bengtson, & Peterson, 1972). The researchers wanted to state specifically why role losses due to disengagement in later life would lead to reduced life satisfaction, and to generate some data supporting the theory.

Lemon, Bengtson, and Peterson reasoned that social role involvements contribute to life satisfaction at all stages of life, because people need interaction with others in order to get reinforcement for their sense of self. Sociologists generally agree that we develop our sense of who we are from others' behavior toward us. Basically, we become who we are because we fill in that information on the basis of the way others treat us. And the process does not stop at the point when our "self" has been created. We need reinforcement of that self throughout life. That reinforcement comes from interactions with other people in our roles. According to the authors of activity theory, if we lose these opportunities for social interaction because we disengage (or are forcibly disengaged), our sense of self is "starved" for reinforcement and our life satisfaction will decline.

To test if this is true, they interviewed people who were planning to move to a retirement community in Southern California. The researchers measured both the levels of activity and life satisfaction among the study subjects. Life satisfaction was measured by an established, thirteen-question scale that had been used in previous research. To measure activity levels, subjects were asked to estimate (from low to high) how much time they spent engaged in each of three types of activity: (1) social activity, such as interaction with friends, neighbors, and relatives, (2) formal activity, such as attendance at church or meetings of community organizations, and (3) solitary activities, such as reading, watching television, and knitting. Of course, the researchers hypothesized that subjects reporting greater activity levels would also have the higher scores for life satisfaction. What they found must have disappointed them.

In general, life satisfaction seemed unrelated to activity levels. For the overall sample, scores for life satisfaction were uncorrelated with formal activity, solitary activity, or social activity. It turned out that there was only one, narrow set of conditions in which higher levels of activity were related to higher life satisfaction: among women in the sample (though not among the men) there was a relationship between life satisfaction and levels of social activity with friends and neighbors (though not with relatives). It seems like a pretty flimsy hook on which to hang a theory, doesn't it? After all, wasn't the theory supposed to apply broadly to social engagement in later life?

Let me give you at least one more chapter in the story. In 1982, ten years after the Lemon, Bengston, and Peterson study, activity theory was retested in a study by Longino and Kart entitled "Explicating Activity Theory: A Formal Replication." They thought that the way the 1972 study was conducted might have prevented it from generating the results that would adequately support activity theory. Longino and Kart improved on the original study in a number of important ways. For example, they used a larger, more representative sample from three regions of the country (not just the one region sampled in the earlier study). Second, the people in their sample were already living in retirement rather than planning on moving into a retirement village (as was the case in the earlier study). The reason this is an improvement is that respondents in the earlier study were reporting on activity levels from before retirement, while respondents in the Longino and Kart study had time to change their patterns after retirement. Third, Longino and Kart improved the measure of activity levels by having respondents keep a diary of what they did with their time. This allowed a more accurate measure of the time spent in each type of activity (social, formal, and solitary) than did the recall method in the earlier study. They used the same sort of measure of life satisfaction as before. This time the results were strong, clear, and somewhat surprising.

First, they did find support for activity theory. In all three samples they found positive relationships between scores for life satisfaction and social activities of all kinds. And this time the positive relationship was found among women *and* men. There was also no relationship found between life satisfaction and solitary activity. Older people who read, watched television, or knitted more than others did not seem to be more satisfied with their lives. This make sense in terms of the reasoning in the original statement of activity theory, because solitary activities do not involve interaction with others and therefore cannot be used by individuals to reinforce their sense of self. But what about formal activity?

Longino and Kart found that formal activity actually was *negatively* related to the life satisfaction of the respondents. That is, the people who engage most in the formal activities, such as attending church or meetings of community organizations, had significantly lower levels of life satisfaction than did those who engaged less in these activities. Why? After all, these kinds of activities involve interaction with others. Why don't they contribute to the sense of self of the participants? Longino and Kart speculated at length about why they found this. I'll tell you my favorite guess, then leave it to you to speculate on some explanations. One possibility is that formal activities might be the only ones available to some people. They want some interaction with others but, for some reason, lack the opportunity to interact socially with friends, neighbors, or relatives. If the reason is that these people are not very pleasant or happy in the first place, then they might also not have choices about activities other than formal ones. If this seems far-fetched, I suggest you come up with more plausible explanations, then conduct the research to test them. That is how a great deal of social research is started: by someone reading a study in which unexpected findings are presented and then doing the research to explain why they appeared.

Application

In American society one of the most important social structures, and one of the largest in terms of numbers of people, is the occupational structure. Using a scale designed to measure the status of a variety of occupations, researchers have collected strong evidence showing that most Americans agree about which occupations are high in prestige and which are low. These national evaluations of occupational prestige have been very stable over the years. Occupations ranked highly in 1947 were ranked in about the same positions in 1963, and this stability was true throughout the original list of ninety prestige rankings (Hodge, Siegel, & Rossi, 1964).

A Survey of Occupational Status

Listed in this application are sixteen occupations from the original ninety. I have also provided a five-point rating scale so that any person can give his or her own opinion of the general standing of that occupation in the society: (1) a poor standing, (2) a slightly below-average standing, (3) an average standing, (4) a slightly above-average standing, or (5) an excellent standing. You are to ask a sample of ten college students to rate each of the sixteen jobs. You should practice reading the introduction and instructions until they sound fairly conversational.

Introduction: I am doing a survey of students' attitudes about the status of a number of occupations. Would you mind answering a few questions? It will take less than a minute.

Instructions: I am going to read you a list of sixteen occupations. For each one, could you respond by saying whether, in your opinion, it generally has (1) a poor standing, (2) a slightly below-average standing, (3) an average standing, (4) a slightly above-average standing, or (5) an excellent standing? OK. Here are the occupations.

(Read the list of occupations, and record the respondent's answers on the Data Recording Sheet. Remember that the answers of a single respondent are recorded from top to bottom.

Analysis

Use the Data Recording Sheet to begin your analysis. Simply add the rating scores for each occupation (across the row, not down a column) and enter the total in the space labeled "Job rating total." Then, transfer these scores to the far left column of the Data Analysis Sheet, provided next. In the next column to the right, divide each job rating total score by ten to get the average scale rating for each occupation from your responses. Then, in the last column on the right, rank-order the average ratings, starting with the highest-rated occupation. For example, if the total rating score for college professor was 91, the average rating would be 9.1. If this were the highest average rating of your responses, it would be the most prestigious occupation and would be rank-ordered as 1. The next most prestigious occupation might have an average rating of 8.7 and would be rank-ordered 2. Using this method, rank-order all the occupations from 1 to 16 in the right-hand column.

Interpreting the Data

The sixteen occupations that your sample of students rated were selected from an original list of ninety occupations. They were chosen to represent the entire range of rankings. Table 13-1 lists these jobs, their ratings, and their rankings from 1947 and 1963. The occupations used in this assignment are underlined, to show their positions clearly distributed throughout the list.

Notice that the ratings and rankings of occupations were highly consistent from the first study to the second, even though sixteen years had passed. Some occupations (such as garbage collector, mail carrier, and Supreme Court justice) remained in exactly the same rank on the list. But a number of occupations changed ranks quite noticeably, although they remained in roughly the same position. Lawyers, for example, moved down seven ranks on the list. (In fact, during the Watergate hearings and Nixon impeachment proceedings, newspaper reports suggested that the occupation of lawyer had moved to the bottom end of the occupational prestige ladder, below some unskilled labor jobs.) Bankers moved up fourteen ranks during the same sixteen years.

How do the rankings of your sample of ten students differ from the data presented in the national samples? If there are obvious differences from your data, what events in the United States in the recent past might explain such differences? For example, is there anything about either the economy or foreign policy that might have influenced the prestige of a particular occupation or set of occupations? Does the prestige of bankers go up or down when interest rates climb quickly?

TABLE 13-1

Distributions of prestige ratings, United States, 1947 and 1963

Occupation	March 1947 NORC[a] Score	March 1947 NORC[a] Rank	June 1963 NORC Score	June 1963 NORC Rank
U.S. Supreme Court justice	96	1	94	1
Physician	93	2.5	93	2
Nuclear physicist	85	18	92	3.5
Scientist	89	8	92	3.5
Government scientist	88	10.5	91	5.5
State governor	93	2.5	91	5.5
Cabinet member in the federal government	92	4.5	90	8
College professor	89	8	90	8
U.S. representative in Congress	89	8	90	8
Chemist	86	18	89	11
Lawyer	86	18	89	11
Diplomat in the U.S. foreign service	92	4.5	89	11
Dentist	86	18	88	14
Architect	86	18	88	14
County judge	87	13	88	14
Psychologist	85	22	87	17.5
Minister	87	13	87	17.5
Member of the board of directors of a large corporation	86	18	87	17.5
Mayor of a large city	90	6	87	17.5
Priest	86	18	86	21.5
Head of a department in a state government	87	13	86	21.5
Civil engineer	84	23	86	21.5
Airline pilot	83	24.5	86	21.5
Banker	88	10.5	85	24.5
Biologist	81	29	85	24.5
Sociologist	82	26.5	83	26
Instructor in public schools	79	34	82	27.5
Captain in the regular army	80	31.5	82	27.5
Accountant for a large business	81	29	81	29.5
Public school teacher	78	36	81	29.5
Owner of a factory that employs about 100 people	82	26.5	80	31.5
Artist who paints pictures that are exhibited in galleries	83	24.5	78	34.5
Musician in a symphony orchestra	81	29	78	34.5
Author of novels	80	31.5	78	34.5
Economist	79	34	78	34.5
Official of an international labor union	75	40.5	77	37
Railroad engineer	77	37.5	76	39
Electrician	73	45	76	39
Owner-operator of a printing shop	74	42.5	75	41.5
Trained machinist	73	45	75	41.5
Farm owner and operator	76	39	74	44
Undertaker	72	47	74	44
Welfare worker for a city government	73	45	74	44
Newspaper columnist	74	42.5	73	46
Police officer	67	55	72	47
Reporter on a daily newspaper	71	48	71	48
Bookkeeper	68	51.5	70	49.5
Tenant farmer—one who owns livestock and machinery and manages the farm	68	51.5	69	51.5
Insurance agent	68	51.5	69	51.5
Carpenter	65	58	68	53
Manager of a small store in a city	69	49	67	54.5
A local official of a labor union	62	62	67	54.5
Mail carrier	66	57	66	57
Railroad conductor	67	55	66	57
Plumber	63	59.5	65	59
Automobile repairman	63	59.5	64	60
Playground director	67	55	63	62.5
Machine operator in a factory	60	64.5	63	62.5
Owner-operator of a lunch stand	62	62	63	62.5
Corporal in the regular army	60	64.5	62	65.5
Garage mechanic	62	62	62	65.5
Truck driver	54	71	59	67
Fisherman who owns his own boat	58	68	58	68
Clerk in a store	58	68	56	70
Milk route man	54	71	56	70
Streetcar motorman	58	68	56	70
Lumberjack	53	73	55	72.5
Restaurant cook	54	71	55	72.5
Singer in a nightclub	52	74.5	54	74
Filling station attendant	52	74.5	51	75
Dockworker	47	81.5	50	77.5
Railroad section hand	48	79.5	50	77.5
Night watchman	47	81.5	50	77.5
Coal miner	49	77.5	50	77.5
Restaurant waiter	48	79.5	49	80.5
Taxi driver	49	77.5	49	80.5
Janitor	44	85.5	48	83
Bartender	44	85.5	48	83
Clothes presser in a laundry	46	83	45	85
Soda fountain clerk	45	84	44	86
Garbage collector	35	88	39	88
Street sweeper	34	89	36	89
Shoe shiner	33	90	34	90

[a] National Opinion Research Center.

Source: Adapted from Robert Hodge, Paul Siegel, and Peter Rossi, "Occupational Prestige in the United States: 1925–1963," *American Journal of Sociology* 70 (1964): 286–302.

Is there something about your region of the country that might make data collected at your college or university differ from those in a national sample? For example, do you go to school in a coal mining region, where miners would be held in particularly high regard? Perhaps you go to school in the bartending capital of the United States.

Lastly, you probably noticed that the data used in this application are thirty years old. Because the social world can change so fast, it is critical to sociology that we update our findings regularly. Not only do we learn what conditions are, but we keep track of rates of social change. Updated data on occupational prestige ratings are only now (as of early 1993) being prepared for publication as a journal article. The National Opinion Research Center in Chicago sent me a copy of an internal paper (meaning only the scholars there have it) that reports on the methods used to update the occupational prestige rankings from 1989 data (Nakao and Treas, 1990). As of this writing the data are not in a form that can be compared directly with the data from the 1964 study by Hodge and his associates. Still, it is possible to determine from the paper that there seem to have been some changes in the rankings for the sixteen occupations underlined in Table 13-1. In each of the following cases, the prestige ratings of occupations in 1989 changed their relative positions from the rankings in 1963. For example, in 1989 the prestige rating for lawyers was slightly higher than for college professors. Unlike the case in 1963, coal miner was rated as a more prestigious occupation than truck driver or restaurant cook. And by 1989 the prestige rating of police officer had passed that of farm owner and operator, electrician, musician in a symphony orchestra, and even banker. It will be interesting to see the data for the entire study when it is published in a journal and made available for public discussion. (Notice the connection between the word *public* and the word *publish*.) Sometimes we have to wait for data we want.

Data Recording Sheet Occupational prestige study

Occupation	Respondent number										Job rating total
	1	2	3	4	5	6	7	8	9	10	
1. Truck driver	___	___	___	___	___	___	___	___	___	___	________
2. Musician in a symphony	___	___	___	___	___	___	___	___	___	___	________
3. Minister	___	___	___	___	___	___	___	___	___	___	________
4. Farm owner and operator	___	___	___	___	___	___	___	___	___	___	________
5. Physician	___	___	___	___	___	___	___	___	___	___	________
6. College professor	___	___	___	___	___	___	___	___	___	___	________
7. Carpenter	___	___	___	___	___	___	___	___	___	___	________
8. Electrician	___	___	___	___	___	___	___	___	___	___	________
9. Restaurant cook	___	___	___	___	___	___	___	___	___	___	________
10. Banker	___	___	___	___	___	___	___	___	___	___	________
11. Lawyer	___	___	___	___	___	___	___	___	___	___	________
12. Garbage collector	___	___	___	___	___	___	___	___	___	___	________
13. Police officer	___	___	___	___	___	___	___	___	___	___	________
14. Bartender	___	___	___	___	___	___	___	___	___	___	________
15. Coal miner	___	___	___	___	___	___	___	___	___	___	________
16. Plumber	___	___	___	___	___	___	___	___	___	___	________

Data Analysis Sheet Occupational prestige study

Occupation	Job rating total	Average scale rating	Occupational rank-order
1. Truck driver	________	________	________
2. Musician in a symphony	________	________	________
3. Minister	________	________	________
4. Farm owner and operator	________	________	________
5. Physician	________	________	________
6. College professor	________	________	________
7. Carpenter	________	________	________
8. Electrician	________	________	________
9. Restaurant cook	________	________	________
10. Banker	________	________	________
11. Lawyer	________	________	________
12. Garbage collector	________	________	________
13. Police officer	________	________	________
14. Bartender	________	________	________
15. Coal miner	________	________	________
16. Plumber	________	________	________

CONCEPT 14

Group: From Small Group to Society

Definition

Group A number of people who share some common interests, interact with one another, accept the rights and obligations of membership (including rules for behavior) within the group, and share a sense of identity and belonging with others in the group.

Society The largest possible group (because it is not a subgroup of any other group). Its broad goal is satisfaction of the basic survival needs of its members; its beliefs and rules for behavior are expressed in a shared culture; and its membership is generally territorially bounded and recruited by the automatic inclusion of the children of members.

My cousin once got stuck in an elevator for five hours with nine other people. They all had gotten in at lunchtime on the tenth floor of the office building in which they worked, and everyone followed normal elevator procedure. That is, even though they all worked for the same company, no one looked at anyone else; they all faced forward to look at the floor indicator. But when the doors closed, nothing happened. After about a minute, someone started pushing buttons and then pushed the call button. When it became apparent that they were really stuck, people started talking to one another. First they talked about their predicament, but later, while the engineers worked to get the elevator started, conversations inside the elevator became more personal. Everyone told what job she or he did and in what department. Office gossip was shared. One woman, who was particularly anxious and uncomfortable, was calmed by others, and bag lunches were passed around. There was no high drama—no baby born or killer apprehended—but by the time the elevator doors opened, friendships had been started, and everyone agreed to meet for lunch once a year to commemorate their "ordeal." In fact, these people began to meet frequently and had parties together as well as an annual lunch.

Before the elevator became stuck, the people inside could have been called an **aggregate,** or even a **category.** An aggregate is just a number of people who happen to be in the same place, such as people at a corner waiting to cross the street. A category is a number of people who share some characteristic, such as mothers with two children. The people in the elevator began as an aggregate (they were all in the same place) and a category (all sharing the characteristic of being on their way to lunch). But it took the experience on the elevator to make them a **group,** in the special, sociological sense of the term.

Group Defined

There are many kinds and sizes of groups, but sociologists generally agree that they all have the following four characteristics:

1. **Members Share Some Common Interests.** In the elevator group, the common interest at first was to get out. Later the common interest became social, as they got to know one another outside the work setting. Groups develop around interests as wide-ranging as stamp collecting, promoting the reputation of a community, making money, and saving a species from extinction.
2. **Members Interact with One Another.** This is not the kind of accidental, nonoriented interaction that occurs within an aggregate. It is the purposeful and structured interaction of people expressing their shared interests and goals, such as a chamber of commerce meeting weekly to discuss how to get more tourists to visit the town or a work group within a corporation meeting to plan a marketing campaign. Categories of people, such as taxpayers, are not really groups, because they rarely interact purposefully.
3. **Members Share the Rights and Obligations of Membership.** Every true group establishes rules for behavior within the group (ways of talking, jobs undertaken in pursuit of group goals, attendance at meetings, dress). These are some obligations of membership. In the elevator group, members took turns giving parties. But rights of membership are also established, such as the right to hold office, to expect help from other group members, and to enjoy the friendship of others. Friendship was the main right of membership in the elevator group.
4. **Members Share a Sense of Identity and Belonging Within the Group.** Members must define themselves as belonging to the group, at least to some extent. Thus, a person can't be a group member without realizing it.

Of course, the feelings of belonging can vary greatly in intensity. If the elevator group had continued to meet only once a year, a minimum amount of group identity might have survived. But, as it met more and more frequently, identity among group members grew. They even made little elevator buttons for group members to wear. In fraternities and other tightly knit groups, belonging can be extremely absorbing and can pervade a person's life to the point at which membership is the most important identity he or she has.

Group Solidarity and Group Boundaries

The stability and permanence of groups depend on the shared sense of belonging that the members have. The group's ability to hold together in the face of obstacles is called its **cohesion,** or **solidarity.** (For a fuller discussion of the concept of group solidarity see the material in Concept 10.) How much solidarity do college classes have? If students heard that the classroom building was locked, would they still make an effort to go? What if the teacher were late or if there were a bomb scare? What degree of difficulty would any group be willing to overcome to continue to meet? This is a measure of the group's solidarity. The greater the sense of belonging among the members, the more they are willing to consider the interests of the group (such as continuing to meet) before considering their own interests (such as personal convenience).

A further indication of the solidarity of a group is the clarity with which its boundaries can be identified. How easy is it for members to distinguish between those who belong to the group ("we") and all others ("they")? Sometimes this process of differentiation is aided by physical symbols of membership, such as pins, jackets, tattoos, or secret handshakes. As a tactic to increase feelings of belonging in a group, outsiders (they) are sometimes described as more threatening or organized than insiders thought. This distortion emphasizes the importance of membership. I have seen this technique used often in local elections: politicians raise the threat of "outsiders telling us how to run our town" and offer themselves as the best insiders for the job. Obviously, the distinction between us and them has some positive consequences for the stability and cohesion of groups but some negative consequences for relationships between groups.

Types of Groups

Small Groups

Groups range in size from the smallest possible (the two-person group, or dyad) to the largest (the society). A **small group** is one with few enough members so that personal contact occurs among all of them. The best known work on the character of small groups was done by the German sociologist Georg Simmel. He recognized that the kind of interaction that takes place in a group is strongly influenced by the number of people who belong to it (Simmel, [1908] 1964). If one person drops out of a dyad, the group is gone. But in a three-person group (triad), the group persists even if one member drops out. Therefore, all other things being equal, and only up to a point, the more people who belong to a group, the more persistent and stable it can be. In addition, in a dyad, disagreements can be overcome only by the personal power or persuasiveness of one member. But in a triad or larger group, social pressure can be brought to bear. A person who is outvoted 2 to 1, or by a greater margin, faces social rather than personal pressures.

Primary and Secondary Groups

Primary Groups One very important example of a small group is the **primary group.** This term, coined by Charles Horton Cooley (1909), refers to small, relatively enduring groups whose relationships are emotion-based, rather than goal-based, and involve virtually every aspect of the lives of the members. The family and the childhood peer group (the group of close friends of like age) illustrate the characteristics of primary groups.

1. **Primary Groups Are Small.** In Cooley's terms, such groups are small enough to allow intimate, face-to-face contact among all the members. Fifteen or twenty members is about the upper limit for most families and childhood peer groups.
2. **Primary Groups Are Enduring.** Small groups that form and disband quickly do not qualify, even if they have all the other characteristics of primary groups. Families and childhood peer groups (your circle of close friends when you were growing up, for instance) last for years.
3. **Primary-Group Relationships Are Diffuse.** No facet of a member's life is beyond the concern of the other members of a primary group. Family or peer group members can legitimately express interest in the work, leisure, medical, financial, or other aspects of other members' lives.
4. **Primary-Group Relationships Are Emotion-Based and Noninstrumental.** Relationships within a primary group rest mainly on how the members feel about one another. The emotional attachments among people who love one another, not the performance of tasks, are the basis of the group's solidarity. Families and peer groups, then, provide members with an emotional refuge in which acceptance is guaranteed in spite of other kinds of failures. For example, a parent or friend will say, "I love you no matter what you do." This kind of sentiment also illustrates the noninstrumental character of primary relationships. An instrumental relationship is one maintained to achieve some goal beyond the relationship. For example, the relationship between a mer-

chant and a customer is instrumental—each is dealing with the other in order to make an exchange of cash for goods or services. A noninstrumental relationship is maintained for no purpose beyond the relationship itself. The bond is the point of it all, the main reward.

Secondary Groups By contrast, **secondary groups** range in size from small to very large, may be short-lived or very enduring, and have relationships that are formal, instrumental, and segmental. The most common examples are work groups. To illustrate the characteristics of secondary groups I'll use college classes:

1. **Secondary Groups Can Be Any Size.** College classes range from three or four students to over a thousand in freshman lecture courses. In the larger sizes there is no way to interact face-to-face with all members.
2. **Secondary Groups Vary in Permanence.** College classes may stay together for ten to fifteen weeks. The work force in a factory or corporation may endure much longer. In either group, however, personnel may change. An important fact is that a secondary group is relatively unaffected by changes in its membership, whereas a primary group would be severely influenced by such changes. Primary-group members are not easily interchangeable.
3. **Secondary-Group Relationships Are Specialized and Segmental.** The relationships within a secondary group concern only a special part, or segment, of the whole person. In a college class, for example, matters pertaining to school work and some small portion of school life outside the class, such as extracurricular activities, are the only legitimate subjects for discussion. A student's personal life, finances, health, and so on are not included in that segment and should not normally be brought up. The same is true for aspects of the teacher's personal life. When such issues do come up, another kind of relationship is being established. It may be an additional secondary relationship, such as that between adviser and advisee, or it may be a new primary relationship, such as friendship.
4. **Secondary Relationships Are Formal and Instrumental.** Relationships in secondary groups are based on rational, intellectual decisions about the advantages and disadvantages that might come from them. Bonds are established to achieve a goal beyond the relationship. An exchange of services or of goods for cash is the most common example. In college classes, teachers want to be paid for their work, and students want to get college credits. Those are the bare bones of the relationship, niceties of education aside. People involved in a secondary relationship may like one another, but this is rarely important enough to overcome the failure of a member to perform as expected. I like the kid who cuts the lawn, but not so much that I'll still pay him if he fails to do the job.

Reference Groups

In every kind of group discussed so far, the quality of membership has been stressed, in terms of either the cohesion of the group or the character of the interaction of the group. But a **reference group** is one that can influence a person even if he or she is not a member. Reference groups are groups to which people compare themselves when they wish to evaluate their own beliefs or performance (Merton, [1949] 1968). For example, when you began college, with whom did you compare your ideas on issues? Did you recall what your parents would think and try to act accordingly? Or did you begin to compare your ideas with those of your new classmates and teachers? Studies of reference groups suggest that people's choices of groups to compare themselves with strongly influence their political ideas, levels of ambition, expectations of success, and a great deal more (Hyman & Singer, 1968; Stouffer, 1949). And when you are evaluating your performances, with whom do you compare yourself? When you get back an exam, for example, do you need to know how others have done in order to evaluate your own grade? On the face of it, you would think that an 83 on a test means you got 83% of the material correct. However, because we are relatively competitive in America, an 83 on an exam usually means a very different thing if it is the highest grade in the class that day than if it is the lowest. And with what kind of people do you tend to compare your performances—people who you guess are likely to do worse, the same, or better? If you compare your grades with the top students in a class, they look worse than if you choose as your comparison group people who typically do poorly.

One important finding in this area is that the extent to which a person feels deprived is related to her or his selection of a reference group. A person earning $10,000 per year may feel quite wealthy if that sum is compared with the smaller amount he or she earned before or with the even smaller amount earned by those in very poor classes or nations. By contrast, a person earning ten times that amount can actually feel more deprived than the first person if he or she chooses as a reference group the millionaires who live in the next neighborhood. This concept of **relative deprivation** explains how ambition can remain high even among objectively successful individuals or how discontent can stay low among people who seem quite poor.

Society: The Largest Group

Society is the largest group possible, because it is not a subgroup of any other group. It fits the definition of a group, except that it is obviously not possible for *all* members of a large society to interact with one another. But all members of a society do have common interests, goals, and rules for behavior. These are expressed in the culture of the society. **Culture** consists of the way of life that is learned and shared by humans and is taught by one generation to another. It includes all the goals we have, the values we hold, and the rules for everyday life by which we operate. (*Society* refers to the structural relationships among a group of people, whereas *culture* refers to their shared values, beliefs, and ways of life. The people referred to are the same, but the meanings of the terms differ.)

The goal of the society is to ensure that the basic needs of its members are satisfied so that the group can be considered self-sufficient. In addition, societies are maintained by the recruitment of the children of all members. Citizenship is automatic. A society also generally coincides with territorial boundaries and often with national names and characteristics.

Types of Societies Societies have been categorized in terms of the way they survive. Throughout history, advances in social organization and in technology have enabled human beings to take increasing advantage of the environment's resources, and the character of their societies has changed with these advances (Lenski & Lenski, 1978).

1. *Hunting-and-gathering societies* depend on the ability of the members to collect and kill food. When game is scarce or wild food sparse, the group must move across large areas to survive. The size of such societies is limited by natural conditions and the need to travel.
2. *Horticultural societies* began to develop about 9,000 years ago with the first controlled cultivation of crops and herding of animals. For these to occur, technological changes were necessary. For example, cultivation required tools for planting, plowing, harvesting, and storing crops. A different form of social organization was needed to deal with the specialized skills that developed. Labor, crafts, and administration began to be differentiated. These changes were necessitated by the greater concentrations of stable populations that could be supported by less land. Horticultural societies have greater control over the exploitation of natural resources than hunting and gathering societies do, but less than agrarian (or agricultural) societies.
3. *Agrarian societies* are different from horticultural societies primarily in degree. The beginning technological advances of horticultural societies were extremely limited compared with those that later multiplied the food-producing capacity of the land. With greatly improved methods of planting, cultivation, irrigation, transport and storage of crops, the ability of populations to live relatively permanently in dense settlements was also greatly improved. At the same time, the complex division of tasks in social structures multiplied. Specialization in society gave rise to military, financial, governmental, religious, educational, and administrative positions in rough proportion to the density of the population served.
4. *Industrial societies* developed out of the Industrial Revolution and have been the dominant form of society in the West through the middle of the twentieth century. These societies became increasingly

complex and densely populated, especially in great urban concentrations. They have been based on new forms of energy (such as steam power), factory mass production of manufactured goods, division of labor (such as found on the assembly lines), and emphasis on the control of the cost of production, including the cost of labor. The urban areas in industrial societies became the largest and most compact settlements in history, and their social order became correspondingly complex.

5. *Postindustrial societies* have developed out of, though they have not fully replaced, the industrial societies that preceded them. They reflect the economic shift from the manufacture of goods to high technology, information control, and the delivery of services. The computer age has greatly changed the way data are stored, transmitted, and manipulated, with consequences for the way work is done and services delivered. While the American economy is clearly more focused on postindustrial tasks such as financial management, it is also true, as of the early 1990s, that there has been a debate about the need to renew and recreate a manufacturing base. It may be reasonable to consider holding off pronouncing the arrival of postindustrial society until we see how high technologies alter the nature of the manufacture of goods.

Summary

A *group* is defined as a number of people who share some common interests, interact with one another, accept the rights and obligations of membership (including rules for behavior) within the group, and share a sense of identity and belonging with others in the group. A family is an example of a group. A group is very different from an aggregate (a number of people who are in the same place, but need have nothing else in common) and from a category (a number of people who have some characteristic in common, but need not be in contact at all).

The *solidarity* (or *cohesion*) of a group is its ability to hold together in spite of obstacles. The greater the solidarity, the greater the obstacles its members are willing to overcome in order to persist as a collective. A further indication of the solidarity of a group is the clarity with which its boundaries can be determined. This is easily seen in the ability of members to distinguish "us" (group members) from "them" (nonmembers).

Groups range in size from the smallest possible (the two-person group, or dyad) to the largest (the society). One very important kind of small group is the *primary group* (so named by Charles Horton Cooley). It is an enduring, small group with diffuse, noninstrumental, emotion-based relationships. By contrast, a *secondary group* is a group of any size or permanence whose relationships are formal, instrumental, and segmental.

Some groups are used as reference, or comparison, points by people, even if they are not members of the group themselves. Reference groups are groups to which people compare themselves when evaluating their own beliefs or performance. Because a person may choose any one of a number of groups for such evaluations it is possible to feel much better or worse off depending on the selection of a reference group. Sociologists use the term *relative deprivation* to refer to a feeling of dissatisfaction arising from the comparison of one's condition with that of a reference group or with some more favorable condition in history or imagination.

Society is defined as the largest possible group (because it is not a subgroup of any other group). Its broad goal is the satisfaction of the basic survival needs of its members; its beliefs and rules for the behavior are expressed in a shared culture; and its membership is generally territorially bounded and recruited by automatic inclusion of the children of members. Societies often are divided into five types according to their methods of survival and production: (1) hunting and gathering societies, (2) horticultural societies, (3) agrarian societies, which have much more ability to cultivate land than horticultural societies, (4) industrial societies, and (5) postindustrial societies.

The Illustration for this concept summarizes a classic study of reference-group behavior of blind people. Helen May Strauss was curious whether the blind would tend to compare their own abilities and beliefs with those of other blind people, as early reference-group theory had suggested, or whether they would use sighted people as their primary points of reference. Her results brought into question not only early assumptions about reference group selections but also assumptions about the self-concepts of blind people.

The Application for this concept is an exercise in specifying how different groups satisfy the four elements of the sociological definition of *group*. You are asked to identify three groups of varying size, frequency of interaction, cohesion, or shared interest and to apply the definition to each.

Illustration

Helen May Strauss, "Reference Group and Social Comparison Processes Among the Totally Blind," pp. 222–37 in Readings in Reference Group Theory and Research, *ed. Herbert H. Hyman and Eleanor Singer (New York: Free Press, 1968)*

Reference groups, as we have seen, are the groups with which people compare themselves when they wish to evaluate their own beliefs or performance. Groups used for the evaluation of beliefs are called **normative reference groups.** Groups used for the comparison of performance are called **comparative reference groups.** But how do people choose their reference groups? A good deal of our understanding of the process of reference group selection is due to studies such as Strauss's classic study of reference-group choices of blind people.

Early in the study of reference-group processes it was proposed that the basic principle in group selection was similarity (Festinger, 1954). That is, people would tend to choose for comparison those who were like themselves. Such a comparison supposedly would be the most useful because the standards of evaluation would be realistic. According to this principle, if a sociology major wanted to compare his or her grade on a math exam with someone else's, he or she would be more likely to choose another sociology major than a math major as a "comparison-other." Strauss's research raised some serious questions about this "similarity hypothesis."

Strauss suggested that there might be forces other than similarity at work in reference-group selection. She suggested that dissimilar comparison-others may be selected if a person wishes to identify with them. She reasoned that blind people would provide an exceptionally clear test of the similarity hypothesis for a number of reasons. First, their reference-group selections could be only blind individuals (similarity) or sighted people (dissimilarity). Second, Strauss noted that blind people were subjected to opposing social pressures. On the one hand, traditional practice taught that blind individuals would benefit most by isolation and protection from the demands of the sighted world. On the other hand, reformers argued that blind people would benefit most from integration into the sighted world, because such contact would lead to the greatest development of their potential. Strauss hypothesized that to the extent that some blind people believed in the traditional approach and some in the alternative approach, the research would uncover clear differences in their reference-group selections.

Testing the Hypothesis

To test this hypothesis, Strauss contacted a representative sample of 197 totally blind adults from across the United States. They were interviewed at length by professional, specially trained interviewers who asked them a number of questions, including several about their reference groups. In these questions the respondents were asked how they evaluated three of their own qualities: (1) their personal appearance, (2) their ability to learn quickly, and (3) their general character—"whether you are a good person." The respondents were asked whether, when thinking of each of these issues, they tended to compare themselves with "people [their] own age who are blind, or who are sighted."

These qualities were chosen because they were thought to vary in the extent to which blindness would make a difference. That is, it was guessed that blindness would be considered unrelated to "goodness," only somewhat related to appearance, and most related to learning speed. If this were the case, then reference-group selections for each issue might differ. But would comparison with similars occur more frequently for issues in which blindness mattered more or for issues in which it was not important?

One last point before I summarize the findings of the study. The interviewers were instructed to try to force the respondents to choose one reference group or the other (blind or sighted) for each issue. Respondents were not given the choice "I use both groups." Nor were they given the choice "I don't compare myself with anyone." A nonstandard answer was to be accepted only if a respondent insisted on giving such an answer even after being urged by the interviewer to choose one of the two reference choices. The effect of this procedure was to make it quite difficult for the respondent to choose "both" or "none" as an answer. In other words, the research design caused the percentages of blind people who were "both-comparers" and "noncomparers" to be diminished.

Results of the Study

One of the most interesting findings of the study was apparently unintentional. Strauss had set out to determine what sort of reference-group selections blind people make; that is why she tried to compel the respondents to choose similar or dissimilar reference groups. She was not originally interested in the rate of "noncomparing." But she found it anyway. Overall, in 20 percent of the cases, blind people said that they made no social-comparison choices. This figure is especially impressive given the barriers that the respondents had to overcome in order to have their answers accepted. It can be assumed that the rate of noncomparison among blind people was much higher than this.

Therefore, this test of the similarity hypothesis of reference-group selection begins by questioning the inevitability of social comparison itself. But why would noncomparing occur? In the case of this study, Strauss concluded that "the absence of social comparisons was found, by many analyses, to be most characteristic of those blind who suffered the greatest social isolation, past or present." Social experience, then, shapes the kind of evaluations a person is capable of making. If you do not know what other people are like (or, in extreme cases, that they exist), how can you compare yourself with them?

However, many of the blind people in Strauss's sample did make reference-group selections. What sorts of selections did they make? A portion of the results are summarized in Table 14-1.*

The first thing that should strike you about the data is the very high proportion of blind people who chose sighted people as their reference group. As you can see in the bottom row of figures, almost two-thirds of the respondents made this choice. And the use of the dissimilar reference group was very consistent across the three issues. Sighted people were most often used as a reference group for judging matters of appearance, but the majority of blind people even evaluated character, which the researchers thought would be unrelated to ability to see, in comparison with sighted people. It is clear, as Strauss pointed out, that these data raise serious questions about the principle of similarity in reference-group selection.

Another interesting finding is that the remaining reference-group selections were predominantly for both blind and sighted people rather than for blind people alone. In fact, an average of only 9 percent of all reference-group selections were for people similar to the respondents. In evaluations of character, in fact, blind people compared themselves with other blind people only 5 percent of the time. Among blind people who did not compare themselves with sighted people, it was much more common for them to say that both groups were used for comparison purposes (25 percent of the time, overall).

How are we to explain these results? According to Strauss, blind people are subjected to social pressures that urge them to live effectively in the sighted world. If blind people are to learn the skills necessary to do this, they cannot afford to adopt standards for performance or belief that are less demanding than those of sighted people. In addition, to the extent that blind people are defined in American society as inferior, comparing oneself with them might damage an individual's self-esteem. This might help to explain the fascinating finding (also, I suspect, unexpected by Strauss) that among the sample of blind individuals, one-third reported that they did not even define themselves as blind. For sociologists interested in the relationship between social interaction and self-concept, the findings of studies like this suggest that reference-group theory can teach us a great deal.

*For the purposes of this illustration I have reported on only a portion of the data from Strauss's original study. Strauss compared the responses of people who had become blind early in life with those who became blind after the age of eighteen. In the table I have combined (or *collapsed,* to use the term employed by researchers) the results. You may also notice that the original sample contained 197 people, but that none of the totals adds up to 197. This is because the responses of the "noncomparers" are not included in the table.

TABLE 14-1
Percentage of blind individuals choosing various reference groups

Issue	Other blind	Both blind and sighted	Sighted	Total number of responses
Learning	15% (24)	24% (38)	62% (100)	162
Appearance	7% (10)	16% (23)	77% (108)	141
Character	5% (7)	34% (53)	62% (97)	157
Totals	9% (41)	25% (114)	66% (305)	460

Application

A *group* is a number of people who share some common interests, interact with one another, accept the rights and obligations of membership (including rules for behavior) in the group, and share a sense of identity and belonging within the group. Most people belong to a number of groups. Some groups may be very important to your everyday life (like your family or the people you work with), while others are more marginal (a specific class you might be taking or a club whose meetings you attend infrequently). It is possible (even likely) for a person to be a member of a group without ever thinking about what makes it a group. This application is designed to allow you to specify how various groups satisfy the four elements of the definition of group.

A Study of the Characteristics of a Group

The first step in completing the application is to think of three groups to analyze. You do not have to be a member of such a group in order to use it in the application, though that would probably help you identify its characteristics. Students in my classes sometimes have used groups to which they once belonged (for example, a Little League team and a rock group from their high school years) and groups about which they were curious (for example, a theater group on campus and a sorority). In choosing groups to examine, try to get some variation in characteristics, such as intensity of membership, frequency of interaction, number of members, and interest shared by the members. For instance, don't choose three groups in which the members are only casually involved, three small groups, or three groups focused on sports. The idea is for you to see the way groups vary in their characteristics, yet manage to satsify the four-part sociological definition of *group*. Also, try to choose at least one group with which you are unfamiliar. This will make things more interesting. It also can make clearer the analytical value of sociological ideas.

Once you have selected the three groups to study, identify each in the space provided on the analysis sheet. Then for each fill in the four group characteristics in the spaces provided. Of course, for the group with which you are unfamiliar, you will have to speak to group members to get this information. The questions to ask are easy for the first three characteristics of groups.

1. *Members of a group share some common interests.* Most of the time the common interests of a group are made clear in its name. "The Beagle Owners Club of Greater Taunton," for instance, is not too complicated to understand. But sometimes the common interest that binds the group members is not so clear, or consists of more than one thing. For example, families are likely to share a number of interests, some of which may be neither obvious nor the same as those in your family. Or what would you think was the interest held in common among the members of the staff of a college newspaper? Is it really just the publication of the news?
2. *Members interact with one another.* This one is easy. How often do the members interact, and for how long each time? Are the interactions scheduled?
3. *Members share the rights and obligations of membership.* Some of these are formally established in written documents, such as dues, clothing styles or uniforms, responsibilities to attend or behave in specific ways, and tasks to be accomplished by members. More often they are informally understood and reinforced. Be sure to include both rights and obligations.
4. *Members share a sense of identity and belonging within the group.* This one might be the most difficult to identify and describe, because group members often experience a sense of membership without ever thinking consciously about it. To what extent do you identify yourself as a member of the groups to which you belong? You might have a strong identity as a fraternity or sorority member but never have considered that you are also a group member in every class you take in college. Remember that groups may differ in the intensity of the group membership (high in fraternities and sororities and relatively low in college classes) and still satisfy all the conditions for the sociological definition as a group. So, when answering this question, you may have to be more imaginative. What are the indicators of the identity of group membership other than membership cards or uniforms?

Data Recording and Analysis Sheet

Group #1 ______________________________ ____________________

Name of Group — Number of Members

1. What are the common interests shared by the group's members?

2. What are the ways in which the members interact with one another? How often do they interact?

3. What rights and obligations are shared by the members?

4. What evidence is there of the sense of identity and belonging among the members of the group?

Group #2 ______________________________ ____________________

Name of Group — Number of Members

1. What are the common interests shared by the group's members?

2. What are the ways in which the members interact with one another? How often do they interact?

3. What rights and obligations are shared by the members?

____

____

____

____

4. What evidence is there of the sense of identity and belonging among the members of the group?

____

____

____

____

Group #3 ______________________________________ ______________________

Name of Group — Number of Members

____

CONCEPT 15

Formal Organization and Bureaucracy

Definition

Formal Organization A group that is organized to achieve some specific goal or goals by coordinated, collective effort.

Bureaucracy A type of formal organization featuring extensive hierarchy of authority and expertise, division of labor, and explicit rules of performance for its members.

All through school my friends and I had great plans or, more accurately, great goals. In the fourth grade, for example, we wanted to build an elaborate hideout and treehouse in some woods nearby. By the sixth grade we were interested in getting rich. We were not much more detailed than that. I remember that we tried to start several businesses. It's too bad we didn't manage to do what the kids in the movies did. They'd be sitting on some porch, bored to tears, when one would exclaim, "Say, let's put on a play!" Before you could buy more Milk Duds they were singing and dancing their little profit-making hearts out on their makeshift stage—and selling tons of tickets. It seemed like so much fun, because Hollywood knew to leave out the boring parts, all the organizational stuff.

When a goal is stated, whether it is to make money, to educate children, or to put on a play, the efforts of a number of people need to be coordinated. The simplest kind of coordination of efforts calls for everyone to do the same thing—for example, five people all pushing a car out of a snow bank. But most of the time the organization of efforts to achieve a specific goal requires the designing of much more complex patterns of human effort in which people typically take on many different, interdependent jobs. Since the Industrial Revolution and urbanization, the complexity and scale of the tasks we need to accomplish have increased greatly. Today we must organize efforts to feed, house, transport, police, and provide countless other services for huge populations living in urban concentrations. We can't expect to get these jobs done using the simpler methods of village life in the eighteenth century. Faced with larger, more complex goals, the organization of human efforts has become increasingly formalized. By assigning people to various tasks (presumably according to their talents and training), we can coordinate diverse efforts efficiently. Such specialized tasks tend to become hierarchical, with a leader at the top of a chain of command and titles for the various jobs within the organization. And formal organizations tend to persist; that is, they survive even if their original members drop out. For example, the group of kids who put on a play might have to replace an actor or musician who has moved out of the neighborhood. The positions within a formal organization are more enduring than the specific people who fill those positions at a given time.

There are many kinds of formal organizations. They can differ in terms of the goals they seek to achieve and the formality of their structures. First, we'll look at goals.

Goals of Formal Organizations

One convenient way to categorize the goals of formal organizations is in terms of the people who benefit from their activities. Blau and Scott (1962) identified four types of formal organizations:

1. *Mutual-benefit organizations* act in the interest of their own members, people who join such groups (usually voluntarily) for their own benefit. Examples include trade and manufacturers' associations, landlord groups, taxpayers' associations, and labor unions. Sometimes we realize that such groups exist only when they manage to publicize their existence—for example, by declaring National Turnip Growers' Week. Business associations and labor unions make up by far the largest number of mutual-benefit associations, and they maintain national organizations for the promotion of their interests.
2. *Businesses* are formal organizations that act in the interest of their owners, whether one person or a large number of stockholders. The goal of businesses in capitalist systems is to make a profit for the owners.
3. *Service organizations* act in the interest of people who need specific services, such as health care, legal advice or representation, jobs, housing, and education. Hospitals, law firms, colleges, employment

agencies, real estate or housing services, and so on are examples of service agencies. (They may also be businesses.)

4. *Commonwealth organizations* are more generally concerned with the service needs of the public as a whole. They meet the most varied goals of the populations they serve and must adapt to changing needs expressed by these populations. Examples include governments at all levels (federal, state, county, and local) and the military.

It may have occurred to you as you read these descriptions that the categories overlap some. It is not always apparent from the officially stated goals of a formal organization who benefits from its activities. For example, although hospitals, colleges, and employment agencies clearly are service organizations focused on meeting the specific needs of individuals, law firms and real estate offices are also profit-oriented businesses. From the officially stated goals of such formal organizations, it is often difficult to untangle the underlying web of benefits. For example, when the National Rifle Association lobbies in Washington, D.C., for legislation in the interest of its voluntary membership, it is acting as a mutual-benefit organization. At the same time, the organization uses arguments that are based in all citizens' constitutional rights to bear arms and to self-defense against crime, thus acting as a commonwealth organization. Lastly, the legislation that they wish to influence typically benefits the profit margins of gun manufacturers, so that the organization is acting in the interests of businesses.

Structures of Formal Organizations

Formal organizations are often divided into two main structural categories, **voluntary associations** and **bureaucracies.** Voluntary associations are groups that develop out of the shared interests of their members. These interests may be recreational (a model airplane club) or service-oriented (a church group formed to send food to victims of a flood). They tend to be less formally organized than bureaucracies, but they vary from very loosely organized associations, such as singles bars and weekend softball teams, to groups that regulate almost every aspect of a member's daily life, such as the U.S. Army and a large urban mental hospital.

The most formally structured of organizations are called bureaucracies. They are characterized by an elaborate hierarchy of authority and expertise, a very specialized division of labor, and specific rules about how each member is to do his or her specific job. Bureaucracies influence many aspects of our daily existence because they have become the dominant form of social organization for the accomplishment of the goals we have set. Bureaucratic organization characterizes almost every level of government, schools, most private businesses, and public utilities (such as the phone company), to name just a few. But bureaucracies have a very bad reputation. In fact, the word *bureaucracy* has become synonymous with inefficiency, lack of concern for individuals, and even arrogance. This is especially interesting because the bureaucracy was originally a utopian concept, considered a solution to the most difficult problems faced by industrial society. To understand where things began to go wrong, let's look at Max Weber's model of this structure.

Weber's Model of Bureaucracy

Weber ([1925] 1946) described bureaucracy in ideal terms; that is, he outlined the characteristics of a form of social organization that theoretically would be capable of accomplishing large, complex goals with efficiency. He was not describing one specific organization but what he called an *ideal type,* or theoretical model for how things should work in such an organization. (For a more detailed discussion of ideal type, see Concept 2.) Here are its elements:

1. **Impersonality.** Positions in the organization become explicitly defined offices, each existing independently of who occupies it. This means that bureaucracies are built of positions rather than specific personalities. They can survive, and even operate efficiently, when people leave the organization. In addition, relationships between offices are supposed to be impersonal. In organizational terms, the head teller deals with the vice president of personnel, rather than Ted talking to Edna.
2. **Rules.** Ways of behaving within the organization, including how one does a job, are set out as specific rules and regulations. In many cases rules are collected in lengthy job descriptions listing every task for which a given officeholder is responsible and the exact ways in which each task must be accomplished. At a large bank in Boston, for example, new employees in the personnel department are given scripts for answering the telephone and are not supposed to deviate from the words on the page.
3. **Specialization.** Tasks within the organization are divided into very small units, each designed to fit into a plan for the coordination of all tasks. The idea is to accomplish the goals of the organization the way an assembly line produces a finished product. In theory, the simpler a task, the more quickly and efficiently it can be done. For example, it should take one person longer to fold a letter, stuff it in an envelope, and close it than it will take three people if the first just folds letters, the second stuffs, and the third closes.
4. **Hierarchy.** Offices become arranged in bureaucracies in a formal chain of command so that each office is subordinate to the layer above. In theory the coordination of complex efforts requires a pyramid of control and responsibility.

5. **Expertise.** In line with the chain of command and responsibility, a hierarchy of expertise must exist in an efficient bureaucracy. Those at the top must have knowledge about the overall operation of the organization to make adjustments to it. Each layer within the organization must be able to make similar changes within the area for which it is responsible.

Weber's model describes a utopian form of organization, because it was intended to achieve goals well beyond the capability of the simple forms of organization that had existed before. Bureaucracy was intended to achieve goals while allowing people to maintain some individuality (everyone doing her or his own special job within the structure). But something obviously has gone at least a little wrong, and the seeds of the trouble can actually be seen in the very theory of bureaucratic structure.

Sources of Bureaucratic Inefficiency

Perhaps bureaucratic inefficiency develops because human beings react to bureaucratic arrangements in ways that are understandable but are not predicted by the model (Peter & Hull, 1969). The following are some human reactions to bureaucratic structure that are anything but efficient from a bureaucratic planner's point of view:

1. *Impersonality* is unlikely, because humans make friends, make enemies, and generally develop very personal relationships in spite of the theory that official relationships should be impersonal. Some people who are friends will then ignore the rules of the bureaucracy to cover for one another's errors. Officials will replace efficient workers they don't like with less efficient friends. Both practices have the same harmful consequences for the operation of the organization.
2. *Rules* that specify what each worker is to do make it possible to have an overall plan, but humans resent such restrictions and can take advantage of them. After all, if everything about a job is supposed to be in the job description, then anything not on the list is not part of the job. A resentful clerk at the motor vehicle bureau can let a person stand forty-five minutes in the wrong line, knowing that it is happening, because the clerk's job description does not specify redirecting lost people to the proper line.
3. *Specialization* may seem like a good idea for raising the efficiency of work until you try to get people to do the same tiny, mindless job all day. It is boring. The worker's attention wanders, and resentment builds, so that errors are made. Sabotage may even occur if enough resentment is felt.
4. *Hierarchy* also can be resented, especially when it is supposedly linked with expertise. These are discussed together next.
5. *Expertise* is supposed to be the basis for authority in a bureaucracy. Yet how often have you noticed that people in positions of authority

know less about the work of their subordinates than the workers know themselves? In theory, decisions in a bureaucracy are to be made by superiors in the organization. In practice, the bosses often do not know how things should be done. In addition, hierarchy of command and expertise is antidemocratic. Americans believe that authority ought to flow upward, from the "grass roots." In representative democracy, executives are supposed literally to execute the will of the electorate. But almost all of government is bureaucratic in structure, with expertise supposed to flow from the top down. There is a basic conflict between the everyday operation of bureaucracies and the American belief in the control and knowledge of the citizens.

Both workers in bureaucracies and victims of their inefficiencies seem to have been robbed of their control and creativity (as you will see in the Illustration). All in all, bureaucracy seems to be something of a failure, in spite of its utopian origins. So what keeps it in place in modern societies?

The Persistence of Bureaucracy

Bureaucracies persist partly because our goals seem to have increased in size and complexity. Now we have to feed, transport, protect, and otherwise serve ever larger, more urban populations and satisfy their ever more specialized tastes. We want to travel to the moon, bank from home, eat Mandarin food, insure lives and stereos, learn about sixteenth-century medicine, and vacation in Nepal. Unless we begin to reduce both the scale and the complexity of our goals (as some people have begun to suggest), only bureaucratic structure seems capable of accomplishing them.

A second reason for the persistence of bureaucracies is their ability to handle information. Computers represent a technological leap in information manipulation (I'll resist calling it information control for the time being). This development can only further entrench bureaucratic structure in industrialized societies.

A third factor in persistence is that bureaucracies can elicit worker compliance. Amitai Etzioni (1975) classified a number of the ways in which they do this. *Coercive* organizations such as the military and mental hospitals can force compliance. *Normative* bureaucracies get compliance by socializing members to believe in the goals and means of the organization. For example, service organizations (hospitals, colleges, and the like) often develop a sense of mission and spirit of cooperation that express normative membership and induce members to work toward organizational goals. *Utilitarian* organizations get compliance by controlling a wide variety of rewards, such as salaries, promotions, and job security. Such rewards are also incentives for compliance in coercive and normative bureaucracies.

As you can guess from this list, there are organizations that increase the likelihood of getting worker compliance by using all three methods.

In the U.S. Army, for example, coercive compliance is enforced by the right of superiors to discipline enlisted personnel, even to the extent of court proceedings that take place within the organization. (Civilian courts do not have jurisdiction.) Normative compliance is accomplished by powerful socialization to army goals and "spirit" during basic training and/or officer training programs. And rewards are carefully distributed by a complex structure of incentives and promotions.

Summary

A formal organization is a group organized to achieve some specific goal or goals by coordinated, collective effort. Formal organizations can be distinguished by the types of goals they pursue. Mutual-benefit organizations act in the interests of their own members. Businesses act in the interests of their owners. Service organizations act in the interests of people who need specific services such as health care. Commonwealth organizations are concerned with the needs of the public at large.

Bureaucracies are special types of formal organizations featuring an extensive hierarchy of authority and expertise, division of labor, and explicit rules of performance for members. Max Weber's model of bureaucracy includes five elements: (1) impersonality in the dealings between offices of the bureaucracy, (2) rules that specify how members of the bureaucracy are to behave within the organization, (3) specialization of bureaucratic tasks in the interest of increased efficiency, (4) a hierarchy of positions in the bureaucracy, forming a chain of control and responsibility, and (5) the investment of expertise in the topmost positions of the organization, so that decisions about daily operation come from above to the line workers.

If bureaucracies have a reputation for inefficiency, it is not because they were designed to be that way. In fact they were rationally designed to deal with the great problems posed by a large, dense, and highly individualistic population. But unplanned components of bureaucracies (such as personal friendships, failure to follow rules, and inability of supposed experts at high levels to actually solve the technical problems of bureaucratic operation) undermine their efficiency and victimize both workers and outsiders dealing with these organizations.

The Illustration for this concept summarizes a study of how bureaucracies (specifically, social welfare agencies) can subvert their intended operation. Michael Lipsky reveals how policies made at the highest levels of the bureaucracy are mediated (and even remade) by the line workers who deal with welfare clients. This subversion of organizational operation is built into the structure of the worker's situation.

The Application for this concept uses the techniques of participant observation to discover some of the rules by which a bureaucracy is intended to run and some of the important ways in which it differs from the plan. Although this exercise can be done in a few hours, it also lends itself to expansion into a semester-long project for an entire class (or several smaller study groups looking at different organizations).

Illustration

Michael Lipsky, Street Level Bureaucracy *(New York: Russell Sage Foundation, 1980)*

In theory, bureaucracies are designed to deal with complex, large-scale tasks by the most efficient organization of human effort that can be designed. High-level administrators and policymakers are charged with responsibility for coordinating the separate activities of a large number of workers so that the work is divided into well-defined, interdependent jobs. The whole effort is supposed to be rational, planned, and efficient. But as anyone who has dealt with bureaucracies can testify, the result is not always so successful. Most of the studies of bureaucratic inefficiency focus on the organization as a whole. But a great deal about bureaucracies can be understood by looking at the process from the point of view of the worker, which is what Lipsky did in his study *Street Level Bureaucracy.*

Lipsky focuses on the experiences of social workers, the street-level bureaucrats whose job is to deliver services to clients of the welfare state. Social workers must mediate between the large bureaucracies within which they work and the clients who need services. After social work agencies hand down policies about how services will be delivered, the workers must *apply* the policies every day. As Lipsky makes clear, the very nature of social work at the street level requires that social workers remake or even ignore higher policies just to survive in the job.

Like police officers, judges, physicians, teachers, and other people who provide human services, social workers tend to enter their profession with a high sense of purpose. They wish to help people in need. They soon realize that between them and the service of their clients stand two major difficulties: (1) the chronic inadequacy of resources and (2) a host of bureaucratic policies. In combination these factors prohibit the practice of social work as they had conceived it.

For one thing, the caseloads are too high to service a case adequately. In Massachusetts, for example, the average caseload for a single worker as of 1990 was 160 cases. Because no worker can hope to actually deal with the needs of so many clients, some shortcuts must be taken. Information about client needs is gathered in less time than agency guidelines suggest. Or less time is devoted to discovering various avenues of help for each client. Training of workers is necessarily reduced, with the result that they never learn all the techniques necessary to help clients. As Lipsky points out, increased resources do not necessarily reduce case overloads. As resources rise, demands rise to match them. There is, apparently, an almost inexhaustible supply of people in need, waiting to seek aid from welfare agencies if resources become available.

Beyond the chronic inadequacy of resources, street-level bureaucrats are required to follow work guidelines that were designed to increase efficiency but that reduce the ability of workers to deal with clients as individual humans with unique sets of needs. People at the upper levels of the bureaucracy set policies that turn the worker into a "people processor," armed with mandatory questionnaires about property, income, work, and even sexual behavior. The rules for eligibility are overwhelming, and the maze of guidelines is impossible to follow entirely. Increasingly specialized jobs in the welfare system prevent workers from dealing with their clients' problems as a whole. Social workers see only segments of a client's needs and expect other specialists within the bureaucracy to deal with other facets of the case.

Another influence on social workers combines the elements of inadequate resources and bureaucratic structure. With the recent budget shortages, additional pressures to cut welfare costs have been applied. Workers have been given incentives to limit eligibility for aid and guidelines for accomplishing this goal. Resources are further reduced, and bureaucratic standards for work increase. The needs of clients, which were at the top of the list of priorities at the beginning of a social worker's career, get pushed further down the list.

How do street-level bureaucrats react to the clash between their ideals as social workers and the twin constrictions of inadequate resources and bureaucratic structure? Many quit, especially the most idealistic workers, who cannot tolerate the compromises they would be forced to make if they stayed in the profession. For those who remain, idealism suffers a number of indignities. Workers must accept the fact that they can serve neither the needs of all their clients nor all the needs of those who are served. The criteria for deciding who gets helped are not even based on the ideal notion of need. Workers keep in mind the reaction of clients. If a client is likely to be troublesome when aid is denied, she or he may receive more attention. This is hardly the distribution of aid according to need.

The routine of work in the bureaucracy takes over the everyday life of the worker. Instead of responding to human demands, workers are forced to adhere to bureaucratic standards of success. In evaluating how effectively workers are doing their jobs, agencies must find ways of measuring successful work. But by its very nature, dealing with human beings who have highly individual problems defies measurement. How can a number be attached to the subjective evaluation by a client of the services he or she receives? So social workers are evaluated for behaviors that *are* measurable, such as payment levels and, especially during recent fiscal crises, reductions in caseload by the removal of "ineligibles." In fact, workers who are successful and energetic in dealing with large caseloads find themselves rewarded only with more cases. According to Lipsky, workers soon realize that the way to keep their workloads manageable is to provide a "consistently inaccessible or inferior product."

Inadequate resources and bureaucratic structure also have consequences for the way social work policy is made. Because of their huge caseloads and endless bureaucratic regulations, social workers have great autonomy and discretion. They are like police officers, who can choose what crimes to notice, what lawbreakers to pursue or arrest, and what work regulations to follow. As a consequence, the street-level bureaucrat, the person who actually delivers the services, becomes the policymaker. The social worker determines who is eligible, for what amount of aid, how the aid will be distributed, and which agency guidelines will be followed. By this process, publicly stated institutional goals are subverted to the needs of the worker who must survive from day to day in an essentially impossible job.

Workers may rationalize the process as the only way a mature person can deal with such contradictory demands or as a practical and political reality—the way the world works. And legislators and high-level bureaucrats may still believe that *they* shape social welfare policy.

Application

The aim of this application is to have you observe an operating bureaucracy to discover some of the roots of its difficulties. Weber's model of bureaucratic structure serves as a convenient way to organize the research. He identified five major components of bureaucracies: (1) impersonality, (2) rules, (3) specialization, (4) hierarchy, and (5) expertise. (You may want to review the discussion of these in the Definition section.) These five elements form the rational basis by which bureaucracies are designed to operate. But they apparently do not live up to the plan.

Participant Observation of a Bureaucracy

In order to see how Weber's model fits a real bureaucracy, you will conduct a bit of research. Choose a bureaucratic organization to study. Consider its accessibility, because you will have to spend a few hours there and will have to talk to some of the people who work there. It is helpful if you know someone who works in the personnel or public relations office or has some higher executive position, but that is not necessary; you can do the study without such a contact inside. Examples of bureaucracies that are interesting to study are hospitals; large corporations such as the phone, electric, or gas company; a university or large college; a government agency or bureau such as the motor vehicle bureau; and a large department store. There are others, but this list should give you an idea. The research consists of three parts, each intended to apply to some component(s) of Weber's model.

Part 1: Organizational Chart

Once you have chosen a bureaucracy to study, obtain its organizational chart. Most bureaucracies print one for themselves, and a personnel or public relations staff member should be able and willing to locate one for you. In government agencies (such as the motor vehicle bureau) and some other organizations, you may have to go to the headquarters rather than to a branch office. If you cannot locate an organizational chart, either try to draw one yourself with the help of someone in personnel or public relations (one organization decided to print one for the first time when my student asked for one), or choose a different organization to study.

Part 2: Job Descriptions

When you ask for an organizational chart, also ask for copies of job descriptions. Most bureaucracies have these as well. In fact, they have too many for you to study them all (use just a selection), and they are bewilderingly detailed. You may have some trouble getting job descriptions for positions at the top of the organization. These jobs are sometimes not delineated. You might try to interview these higher-ups about the responsibilities of their jobs. They often like to talk about their work.

Part 3: Participant Observation

Conduct an observation of the actual daily operation of the bureaucracy you have chosen. For example, go to a busy office of the motor vehicle bureau, and go through the process of registering a car. Fill out all the forms. Ask for help and instructions. Observe other people who are clients of the bureaucracy, and observe the people who work there. In most cases, it is quite easy to stand aside and simply observe without drawing anyone's attention. Usually there is too much going on for people to notice. How do the workers treat the clients? How do the clients react? What about clients who are particularly at a loss about what to do? Do they get especially helpful treatment, or the opposite?

Collect as many of the forms as you can. Read them and try to figure out what they require. How regimented are the procedures? How clear? How narrowly defined are the steps? How specialized are the responses of the workers?

In acting as a client of the bureaucracy, try to follow the procedures as exactly as you can, but also try doing a few unusual things. For example, smile at a line worker and be friendly and warm. Notice the reaction. Try asking a particularly difficult question, or come up with a problem unlikely to be covered by standard procedures. For example, you might ask at a motor vehicle office whether to register a three-wheeled vehicle as a car or a motorcycle, or you might inquire how to make a citizen's arrest for a registry violation. Again, notice the reactions. Sometimes, breaking the rules (even slightly) helps to highlight them.

Analysis and Interpretation of the Results

Try to apply your data to the five elements of bureaucracy in Weber's model. You should be able to fit data from each part of your research into the model. Here are some suggestions to start the process:

1. **Impersonality.** In your observations, did workers deal with one another on a strictly professional basis? Were the relations with the clients also impersonal, or were they *more* than impersonal; that is, were workers rude and cold? How did they react to your friendliness? Of the five elements in the model, this is the most difficult to evaluate without being an insider. You might consider talking to some friends who have worked in bureaucracies to discover what relationships among workers are like.
2. **Rules.** The job descriptions in a bureaucracy are basically lists of rules for behavior in every position. How detailed are they? Do they seem possible to obey? The forms and instructions that regulate the interactions of clients with the bureaucracy are another source of information about the rules for bureaucratic behavior. How specific are they? How clear? How possible to obey? In your observations, were rules actually obeyed? To what extent? To excess in some cases? (A worker at the motor vehicle office once told me that I was not standing in line properly. I was off to one side and was "confusing things.") What happened when you presented a problem for which there was no rule? Were there apparent standards of dress? How strict were they? How strictly were they followed? This could be especially important in private corporations.
3. **Specialization of Tasks.** In the job descriptions, how specialized were the tasks described? From the organizational chart could you draw some conclusion about the extent of the division of labor? In your observations did the jobs seem very narrowly specialized? (One woman at the motor vehicle bureau told me that she stamped only a particular form and was not allowed to give me directions about where to go next; that job was assigned to the information officer on the other side of the room.)

4. **Hierarchy.** Is the hierarchy of the organization clear from the organizational chart and the job descriptions? Job descriptions often specify the people to whom and for whom each individual is responsible. In your observations did you notice anyone walking about "supervising" subordinates? What was the physical arrangement of superiors and inferiors? Did office size and placement separate positions in hierarchical terms? How about clothing differences such as uniforms? Were there other signs of hierarchy?
5. **Expertise.** Did the job descriptions refer to knowledge held only by higher-level workers, knowledge used to determine the tasks of lower-level workers? Did you notice in your observations any line workers going to a superior for information? If you managed to ask a particularly difficult question or presented a problem that was somewhat outside organizational procedure, did it cause the line worker to have to go to a superior for information or a ruling?

The results of your research can be presented in a variety of formats. Using the Data Summary Sheet supplied, you could arrange your findings according to Weber's five points, placing rational and then irrational elements under each. Or you could write a summary of your findings, dividing the information into two categories: (1) information supporting the rational organization of bureaucratic efforts, and (2) information exposing the irrational or inefficient components of bureaucracy. However you present the findings, try to keep in mind the gap between the rational plan by which bureaucracies are supposed to operate and the actual operation you discovered.

Data Summary Sheet Analysis of bureaucracy

Bureaucratic component	Examples of rationality (bureaucratic model adhered to)	Examples of irrationality (bureaucratic model violated)
1. Impersonality		
2. Rules		
3. Specialization of tasks		
4. Hierarchy		
5. Expertise		

CONCEPT 16

Institutions

Definition

Institution A set of rules that establishes how group members agree to accomplish universal issues of survival, such as procreation, socialization, and care of the young or distribution of power.

In William Golding's novel *Lord of the Flies* (1954), English schoolchildren are stranded on a small island after a plane crash. There are no adult survivors, so the children have to make do for themselves. How to proceed? The first thing they do is to decide who will be in control. Ralph, who has called them together for the first meeting by sounding a sea conch like a trumpet, starts things off:

> *"Seems to me we ought to have a chief to decide things."*
> *"A chief! A chief!"*
> *"I ought to be chief," said Jack with simple arrogance, "because I'm chapter chorister and head boy. I can sing C sharp."*

Jack's logic, a holdover from their former lives, does not win out. An election is held, and Ralph, who has the conch and has called the meeting, is elected chief. Jack is made leader of the hunters (because the children realize they will need food), and in rapid succession a variety of other tasks is established. A map of the island is ordered, and a signal fire set up. Shelters are proposed, and so is a system of punishment for disobedience:

"We'll have rules!" he cried excitedly. "Lots of rules! Then when anyone breaks 'em—"

"Whee-oh!"

"Wacco!"

"Bong!"

Tales of people stranded on islands or in blizzards, even real-life accounts, all have this process in common. Once the shock or the thrill of the situation wears off, the people involved realize that there are things they must do if they are to survive. And they recognize that, if they don't come to a general agreement about how things will be done, there will be internal conflicts that could threaten their survival. Usually such stories focus on the special difficulties the group must overcome, such as caring for the injured after a crash, getting rescued, or getting enough food and shelter. And the stories usually end when the major difficulties are resolved by rescue, or death.

But these same tasks face the members of much larger and more stable societal orders, including modern industrial and postindustrial societies. In order to survive, we must face the same **universal issues** and agree about how we are going to deal with each of them. The sets of rules that group members establish about how such universal issues will be accomplished are called **institutions.** Unlike Golding's survivors, who were eventually rescued, members of large-scale, more permanent societies establish long-standing, stable institutional patterns that become ingrained in the structure of the society.

Institutions and Universal Issues

Any society must develop institutions to deal with at least the following universal issues:

1. **Procreation.** New generations of the young must be born.
2. **Sexual Access.** People who are allowed to engage in sex by societal agreement must be provided approved institutions in which such access is considered legitimate. Otherwise, sexual behavior could become predatory and destructive of social order.
3. **Care of the Young.** Children must be fed, clothed, and protected until they can fend for themselves.
4. **Socialization.** The way of life that is shared in a culture must be taught by one generation to the next.
5. **Education.** The body of information accumulated by a society can be maintained only by teaching it to the next generation.
6. **Religion.** All societies must deal with questions about the origins and meaning of human life.
7. **Distribution of Power.** Social groupings are dependent on the willingness of their members to subject themselves to the will of the group. This

process involves the agreement that some person or persons will exercise the accumulated authority in the name of those who belong to the group. Some decisions must be made about how that power will be distributed among group members.

8. **Production, Distribution, and Consumption of Goods.** The group must provide for the material needs of members. Decisions must be made about how such needed goods will be produced (grown, manufactured, collected, and so on), who will do this, how the goods will be distributed among the members, and who will consume what goods in what proportions.

9. **Social Control.** Whatever agreements are reached by the members of a society there is always the danger that individual members will disagree with them. If social order and the agreements that support it are to remain stable, deviations from prevailing social patterns must be controlled. Legitimate forums for such disagreements may be provided, and punishments or the threats of punishment are usually specified.

American Institutions

The institutions that develop around these universal issues rarely deal with just one issue. Institutions tend to have multiple functions and to overlap somewhat with one another. Some major American institutions and the issues with which they are concerned are discussed in the sections that follow.

The Family

The family is probably the most basic of institutions in American society (and many others), because it deals with such a wide variety of universal issues (Bernard, 1972). It provides the structure for legitimate sexual access between adults and for legitimate procreation. (We even label children born out of marriage *illegitimate.*) The family is also the prime socializer and educator of the very young, and it is responsible for their care and social control until a certain age. The family interprets many issues of religious belief. It was an important unit of production when family farms were more common than they are now, and it is still an important unit of consumption within the American economy.

The Government (or Polity)

The primary function of government is to provide a structure for the legitimate distribution of power in society. In the United States the form of government is representative democracy. Because we elect officials to make decisions in our interests on a wide variety of issues, all the other universal issues are at one time or another of concern in the legislation these officials consider and enforce. What the government does influences patterns of social control, economic behavior, education, socialization, and even procreation and sexual behavior (Orum, 1978). When changes are proposed in any institutional structure of society, the conflict over whether to allow such change is generally played out within the rules of government.

The Economy

The production, distribution, and consumption of goods are functions of the economic institutions of society (Smelser, 1975). The American forms are capitalism in combination (and sometimes in conflict) with labor. Many issues of concern to the economic institutions are made by government in its role as representative of the general population (regulations concerning the manufacture and transport of goods, taxes and import controls, conditions of work, and so on). Economic and governmental institutions tend to get quite tangled up in one another's business in the United States.

Education

The maintenance and teaching of the accumulated knowledge of the society is the primary concern of the educational institutions (Parelius & Parelius, 1978). In the United States we have an extensive public school system and parallel (although smaller) private and religious school systems. The differences between these forms have recently become very significant as educational institutions compete for shares of the shrinking financial support for education and as controversies increase about what ideas and beliefs ought to be taught in public schools.

Religion

Every society must establish legitimate ways for its members to deal with questions about the origins and meaning of life (Glock, 1973). In most cases such institutions are religious as opposed to secular. In the United States the guiding principle is freedom of choice in the pursuit of such questions, but established Western religions dominate overwhelmingly in the society.

Institutional Differences

Differences Across Time

Institutions reflect the most deeply held values of the society. For example, our capitalist economy reflects values for progress, independent action, competition, and material accumulation. Our democratic form of government reflects values for representativeness, freedom, and majority rule. The educational structure reflects a value for rationality.

In every case the institutions that develop must be consistent with the values of the society. But institutions can change over time within one society when (1) values change (which usually happens very, very slowly) or (2) conditions in the society change (as long as the essential values continue to be reflected in the new institutional arrangements). For example, the family has been changing quite rapidly. Part of this change is a response to economic changes, which increasingly require two incomes to maintain a family's style of life. Part seems to be a result of the independence that a full-time career makes available to women. It is no surprise that people can become very upset when they recognize that institutions such as the family are threatened by change. They perceive changes as attacks on the very basic values of the society. And it is in terms of values that debates about institutional changes are stated. The fundamentalist groups that oppose recent changes in the traditional American family have focused on the "loss of our basic moral guidelines" and suggest that changes in the family both reflect and accelerate that loss. From a sociological point of view it remains to be seen whether the changes in the family structure at the center of the controversy are actually adjustments to the conditions in which we live (changes that continue to reflect the same strong values) or whether they actually indicate value changes at the cultural level.

Differences Across Cultures

Institutions can be different from one time to another within a society, and they can differ between societies. In fact, one of the main ways to distinguish one society from another is to compare the structures they have developed to accomplish universal issues. For example, in the United States the institution that deals with the distribution of power and social control (democratic government) is at least somewhat separate from the institutions dealing with economic issues (capitalism and labor). It is true that they overlap somewhat, but their goals are at least somewhat in conflict with one another. This fact becomes evident when governmental agencies regulate the conduct of business in the private sector. Businesses frequently try to diminish the amount of regulation by lobbying government, to influence legislation, and by working for the election of candidates less sympathetic to government regulation. By comparison, in the People's Republic of China the economic, social control, and power issues are all dealt with by the same institution, the Communist party. Despite the 1989 protests in Beijing's Tiananmen Square against this total control by the institution of communism, it remains the only instrument for the distribution of power, the control of deviance, and all decisions about the production, distribution, and consumption of goods. The contrast between American and Chinese institutions seems quite stark, especially in comparison to the contrast between American institutions and those of other

Western societies. Canada, for example, seems quite similar to the United States except for its parliamentary system of government (which is also a form of representative democracy). England varies largely in having a monarchy (within another parliamentary form of representative democracy) and a somewhat more socialist economic structure than ours (although still dominated by capitalist arrangements). When comparisons are made with Eastern societies, the differences are more extreme. For example, Iran has what amounts to a religious government (theocracy) in which decisions about virtually every phase of life—educational, sexual, economic, and other—are made by the centralized religious authorities, the ayatollahs. Now, that *is* a different institutional arrangement from our own.

Summary

Every society faces certain universal issues, tasks that must be accomplished if its members are to survive. For example, future generations must be born, cared for, and taught the knowledge and beliefs of the society. Power must be distributed by some agreed-on method, and the goods of the society must be produced, distributed, and consumed. To accomplish these aims, societies create and maintain institutions, sets of rules that establish how group members agree to accomplish the universal issues of survival.

American institutions include (1) the family, which provides for legitimate sexual access between adults; provides for procreation, care, socialization, and some education of the young; and serves as a unit of consumption within the economic system; (2) the government (or polity), which is the primary institution for the distribution and exercise of power; (3) the economy, which is responsible for the production, distribution, and consumption of goods; (4) schools, which pass on much of the accumulated knowledge of the society (along with families); and (5) religious institutions, which deal with questions of the origins and meaning of life.

There is often an overlap between institutions dealing with the same universal issues. For example, the family and schools both deal with the universal issue of education of the young. The family is an especially important institution, because it deals with a number of universal issues. But, as important as it is, it can change (as can any other institution), and it is not the only institutional arrangement for dealing with those universal issues. Institutions change when the members of society differ about how the issues are to be accomplished. For example, economic pressures (among other forces) have compelled many women to seek full-time work, with inevitable consequences for the traditional family. Care of the children, then, must be dealt with by new institutions, such as day-care programs. In addition, different cultures have different institutional arrangements for dealing with the same universal issues. These include different forms of governments, economies, educational structures, and so on.

The Illustration for this concept focuses on the family. Specifically, it summarizes a recent study that determined factors influencing divorce rate and examined whether these factors had the same effects during different stages of a marriage. The study utilized a longitudinal survey technique in which the same individuals were interviewed several times over a span of nine years. Among the factors found to influence the divorce rate were race, the labor-force participation of each partner, urban residence, and the wife's level of education.

The Application for this concept is designed to discover what institutions people use in dealing with existential questions such as the meaning of life and death. Is religion the institution to which people turn first when confronted by such questions? Do younger people turn more to secular institutions? This brief survey should indicate what people in your area are like in this regard.

Illustration

Scott J. South and Glenna Spitze, "Determinants of Divorce over the Marital Life Course," American Sociological Review *51 (1986): 583–90*

The attention of social researchers typically is drawn to areas of our social life, including the operation of our social institutions, that seem to be disrupted or changing. In the 1920s, for example, sociologists began to study urban life when the decay of inner cities became apparent. Studies on the subject of aging were spurred by evidence of increased problems faced by older Americans and by the increase in the percentage of the population over the age of sixty-five. And recently, attention to the family has been prompted by persistent evidence of change in its structure. For example, the data show that compared with American families of just twenty years ago, today many more children grow up in single-parent homes, a much higher percentage of married women work outside the home, and the number of children born in the average family is significantly lower. For most people, however, the most obvious indicator of change in the family is the rise in the American divorce rate.

Data collected by the U.S. Bureau of the Census show that the divorce rate has approximately tripled in the last sixty years. In the 1920s and 1930s the rate of divorce remained fairly steady, at 1.5 divorces per 1,000 Americans. The rate began to rise sharply in the mid-1960s, and it leveled off in the early 1980s at a rate of approximately 5 divorces per 1,000

Americans, where it remains today. The point can be made more dramatically by noting the often-cited statistic of a 50 percent divorce rate in the United States. This rate does not mean that half of the marriages performed in a given year will end in divorce. We won't know that outcome for many years. Rather, a 50 percent divorce rate means that the number of divorces granted in a year are half the number of marriages performed in the same year. Measured this way, the United States has had nearly a 50 percent divorce rate since 1975. Whatever the actual rate of divorce today, we know that it is much higher than it used to be.

Clearly, such trends are likely to change the way the American family functions. Given the wide range of universal issues for which the family has traditionally been responsible, it is no surprise that there have been strong and varied reactions to evidence of apparent changes in its structure. There are those who think that changes in the family structure threaten the stability of the social order and those who think that change is desirable; however, if advocacy is to make any sense, the sociologist must first gather verifiable information about the issue. The first stage of such research is to document that changes are really occurring. Once change is confirmed, the next step is to try to identify and examine its causes, which brings us to the research that is the subject of this illustration.

South and Spitze's study focused on divorce. They wanted to find out what factors were related to the divorce rate and whether these factors had the same effects during different stages of a marriage. The problem was in how to accomplish this. Many studies of divorce have been **cross-sectional studies;** that is, they have compared data collected from divorced and nondivorced contemporaries on a range of issues thought to be related to likelihood of divorce. (Examples of such variables include religion, home ownership, and levels of income and education.) The assumption has been that, if two groups differ on some variable, that variable may be a factor in divorce rates. Cross-sectional studies, however, suffer from the problem of establishing causality. For example, even if it is demonstrated that people who have remained married are more religious than those who have divorced, it is still impossible to determine whether religiousness causes people to remain married or whether divorce causes people to become less religious.

To deal with this problem, South and Spitze analyzed data from the National Longitudinal Surveys of Young and Mature Women. This study had collected information about women between the ages of fourteen and fifty-three. These women were periodically reinterviewed over the nine years between 1967 and 1976. **Longitudinal studies** use data collected from a given group of people over a period of years. Though time-consuming and expensive, this technique allows the researcher to establish whether differences in variables, such as religiousness or level of education, occurred before any changes in marital status. In this way, those factors found to be related to the likelihood of divorce can be more clearly identified as causal.

Here is what the data showed. Previous studies had found a number of variables to be related to the divorce rate, and this study confirmed those findings. The variables included (1) the wife's level of education, (2) race, (3) urban residence, and (4) the labor-force participation of each of the partners. Of these, only the first—the wife's level of education—was clearly shown to have different effects on the likelihood of divorce at different stages of a marriage.

Let's begin with those factors that were found to influence the likelihood of divorce but whose influence was not different at different stages of a marriage. That is, the effect of each of these factors was the same in newer marriages as in marriages of longer duration.

1. **Race.** It was confirmed, as had been shown in other studies, that race is an important predictor of divorce rate. American blacks are significantly more likely to be divorced than are whites. One study reports a 50 percent greater divorce rate for blacks than for whites (Glenn & Supancic, 1984). The effect of race was consistently independent of marital duration. That is, blacks had higher divorce rates than whites in marriages of short or long duration.
2. **Labor-Force Participation of Each Partner.** The number of hours per week that a wife worked was found to be a predictor of divorce; the greater the number of hours the wife worked, the higher the probability of divorce or separation. By contrast, in marriages in which the husband worked more during the entire previous year, the divorce rate was significantly lower than in those in which the husband worked less. Again, the effects of these variables were consistent regardless of marital duration. The effects of the wife's labor-force participation and the husband's employment were the same in early years of marriage as in later years.
3. **Urban Residence.** As you might expect, marriages in urban areas were found to be more likely to end in divorce than those in less densely settled areas. Again, this effect did not vary by marital duration.
4. **Wife's Level of Education.** This is the only instance in which a variable was clearly found to affect a marriage differently in the early years than in the later years. More specifically, for marriages lasting less than five years, it was found that in those marriages in which the wife had completed college, the likelihood of divorce was significantly lower than in those in which the wife had less education. However, the wife's level of education was found to have the opposite effect in marriages of longer duration. That is, in those marriages in which the wife was more educated, the likelihood of divorce was significantly *greater* in marriages of longer duration. As South and Spitze put it,

"early in marriage, wife's education appears to deter divorce, but later in marriage it is associated with a higher probability of dissolution" (1986:587).

It is interesting to speculate why this is the case. The researchers suggest that women who are more educated may enter marriage better prepared in some sense than do less educated women. Thus, the divorce rate is especially low in the early years of their marriage. However, after some years of marriage, the wife's college education has a different effect. According to the researchers, it may be that among women who are dissatisfied with their marriages, those who have had a college education (and a career during marriage) have many more alternatives than do women with less education—thus the higher divorce rate. Whatever logic you use to explain the findings of a study like this, it is fascinating to consider what forces are at work in changing a social institution such as the family.

Application

One of the American institutions discussed in this concept is religion. I suggested that religion develops around the universal issue of explaining the origin and meaning of life. But is religion the only institution that deals with such questions? Sociologists have for three decades been studying the apparent decline of religion in the United States, a process usually called *secularization*. Undoubtedly, religion is still important in American life. In recent national surveys as many as 90 percent of the respondents said they believed in God, and as of the late 1980s rates of membership in established religions were around 60 percent of the population. But clearly, some people do not turn to religion when they confront questions about why we exist, why there is death, and what life is for. For example, if a person does not believe in God and does not belong to a formally established religion, he or she may turn to other institutions such as the family (asking parents or spouse), education (asking teachers), or even science (asking professional counselors or technical experts) for answers to such questions.

A Survey of Institutional Sources of Help on Universal Questions

This application calls for you to interview a small sample of younger and older individuals to find out how they would deal with these questions. What people might they turn to, if any at all? Would they turn to the institution of religion or some other? And as a modest test of the assumption that religion has declined in importance in the United States, you will be able to compare the percentage of older and younger respondents who choose to turn to religion for their answers. That is, it may be that older respondents grew up in a time when religion was more commonly the institution used to deal with existential questions and that they still turn to religion for help. By comparison, younger respondents may turn to other institutions. Lastly, using some data about regional differences in religious membership in the country, you should be able to ground the data you uncover in the information for your region.

The main questions of the interview are intended to discover what a respondent would do if he or she suddenly needed to grapple with questions such as the meaning or origin of life. The idea is to find out what percentage of your respondents turn to religion for help with these questions and what percentage turn to some other available institutions, such as family, school, or medical and professional counseling. A range of existential questions could be used, including questions about why life exists, where life comes from, why we die, and whether there is life after death. The study could get very complicated if we tried to deal with all of them. For this application I have included only questions about the meaning of life and about the origin of life. I wanted to limit the scale of the study but felt that I had to include both questions because of the recent debate between the religious and scientific communities over biblical and evolutionary theories. It is possible that this debate may cause people to respond very differently to the two questions. That is, a person may want to turn to religion for help with one question but not for help with the other.

The interview consists of three questions. The first two are about what the respondent would do if confronted with one of these existential questions, and the third is to determine the respondent's age. For some people, questions about religion and age will all be rather personal, so it is important to both justify why you are asking the questions and to reassure respondents that their names are not needed. Anonymity is our best assurance that respondents will feel free to answer our research questions honestly. Also note that the question about age does *not* ask respondents simply to say how old they are. Rather, it asks for a year of birth. It is usually easier for a person who does not want to reveal his or her age to give a year of birth. You can do the math during the analysis phase to determine whether a respondent belongs in the "younger" or "older" category of respondents. Try to interview twenty people, ten of whom are younger (between the ages of eighteen and thirty should do it) and ten of whom are older (over the age of fifty would be best). Use your judgment in approaching individuals for your sample. If you think you are interviewing someone who is clearly in one age category or the other but who turns out to be in the middle (for example, thirty-five years of age), you will have to throw out those responses and replace that interview. If we are to find an age difference in the responses, it is best to make the ages of the group represent strong differences in generational experience.

For the first two questions be sure to read the possible responses. The respondent is asked to choose just one response. If respondents insist on choosing more than one response, ask which would be their first choice if they could have just one, and then record it as the response. If you make twenty copies of the Survey Sheet, you can read the questions to each respondent and record the answers in the spaces provided. Then transfer them to the Data Analysis Sheet. Otherwise, record the responses in the spaces provided on the Data Analysis Sheet as you do each interview.

Here is the introduction that you should read to respondents. Practice reading it beforehand, so that it sounds fairly conversational.

Introduction: I'm doing a survey for a sociology class. It's a study of who people usually get advice from when they need it. There are three questions, and the survey takes less than two minutes to answer. Also, the study is completely anonymous. No one's name is asked.

Analysis

Once you have recorded the data, you summarize the results by adding each column and putting at the bottom the total number of respondents who would turn to a particular institution for help in dealing with the existential question. For example, dealing first with the younger sample, how many responded that they would turn to a religious person such as a minister, priest, or rabbi for help in dealing with the meaning of life (question A)? Add up the number of such responses, and put the figure in the space for Totals at the bottom of the column marked Religion. Do the same for the totals of the other columns for Family, School, Professionals, and Friends. Repeat this procedure for the answers provided by younger respondents to question B (about the origins of life) and for the older respondents' answers to both questions.

The next step is to calculate the percentage of each age group that chose each institution for help. To do this, simply multiply the figure in the Totals line by 10. So if six of the ten younger respondents said they would turn to a religious person for help with questions about the meaning of life, then that becomes a 60 percent selection of religious help. Fill in these figures on the next line down, which is marked Percentage. Finally, of all twenty respondents (in both the younger and older samples), what percentage said they would ask a religious person for help with question A, about the meaning of life? Add up the total who chose religion, divide that total by twenty, and place the figure in the next-to-last space on the Data Analysis Sheet, marked Total percentage choosing religious help for question A. Last, in the same way calculate the total percentage of the twenty respondents who chose religious help for question B, and put that figure in the last space on the Data Analysis Sheet.

Survey Sheet

A. Sometimes things happen, such as the death of a loved one, that make a person think about difficult questions such as what the meaning of life is. If you suddenly found that you did not want to deal with such a situation alone and that you needed help in thinking about this question, whose help would you be most likely to seek?

_____ 1. A religious person, such as a minister, priest, or rabbi

_____ 2. A family member, such as a parent or grandparent

_____ 3. A person at school, such as a professor of science or philosophy

_____ 4. A person who advises people for a living, such as a therapist, counselor, or physician

_____ 5. A friend I can talk to

_____ 6. Someone else who does not fit in any of the above categories

B. What if the question you needed help with was not about the meaning of life but about how life began? Would your choice of whose help to seek be the same as before? Yes _____ No _____ (If the response is no, ask: Whose help would you be most likely to seek for this question?)

_____ 1. A religious person, such as a minister, priest, or rabbi

_____ 2. A family member, such as a parent or grandparent

_____ 3. A person at school, such as a professor of science or philosophy

_____ 4. A person who advises people for a living, such as a therapist, counselor, or physician

_____ 5. A friend I can talk to

_____ 6. Someone else who does not fit in any one of the above categories

C. Last, could you please tell me the year you were born? _______ (Year)

Interpreting the Results

First of all, do you find any differences between the choices of institutional help made by younger respondents and those made by older people? Are older respondents more likely to turn to religious sources for help in dealing with these questions? If you found this sort of difference, it may be due to the cohort differences I mentioned earlier. That is, people raised in a time when religion was generally accepted as the authority in a wide range of moral and philosophical issues may have carried that belief into middle age. By comparison, younger respondents may have grown up in a time when religious institutions have been partially replaced by others

TABLE 16-1

Church membership in the United States (percentage of population)

Region and state	Christian church membership, 1980	Jewish membership, 1990
United States	49.3%	2.4%
New England	59.8	3.3
Maine	41.0	0.7
New Hampshire	44.2	0.6
Vermont	47.7	0.9
Massachusetts	64.0	4.7
Rhode Island	75.0	1.6
Connecticut	60.8	3.5
Middle Atlantic	53.4	6.9
New York	48.7	10.3
New Jersey	53.3	5.6
Pennsylvania	60.5	2.8
East North Central	50.3	1.3
Ohio	49.1	1.2
Indiana	44.6	0.3
Illinois	54.7	2.2
Michigan	42.5	1.2
Wisconsin	64.4	0.7
West North Central	59.6	0.7
Minnesota	64.9	0.7
Iowa	61.1	0.2
Missouri	53.1	1.2
North Dakota	73.8	0.1
South Dakota	66.9	0.1
Nebraska	63.1	0.5
Kansas	53.4	0.6
South Atlantic	44.0	2.3
Delaware	40.1	1.4
Maryland	39.7	4.6
District of Columbia	47.6	4.2
Virginia	41.5	1.1
West Virginia	39.6	0.1
North Carolina	53.9	0.3
South Carolina	51.3	0.3
Georgia	46.9	1.1
Florida	38.0	4.6
East South Central	55.1	0.3
Kentucky	54.1	0.3
Tennessee	54.1	0.4
Alabama	57.3	0.2
Mississippi	54.9	0.1
West South Central	55.5	0.5
Arkansas	56.1	0.1
Louisiana	57.2	0.4
Oklahoma	57.9	0.2
Texas	54.5	0.6

TABLE 16-1 *(continued)*

Region and state	Christian church membership, 1980	Jewish membership, 1990
Mountain	46.1	1.1
Montana	44.2	0.1
Idaho	50.0	0.1
Wyoming	44.0	0.1
Colorado	36.4	1.5
New Mexico	58.9	0.4
Arizona	39.2	2.1
Utah	75.1	0.2
Nevada	29.1	1.9
Pacific	33.8	2.6
Washington	30.9	0.7
Oregon	35.9	0.5
California	34.1	3.2
Alaska	30.6	0.5
Hawaii	33.1	0.6

as sources of authority. In addition, much research about the life course, especially in the study of social forces in later life (social gerontology), consistently shows increased religiousness in later life. It is suggested that, as health declines and the likelihood of death increases, individuals feel the need to come to terms with these existential issues as part of what has been called the "life review." But at what age does it occur? Does the age of your respondents allow for this effect to be reflected in your data?

What about the two questions? Do you see different tendencies to turn to religion when the question is about the meaning of life than when it is about the origin of life? Is religious authority weaker in the area of origin of life? It may depend on where you live. Some sections of the country have very high rates of church membership compared with others. Also, as you probably know, the debate over what should be taught in schools on this question has been focused in certain areas of the country. Listed in Table 16-1 are data taken from the U.S. Bureau of the Census for membership in Christian churches in 1980 and for Jewish membership in 1990 (U.S. Bureau of the Census, 1992:60). Does it help you understand the rate at which your respondents reported an intention to seek religious institutional help?

Data Analysis Sheet Survey of institutional supports for existential questions

Younger Sample

Respondent Number	Question A (meaning of life)					Question B (origin of life)				
	Religion	Family	School	Profes-sionals	Friends	Religion	Family	School	Profes-sionals	Friends
1	____	____	____	____	____	____	____	____	____	____
2	____	____	____	____	____	____	____	____	____	____
3	____	____	____	____	____	____	____	____	____	____
4	____	____	____	____	____	____	____	____	____	____
5	____	____	____	____	____	____	____	____	____	____
6	____	____	____	____	____	____	____	____	____	____
7	____	____	____	____	____	____	____	____	____	____
8	____	____	____	____	____	____	____	____	____	____
9	____	____	____	____	____	____	____	____	____	____
10	____	____	____	____	____	____	____	____	____	____
Totals	____	____	____	____	____	____	____	____	____	____
Percentage (Total × 10)	____	____	____	____	____	____	____	____	____	____

Older Sample

Respondent Number	Question A (meaning of life)					Question B (origin of life)				
	Religion	Family	School	Profes-sionals	Friends	Religion	Family	School	Profes-sionals	Friends
1	____	____	____	____	____	____	____	____	____	____
2	____	____	____	____	____	____	____	____	____	____
3	____	____	____	____	____	____	____	____	____	____
4	____	____	____	____	____	____	____	____	____	____
5	____	____	____	____	____	____	____	____	____	____
6	____	____	____	____	____	____	____	____	____	____

7	____	____	____	____	____	____	____	____	____	____
8	____	____	____	____	____	____	____	____	____	____
9	____	____	____	____	____	____	____	____	____	____
10	____	____	____	____	____	____	____	____	____	____
Totals	____	____	____	____	____	____	____	____	____	____
Percentage (Total × 10)	____	____	____	____	____	____	____	____	____	____

Total percentage choosing religious help for question A ________%

Total percentage choosing religious help for question B ________%

PART FIVE

Inequality, Change, and Social Disorder

Sociology is concerned with a search for the principles underlying social order. Even a cursory examination of social life reveals that society is not like some great, unchanging, orderly machine with all parts operating smoothly. Society changes, it is shot through with conflict and inequality, and many of its members suffer feelings of confusion about what they should do and why. This part discusses the issues of inequality, change, and social disorder, emphasizing that they are not merely problems for society (or sociologists), because they provide opportunities to understand much about the operation of society.

Concept 17 examines social stratification, the system of ranking individuals in terms of their access to and possession of the things that are valued by a society. All societies are stratified to some extent, and various systems of ranking are discussed in this concept.

Concept 18 focuses on the processes by which categories of people (called minority groups) are targeted for unequal treatment by the more powerful members of a society (a process called discrimination) and subjected to negative beliefs and feelings about them (prejudice).

Concept 19 deals with the processes by which society changes, an especially important topic for a time such as ours in which change is rapid and pervasive. Various theories about how change occurs and the causes of social change are introduced, and the relationship between change and social order is discussed.

Concept 20 examines deviance, behavior that does not conform to society's norms. Theories about the origins and consequences of deviance are divided into those that focus on characteristics of the deviants and those that focus on the way social forces define and create deviant behavior.

Concept 21 treats anomie and alienation. Anomie is a condition of social ambiguity in which clear rules for social behavior are lacking. Alienation is a condition of social ambiguity in which social participation lacks meaning. Both are serious issues in our rapidly changing, extremely complex social world.

CONCEPT 17

Stratification and Social Class

Definition

Stratification A system of ranking individuals in terms of their access to, and possession of, the things valued by their society.

Social Class A category of people within a system of stratification who share a similar style of life and socioeconomic status.

My mother went to high school with actor-comedian Art Carney. (If you don't know who he is, then imagine someone very famous.) As she recalls it, he was funny even then, and he was also the biggest success she ever knew personally. For the most part the kids she went to school with are doing about what their parents did. They are working in businesses, driving buses, teaching school; the children of physicians are now professionals themselves. The unusual stories, such as Carney's, are very rare. Here and there the son of a car dealer has made a load of money in real estate, the daughter of a florist is practicing law successfully, and a businessman's son is in jail for embezzlement. But such movements up or down the social class scale are the exception, not the rule, among her high school classmates.

It is obvious, from just looking around, that people in American society differ in terms of the amounts of money, power, and prestige they enjoy. But inequality is not just an individual experience. Inequality in a society can be seen in terms of layers, or strata, of people. Within each layer people share similar styles of life and access to the things that are valued by the society. In American society, each layer is called a **social class,** and the overall system of inequality is called **stratification.** Every society has some form of stratification. But what are the causes of inequality? Why is inequality so stable? That is, why do family members tend to stay at approximately the same position in the system of stratification from one generation to the next? When there is movement up or down the system, how does it occur? These are the questions we will deal with in this concept.

Individual Versus Social Sources of Inequality

Adventurousness and a "pioneer spirit" helped launch and sustain the American enterprise, from the exploration and exploitation of the land (no matter who was already on it) to the rapid expansion of industrial capitalism in the nineteenth century. That spirit is still with us in the American values for individual responsibility and progress. We wish to succeed at whatever we attempt, and we wish to do so as a result of our own efforts and talent. When people make a great deal of money or gain power or prestige, it is common for their success to be attributed to their talent or hard work.

Certainly, that sometimes happens. But it is also clear that *opportunities* to achieve success are not evenly distributed throughout the society. That is, even if everyone had the same amount of talent and worked equally hard, not everyone would succeed to the same degree. Some people face barriers in the struggle to succeed, and some people face no such barriers.

Opportunities to apply one's talents and efforts are called **life-chances.** A person with many life-chances gets to develop her or his intellectual potential through education, to learn skills at home and in school, to use those skills on the job, and to learn about chances to get ahead that are provided by the legal and political institutions of society. I'm not even talking about the person who is handed everything, by inheriting the family fortune, for example. I'm just describing the person who has the opportunity to compete without facing special barriers.

By contrast, large numbers of people are denied adequate diets, health care, education, or the chance to compete with others. They suffer reduced life-chances. People in positions of power do not believe them capable of competing, so these deprived groups never get the opportunity to qualify for jobs that could lead to other rewards. All other individual factors being equal, categories of people such as blacks, women, Hispanic Americans, Native Americans, and the poor are systematically denied equal opportunities to compete for the valued goods of the society. The result is a system of stratification in which the unequal distribution of money, power, and status is often incorrectly attributed entirely to the individual qualities and efforts of members.

The system of stratification is stable, because individual differences in the talents and efforts of societal members are overwhelmed by the uneven distribution of life-chances. On average, women earn about seventy cents for each dollar earned by men. (This figure has risen from just over sixty cents in the early 1980s.) For example, as of 1990 the median income of women living in single-person households in the United States was $12,548, or 63 percent of the $19,964 average for men in single-person households (U.S. Bureau of the Census, 1992:447). Blacks earn about 60 percent of what whites earn and suffer about twice the rate of unemployment. Native Americans experience many times the national average rates for unemployment, suicide, alcoholism, and tuberculosis.

Theories of Stratification

I have argued that, beyond individual differences, social forces work to systematically distribute opportunities for success in a society. Now I would like to discuss two theories about what those forces are and how they operate. The two main theories of stratification, the conflict and structural functionalist views, explain stratification from almost directly opposite perspectives. After explaining each, I'll summarize one attempt to reconcile the two views.

The Conflict View

The **conflict view of stratification** begins with the work of Karl Marx. Marx saw the dreadful consequences of the totally unchecked growth of industrial capitalism in the nineteenth century (Marx and Engels, [1848] 1964). Vast populations of people worked dawn-to-dark hours in dangerous, monotonous work conditions for near-starvation wages. They were packed into dreary company-controlled housing or worker ghettos. Meanwhile, fueled largely by this cheap labor, the economy expanded rapidly, and great industrial fortunes were made.

Marx believed that the circumstances of a person's life stemmed from his or her relationship to the means of production and that inequalities are rooted in society's economic system. In his analysis of capitalism, Marx identified two major economic positions: capitalists, who own the means of production, and workers, who must sell their labor to capitalists to make a living. Marx argued that these two basic relationships to how goods are produced in capitalist economies created two social classes whose interests conflict. Capitalists could earn as much as profits would allow, whereas workers were limited to the amount that capitalists would pay them. The problem for workers, according to Marx, was that profits depended on the ability of capitalists to keep wages low, and capitalists typically controlled both the price charged for the goods produced and the wages paid to workers. Inequality, then, resulted from domination of the life-chances of one group of people (workers) by the actions of another group (capitalists).

Marx predicted that workers would eventually realize that they were being exploited and that the only way to end their exploitation was to seize the means of production by armed revolt. This is just what happened in the Soviet Union and the People's Republic of China. But nineteenth-century Marxism did not apply so well to twentieth-century capitalist societies with their vast middle classes, whose members often share ownership of the means of production through the stock market. Updated versions of conflict theory (often called *neo-Marxism*) still contend that inequality is a consequence of the struggle between social classes; but the number and character of the contending forces and the forms the conflict can take have become extremely varied. Ralf Dahrendorf (1959), for example, has updated Marx's conflict theory to apply to forms of class conflict other than mass revolution. For example, ghetto riots, union strikes, and congressional battles are some of the many forms of conflict that underlie stratification.

The Structural Functionalist View

The **structural functionalist view of stratification** begins with the assumption that society is a large, complex system with many interrelated parts in some kind of balance with one another. The overall operation of the system requires that each part do its job efficiently. The functionalist idea is that any action, idea, or object can be understood only in terms of what it does for the system of which it is an element. If an object, action, or idea contributes to the smooth operation of some system, it is termed *functional.*

In addition, structural functionalism argues that some parts of an operating system are more important for its stable operation than are others. For example, rust-proofing contributes to the operation of the automobile as an operating system (it is functional), because it helps protect the body of the car. But the engine is clearly much more functionally vital to the operation of a car. Without it, the system stops operating altogether. In viewing the society as a system, structural functionalists argue that stratification reflects the variations in functional importance of the many activities that keep society operating.

The founder of structural functionalism, Talcott Parsons, developed it as a general theory for explaining social order (Parsons, 1937, 1954, 1971; Parsons & Shils, 1951). Two of Parsons's students, Kingsley Davis and Wilbert Moore (1945), applied it to the issue of stratification. They argued three major points: (1) Some positions in society are more important for its operation than others. (2) Such positions are very difficult to fill, because they require special talents, demand a great deal of training, and are often very unpleasant or taxing to perform. (3) To fill these positions with the most qualified people and to induce such people to undergo the training and demands of the positions, society rewards the important positions disproportionately with power, prestige, or material compensations. So, for example, physicians are paid well and are respected because society's smooth operation depends on them more than on the people doing many other jobs. The job of physician requires special talents, long training, and arduous efforts. Thus, it is heavily rewarded.

According to structural functionalism, then, systems of social stratification contribute to the smooth operation of society. Inequality is functional, because even those who are not very well compensated benefit from the stability of society. In comparison with conflict theory, structural functionalism is often called *conservative,* because it argues that inequality is the natural outgrowth of the way the system works best. By this logic, it makes sense to allow inequality to persist (to be "conserved"). Conflict theory finds inequality neither just nor inevitable. In fact, conflict theorists assert that inequality is harmful to the stability of society and the quality of life within it. They argue for changes in the system of stratification.

One Attempt to Reconcile the Two Views

Gerhard Lenski (1966) proposed a synthesis of the two competing views of stratification. He claimed that principles from each view come into play in the distribution of the valued goods of society, but under different conditions. Lenski stated two "distributive laws": (1) Until the survival and productivity of everyone in a group is assured, all goods are distributed according to functional principles. That is, the common welfare demands that everyone who can contribute to the operation of the social system be rewarded adequately to allow him or her to do so. (2) Any goods that are available beyond that are distributed by principles of conflict. Surplus, then, is any resource not necessary for the basic operation of the social system. Lenski argued that it is this surplus that groups in society contend for on other than functional terms.

Figure 17-1 illustrates how the distributive laws might work. In Society A there is no surplus, so no resources are distributed by conflict principles. In Society B all the resources above the line are subject to distribution by conflict principles. Lenski also reconciles conflict and functionalist views by noting that functionalist patterns of distribution can continue long after they have stopped being functional. As conflict theorists have pointed out, once a group is in control of rewards, it can do a great deal to solidify (or institutionalize) its advantaged position.

As I am writing this, events in Somalia fly in the face of Lenski's distributive laws. In that country there has been deep and persistent starvation among the civilian population while warring factions struggle for control of the country. There have been countless reports of fighters from

FIGURE 17-1
Lenski's distributive laws

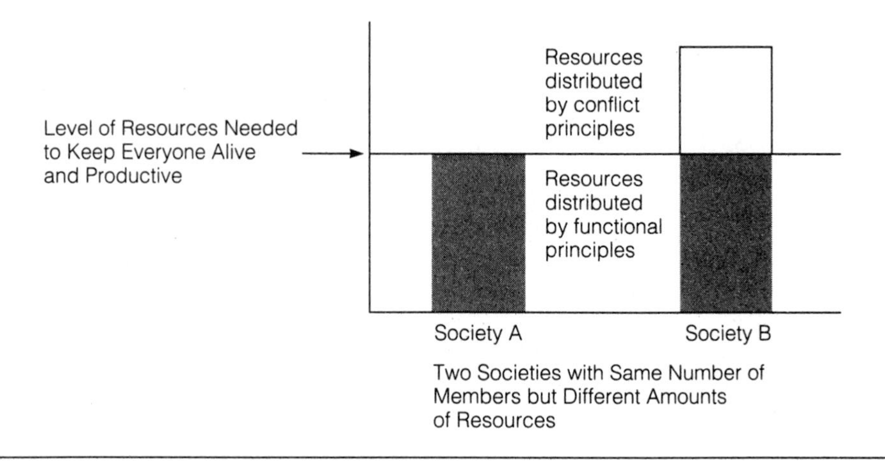

each of the factions, and roving bands of heavily armed individuals with no apparent factional allegiances, stealing shipments of food sent by relief organizations for famine victims. In this case, though there is no surplus, food clearly is being distributed not by functional principles but by conflict.

Some Systems of Stratification

Throughout history and across present-day societies, systems of stratification have varied. I will discuss three major types.

Caste

The caste system is generally regarded as the most rigid form of stratification, because virtually no mobility between levels is possible. It is a system with a number of layers (castes) into which people are born. Adherence to one's caste is supported by custom, religion, and law. The Indian Hindu caste system is perhaps the clearest example, especially as it existed before reforms to reduce its hold on national life. At the bottom of the hierarchy were the Untouchables, who were assigned the most unpleasant tasks, such as dealing with taboo cows. Above them were the Sudra caste, who did unskilled labor. Then came the Vaisya caste, who farmed, ran small shops, and did skilled work. Next to the top were Kshatriya caste members, who were lower- and middle-level military and government officials. At the top were Brahmans, who were the religious and educational elites. This system has been yielding to reforms, but it is so bound up in the religious and occupational life of the country that it is difficult to change entirely.

Estate

People who live within an estate system of stratification are born into the level they occupy in the society, like those in castes. But unlike the caste system, the estate system allows a small degree of mobility between estates (for instance, a serf can be freed by a noble) and is supported by custom and law rather than religion.

The estate system dominated feudal Europe until the rise of the middle class and the Industrial Revolution. At the top of the system were the aristocrats, or nobles, who owned land and controlled the social order. Below them were the merchants, or burghers, who were engaged in commerce and later developed into the middle class. At the bottom was the vast group of peasants, or serfs, who served as labor.

Social Class

In the social class system, mobility between levels is greater than in either the caste or the estate system. This is largely because social class positions are not officially supported by law or religion. Various class theories identify various numbers of layers (or classes).

The first description of the class system was stated by Marx and was discussed earlier in this concept. Marx named two classes, capitalists and workers. He felt that all circumstances of stratification flowed from economic relationships, so only these two classes could develop. After all, a person either is or is not an owner of the means of production.

Max Weber ([1925] 1946) agreed with Marx that social class position is influenced by economic factors, which he called *class*. But Weber also felt that social class can be influenced by one's power (which he called *party*) and by the esteem in which one is held (which he called *status*). A person who has all three—money, power, and status—is certainly a member of the highest social class. Because each of these assets can be used as a tool to gain the others, they are usually found together. But it is possible for a person to have a great deal of status but little money or power (for example, a teacher or priest), or a great deal of money but little power or status (for example, a prostitute or contract murderer). This condition, in which a person has a high level of one measure of social class but a low level of others, has been called **status inconsistency.**

How Many Social Classes?

The social class system is not supported officially in law, so no clear delineation of classes really exists. There are laws that tend to perpetuate positions within the system of stratification, such as laws allowing inheritance of property or the deduction of certain expenses from income taxes. But these are constantly subject to change, as various segments of the class system struggle for access to the resources of the society. So how can we meaningfully talk about the number of classes in any social class system?

One convenient way is to look at the way people earn money. Because the amount of money made often correlates strongly with the way the money is earned, this method can be quite useful. Using it, we can identify six classes: (1) *business owners and entrepreneurs*—those who invest money and stand to gain from the profits made by themselves or others; (2) *professionals and managers*—those who operate at the highest levels of business, government, the military, and so on and who get large salaries for doing so; (3) *skilled and organized labor*—upper level blue-collar workers and those whose job security and relatively high wages are normally reinforced by union membership; (4) *lower level blue-collar workers*—those whose jobs are less secure and who are paid lower wages; (5) *part-time and seasonal workers*—those who are paid less than lower level blue-collar workers and rarely qualify for any job benefits, such as unemployment compensation and medical coverage; and (6) *nonworking, dependent poor*—those who are recipients of support programs such as welfare and public housing and who cannot get into the job market at all.

A serious problem with this system is that there are some confusing overlaps between levels. For example, it is common for professionals and managers to earn much more money and have higher prestige than many owners of small businesses. Some professionals (such as teachers) enjoy higher status but earn a good deal less money than skilled or even unskilled laborers. In addition, members of any class can, technically, be considered entrepreneurs if they invest in any stocks. So other systems of identifying social classes have been devised.

Social classes are sometimes designated (1) upper, (2) upper middle, (3) middle, (4) lower middle, (5) upper lower, (6) lower lower. Many factors can be considered in deciding what social class a person belongs in—income, material possessions, occupational prestige, family history, religion, race, age, sex, club memberships, educational level, schools attended, and so on.

Summary

Every society has some degree of inequality among its members. Sociologists use the term *stratification* to refer to the system of ranking individuals in terms of their access to, and possession of, the things that are valued by the society. The opportunities people get to use their talents and energies (called *life-chances*) vary among individuals (a particularly talented versus a limited person) and, more significantly, among groups or categories of people (blacks or women versus white men). Most often, discussions of social inequality divide modern industrial societies into layers called *social classes.* A social class is a category of people within a system of stratification who share a similar style of life and socioeconomic status.

Two social theories dominate the debate about the sources of stratification. The conflict view, most often associated with the work of Karl Marx, argues that social inequality is the consequence of a struggle between contending groups for the things that are valued in a society. According to Marx, the major contending classes in capitalistic societies are workers and owners of the means of production (entrepreneurs). Marx predicted (and advocated) the overthrow of the owners by a revolution of the workers. More recently, Ralf Dahrendorf reformulated Marxist conflict theory to apply to forms of class struggle other than worker revolution.

By contrast, the structural functionalist theory of stratification argues that social inequality is a consequence of society's need to reward heavily those positions that are most important for the smooth operation of the society as a whole. This view, most often associated with the work of Talcott Parsons, tends to justify inequality on the ground that it aids the operation of society. Thus, changes in the distribution of rewards would only diminish the stability of the social order.

In an attempt to reconcile the conflict and structural functionalist views, Gerhard Lenski proposed that all rewards in a society that are necessary for its survival and the productivity of its members are distributed according to merit (structural functionalist principles), whereas any surplus goods are distributed according to power (conflict principles).

Three main systems of stratification have been identified. (1) In the caste system, a society is divided into rigid layers into which people are born. There is virtually no mobility among castes, and the system is maintained by law, custom, and religious rules. (2) Members of an estate system are also born to their position in the stratification, but some mobility between estates is possible. Estate systems are enforced by law and custom, but not by religious rules. (3) In the class system, mobility between classes is possible and considered desirable. Thus, class position is not supposed to be fixed at birth, nor is law, custom, or religion supposed to support class divisions. However, it is evident in the operation of class structures that classes in advantaged positions do try to protect their advantages by manipulation of law, custom, and religion.

The Illustration for this concept summarizes a 1988 study of the ways people identify their own social class membership. The data presented by Nancy Davis and Robert Robinson make it clear that men and women use very different information to decide in what class to place themselves.

The Application for this concept gives you a firsthand look at some evidence of inequality in the United States. By using data from the most recent U.S. census (collected in a reference called *Statistical Abstracts of the United States*), you can compare incomes for various races, sexes, and age groups in the country generally and in your region or state. The

Application specifies the tables to consult in this very handy and readily available reference source.

Illustration

Nancy J. Davis and Robert V. Robinson, "Class Identification of Men and Women in the 1970s and 1980s," American Sociological Review *53 (1988): 103–12*

One of the elements of sociology most familiar to the American population is stratification. In news reports, magazine articles, and everyday conversation social class descriptions of individuals and groups regularly use terms like *middle class* and *working class.* In less than five minutes of skimming a week's newspapers I found the following: (1) the cost of housing in the Boston area was reported to be still largely beyond the means of most "blue-collar" families (*working class* is the more common term); (2) Charles Stewart, a Boston man suspected of having killed his pregnant wife for insurance money, was described as an apparently upstanding "middle-class" guy; and (3) homelessness in Boston was described as a phenomenon of the "underclass," a category of people described as permanently unable to enter the labor market.

Apparently, social class is a common idea in the culture, but what, exactly, do we mean by the term *social class?* How do we decide what social class a person belongs to, and do we all use the same technique? This illustration summarizes a study in which it is made clear that different people use very different techniques to identify social class in our culture and that the process of class identification can change over time. In order to understand the article better, it will be necessary to expand a bit on one of the ideas from the Definition section of the concept.

Some Elements of Measuring Social Class

In the Definition I mentioned that an individual's social position could be influenced by many factors, including income, material possessions, occupational prestige, age, sex, educational level, and so on. If you could measure each of the variables that influence a person's social class, it would seem to be a simple matter to place an individual on a scale of stratification by adding up all the person's important values and combining them. In fact, that seems to be pretty much what people do when they tell you what their social class is, even if they are unaware that they are doing it this way. Using a fairly systematic version of this procedure, a given college teacher might give herself 30 points for yearly income, 5 more points for the value of her material possessions, 10 points for occupational prestige, and so on. But the task is obviously more complex than this. Ultimately, we must try to (1) include in our measure the most important factors, (2) measure each accurately, and (3) come up with a way of combining them by a formula that reflects their relative importance to social class in the United States. (Income, for example, is apparently many times more important to social class than is educational prestige.)

Assuming that we can accomplish all that (and many measures of social class do pretty well with these tasks), we face an additional, less apparent problem. It is often difficult to separate the characteristics of individuals from those of the groups to which they belong. To illustrate this dilemma, take a few moments to consider your own social position. Specifically, what sort of influence would you say that income and material possessions have on your social class? Full-time students between the ages of eighteen and twenty-one often use their parents' levels of income and material possessions in calculating their social class. Such students typically have not started their own careers and still feel more associated with the social class of their parents. For them, their social class as college students seems too temporary or transitional to replace the class they had as they grew up.

In short, when we ask people what their social class is, we cannot be certain what factors they are using in their evaluation or whether they are using their own characteristics or those of others in arriving at an answer. The research conducted by Davis and Robinson focuses on these questions as they apply to the social class evaluations of men and women.

A Study of Class-Identification Procedures Among Men and Women

Davis and Robinson used data from the National Opinion Research Center's General Social Surveys to study class-identification procedures in the 1970s and 1980s. This information is collected every year from a national sample of Americans and covers a wide range of topics. The researchers merely selected from the broad range of data in the surveys the information that had to do with the social class identifications of survey respondents in the 1970s (using data from 1974 to 1978) and in the 1980s (data from 1980 to 1985). They focused on the different ways that men and women decided on their own class membership (lower, working, middle, or upper). Specifically, they examined which of the following factors were included in their social class decisions:

1. **Education.** The respondent's years of schooling
2. **Occupational Prestige.** Calculated on a scale from the lowest to highest prestige
3. **Ownership.** Whether they owned a business

4. **Income.** Measured in thousands of dollars a year
5. **Spouse's Income.** Measured in thousands of dollars a year
6. **Spouse's Education**
7. **Spouse's Ownership**
8. **Spouse's Occupational Prestige**
9. **Father's Education**
10. **Father's Occupational Prestige**
11. **Father's Ownership**
12. **Mother's Education**
13. **Mother's Employment.** Whether the mother was employed for at least a year after marriage

Notice that only items 1 to 4 are characteristics of the respondents' answers. All the rest are characteristics of family members (either a spouse or parent), which the researchers thought might have been used by respondents in deciding on their own social class membership. To discover if this was happening, Davis and Robinson used a statistical technique (called multiple regression) in which the responses of each individual to the above questions were correlated with their social class identification. Thus, if certain kinds of people were using their own characteristics in deciding on their social class membership, their answers to questions 1–4 would be found to correlate with their social class identification; the other factors would not be correlated. If, however, the characteristics of a spouse or parent were being used, those items would be found to correlate with social class decisions. This technique allowed Davis and Robinson to compare various categories of men and women to discover on what factors they based their social class positions in both the 1970s and 1980s.

Results of the Study

The authors examined three categories of men and women: (1) those in marriages in which the husband was employed in a full-time job and the wife was not employed outside the home, (2) those in marriages in which both husband and wife were employed full time, and (3) single individuals who were employed full time. Tables 17-1 to 17-3 present the results for these groups in the 1970s and 1980s. Each table is followed by a discussion of its data.

In the first category, couples in which the man was employed and the woman was not, a comparison shows clear differences in the way men and women decided on their social class memberships, especially in the 1970s (Table 17-1). Notice that in the 1970s these men relied largely on their own accomplishments (specifically, their own income and occupational prestige) in determining their social class, though they did

TABLE 17-1

Factors correlated with the social class identifications of individuals in male-employed, female-unemployed marriages

Men		Women	
1970s	1980s	1970s	1980s
Own occupational prestige Own income Spouse's education	Own education Own occupational prestige Own income	Spouse's education Spouse's occupational prestige Spouse's ownership Spouse's income	Own education Spouse's education Spouse's occupational prestige Spouse's ownership Spouse's income

include their spouse's education as a factor. By the 1980s, however, men in such marriages relied entirely on their own characteristics (education, occupational prestige, and income) to determine their social class. By comparison, in the 1970s women in this type of marriage used their husband's characteristics exclusively (education, occupational prestige, business ownership, and income) in deciding what social class they belonged to. Their own characteristics had no effect. By the 1980s, the social class identifications of women in these marriages were still dominated by their husband's characteristics, though these women now included their own education as the single factor about themselves that mattered.

In discussing these results, Davis and Robinson identify three techniques of social class identification: (1) In the "independence" model, people use only their own characteristics in deciding on their social class. (2) In the "borrowing" model, people use only the characteristics of other people in making class decisions. (3) In the "sharing" model, people use some combination of their own characteristics along with those of others.

Davis and Robinson suggest that these data show these men to have moved from a largely "independence" model of social class with some "sharing" of the characteristics of others (wife's education) in the 1970s to an entirely "independence" model in the 1980s. They describe the social class model used by these women in the 1970s as being entirely "borrowed," that is, depending on the characteristics of the husband. By the 1980s, these women had moved slightly toward the "sharing" model, now including their own education as a factor, though four of the five factors they used were still about their husbands.

When both members of a couple worked (Table 17-2), the men used three of their own characteristics (education, occupational prestige, and income) and one characteristic of their wives (her income) in the 1970s. By the 1980s even that small amount of sharing had been abandoned, and the "independence" model dominated, based on their own education, occupational prestige, and income. If you look back at the data in

TABLE 17-2

Factors correlated with the social class identifications of individuals in male-employed, female-employed marriages

Men		Women	
1970s	1980s	1970s	1980s
Own education Own occupational prestige Own income Spouse's income	Own education Own occupational prestige Own income	Spouse's education Spouse's ownership Spouse's income	Own education Own income Spouse's income

TABLE 17-3

Factors correlated with the social class identifications of employed single individuals

Men		Women	
1970s	1980s	1970s	1980s
Own education Own occupational prestige Father's ownership	Own education Own ownership Own income	Own income Father's education	Own income

Table 17-1, you can see that the shift from the largely independence model in the 1970s to the entirely independence model in the 1980s occurred among married men whether their wives were employed outside the home or not.

Looking at the data for women, in the 1970s married women who were employed based their social class decisions entirely on their husband's characteristics (the borrowing model). Notice that these data are very much like those for the unemployed married women shown in Table 17-1. By the 1980s, however, the employed married women seem to have been thinking differently from their unemployed married counterparts. Specifically, working women had largely abandoned the use of their husband's characteristics in favor of their own. Only their spouse's income remained as a factor in determining social class, and their own education and income had become important.

The data in Table 17-3 again show a movement among men (in this case, employed single men) from a 1970s model dominated by independence but with some sharing (father's business ownership) to a 1980s model of total independence. Among employed single women, the 1970s model was not the same as it was for married women, who, whether they were working or not, used only the characteristics of others (spouse) for their social class identifications. Single women in the 1970s, in contrast, used a sharing rather than a borrowing model, with their own income and their father's education as the factors used. By the 1980s single women, like all three categories of men in the 1980s, had switched entirely to an independence model. For these women, the only characteristic correlated with social class identification was their own income level.

Summary

After working our way through all this information, we can have the fun of trying to speculate why these results were discovered. The researchers suggest, for example, that the increasing rate of female employment and the effects of the feminist movement may have allowed women to increasingly base their social class identifications on their own accomplishments. But beyond the apparent difference between men and women in these data, Davis and Robinson point to a broader pattern at work. They point out that both men and women, in all categories, moved toward greater reliance on their own characteristics in making social class identifications. They suggest that these results reflect a general increase in individualism in American culture. Although men may have had less room to move toward the independence model than did women (men were almost entirely there by the 1970s), the "me" decade pushed everyone in the same direction, if not to the same degree. Perhaps research conducted on data for the end of the 1980s and early 1990s will reveal that women had become just like men by then in that no sharing of characteristics of others was occurring in the social class identification of either sex.

Application

Stratification, as we have seen, is a system of ranking individuals in terms of their access to the things that are valued by their society. Most often we think of stratification in terms of social class differences. The lowest social classes have the least access to power, money, or status in a society. It is of intense interest to sociologists that social class membership is so closely related to membership in other kinds of social groups. For example, race, sex, and age are highly correlated with social class. Census data consistently document the lower earning capacity of blacks, women, and the elderly in the United States.

Statistics on U.S. Stratification

The purpose of this application is to give you a firsthand look at the documentation of stratification by race, sex, and age in the United States. In addition, you can compare the average figures for the entire country with those for your region or state. You can use one readily available source, the *Statistical Abstract of the United States,* published in yearly updated editions by the U.S. Bureau of the Census. This application was prepared using the 1992 (112th) edition. It is available in virtually every library's reference section, often with supplementary census publications that focus on specific groups, such as black people and the elderly.

Data collected during the 1990 census of the United States are summarized in *Statistical Abstract* in many ways. For example, income averages are calculated for individuals, families, households, and so on. They are further calculated for the entire country, various regions, all the states, races, sexes, age groups, occupations, and other categories. This application focuses on data for race (white, black, and Hispanic origin), sex, and age groups.

For the categories of data specified in the worksheets to follow, look up the appropriate chart in the *Statistical Abstract,* and copy out the data. For most of the information requested, you can also compare the data for your region or state with the data for the country. By the time you have answered all the questions, you should have a strong feeling for the unequal distribution of wealth in the country and the extent to which your state or region reflects that inequality. These are only a few of the indications of inequality available in the data, but they should be adequate to demonstrate stratification. If you would like to extend your analysis, it is a simple matter to look at more data in the same volume (or in the supplementary publications mentioned earlier) for other evidence of stratification.

Income of Households

The amount of money that people earn in the United States is calculated for households. For example, if two adults in the same household both earn income, their incomes are combined for this statistic. Tables for these data are in the section called "Income, Expenditures and Wealth" (look in the table of contents to find the section). First refer to the table entitled "Money Income of Households—Percent Distribution, by Income Level and Selected Characteristics: 1990." In the 1992 edition this is Table 697 on page 446. It reports that as of 1990 the median income for all households in the United States was $29,943. What was the median income for households in each of the following categories?

Median income for households

1. Race: White ____________ Black ____________ Hispanic ____________

2. Family Households:

Married-couple families ____________

Male householder, wife absent ____________

Female householder, husband absent ____________

3. Age: Householders over sixty-five ___________

Householders forty-five to fifty-four ___________
(or similar middle-age group in your edition)

4. Region: Northeast ___________ Midwest ___________

South ___________ West ___________

5. Educational attainment of householder:

Eight years or less ___________
One to three years of high school ___________

Four years of high school ___________
One to three years of college ___________

Four years or more of college ___________

What differences do you find? Are median incomes for households different for the races, sexes, and ages? By how much? Are there differences for the regions of the country as well? What about income in the region where you live?

Note that this table uses the *median* to measure average household income. That is the point on a scale of incomes below which half the scores fall and above which half the scores fall. It does not matter what the highest score on the scale is. It could be $35,000 or $35,000,000, for example; the median would remain the same. So, the median is a measure of central tendency that is not affected by extreme values.

Another table in the section on "Money Income of Households," however, calculates average household incomes as an arithmetic average (called the *mean*). Because the mean adds all incomes together and divides by the number of incomes, it is strongly influenced by very high scores. (Think of how much a single household with an income of $2 million a year would raise the mean income figure for the households on an average street.) Using the mean as a measure of average household income, if a category has some people with very high incomes, group differences will be even more dramatic than when using the median.

Just such a table, comparing mean household incomes, is entitled "Money Income of Households—Aggregate and Mean Income by Race and Hispanic Origin of Householder: 1990." In the 1992 edition this is Table 698 on page 447. It reports that the mean income for all households in the United States in 1990 was $37,403. (Notice that this is higher than the median income reported in the last table. It should be, because the mean is influenced by extreme scores.) Now compare the mean incomes for the three racial categories in the country generally and in the four regions.

Mean income for households

1. Race: White ___________ Black ___________ Hispanic ___________

2. Race by region of the country:

White:	Northeast ___________	Midwest ___________
	South ___________	West ___________
Black:	Northeast ___________	Midwest ___________
	South ___________	West ___________
Hispanic origin:	Northeast ___________	Midwest ___________
	South ___________	West ___________

Focusing on mean incomes for the races, do you find big differences among them? Are they greater than the differences shown in median-income measures? How do the regions of the country reflect such differences? Are they greater or less where you live?

You can see that both the mean and the median are figures that can quickly summarize a great deal of information. Sometimes they also hide some information. For example, we may know the mean or median income of families without knowing how incomes vary around that central figure. It is possible for one country with a mean family income of $100,000 to have every family earning around $100,000 every year; another country with the same high average income may have 95 percent of the families earning essentially nothing (perhaps $1,000 per year) and the other 5 percent earning millions a year. These are very different income distributions. To find out a bit more about how income is distributed in the United States, turn to the table entitled "Money Income of Households—Percent Distribution by Income Level in Constant (1990) Dollars, by Race and Hispanic Origin of Householder: 1970 to 1990." In the 1992 edition this is Table 702 on page 449.

This table records the percentage of the population with household incomes within certain ranges and does this for data collected between 1970 and 1990. Thus, you can see (at the upper left corner of the table) that in 1970, 8.7 percent of all households in the United States had incomes below $10,000. Following down that column, you can also see that by 1990 the percentage of American households with incomes below $10,000 had increased slightly to 9.4 percent. To see how income was distributed around the average figures given in the previous two tables and to see how income distributions have changed over time, look up the following data also:

1. Percentage of all households with incomes of $75,000 and over:

   1970 ___________ 1990 ___________

2. Percentage of white households with incomes under $5,000:

   1970 ___________ 1990 ___________

3. Percentage of black households with incomes under $5,000:

1970 __________ 1990 __________

4. Percentage of white households with incomes of $75,000 and over:

1970 __________ 1990 __________

5. Percentage of black households with incomes of $75,000 and over:

1970 __________ 1990 __________

What is the household income like at the far ends of the distribution, away from the mean-income figures you were looking at before? Does the United States have equal numbers of very poor and very wealthy people? You should realize that looking only at the top and bottom of the distribution is still a very incomplete search. You can get a better picture if you do a graph for the percentage of families in each income level for a given year (see Figure 17-2).

Lastly, what happened to household incomes at the top and bottom of the distribution over the years between 1970 and 1990? Clearly, the percentage earning very little (under $10,000) increased overall, and the percentage earning a great deal (over $75,000) increased a great deal more. Were these changes true for both blacks and whites in America over these years?

FIGURE 17-2

Distribution of household incomes for all households (1987 data)

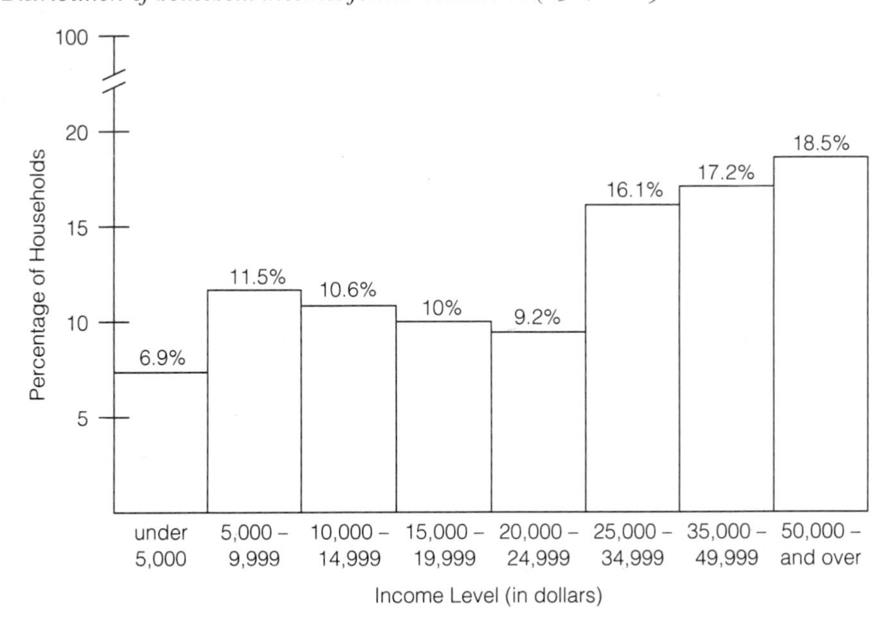

Poverty Level

Another measure of well-being is the percentage of a population that falls below the official poverty level for the country. Also in the section on "Income, Expenditures and Wealth" is a table entitled "Persons Below Poverty Level by Race, Hispanic Origin, Age and Region: 1990." In the 1993 edition this is Table 720 on page 457. It reports that 13.5 percent of all Americans were below the official poverty level in 1990. Compare this figure with the percentages for the races identified in the table, for the country as a whole and by region.

Percentage below poverty level

1. Race: White __________ Black __________

2. Race by region:
   Northeast: White __________ Black __________

   Midwest: White __________ Black __________

   South: White __________ Black __________

   West: White __________ Black __________

What differences do you find in the figures for poverty levels in the various groups? Are these differences greater or less in your region of the country? To see how these figures for percentage of persons below the poverty level have changed over time, look at the table entitled "Persons Below Poverty Level, by Race of Householder and Family Status: 1979 to 1990." In the 1992 edition of the *Statistical Abstract* this is Table 722 on page 458. In the first row of figures in the table you can see that the percentage of Americans below the poverty level increased from 11.7% in 1979 to 13.5% in 1990. What was the percentage of each category of Americans who were under the poverty line for each of the following years?

All persons below the poverty level:

1979 __________ 1985 __________ 1990 __________

White Americans below the poverty level:

1979 __________ 1985 __________ 1990 __________

Black Americans below the poverty level:

1979 __________ 1985 __________ 1990 __________

Extending the Analysis

As you can probably tell from using the census data, a great deal more information is available to document inequality in the United States than is used in this application. To extend your analysis and at the same time become more familiar with the use of this very handy directory of information try finding other measures of well-being, and make comparisons among the various minority groups and regions of the country. You could, for example, look at expenditures for education in the states or rates of unemployment. The further you look, the more evident it will become that American society is highly stratified and that inequality closely follows lines of social groupings for race, sex, and age.

CONCEPT 18

Discrimination and Prejudice

Definition

Discrimination Negative treatment of people on the basis of their group membership rather than their individual qualities.

Prejudice Negative beliefs and feelings about a group of people that serve to justify and reinforce discrimination against them.

I was twelve years old before I ever heard the word *kike*. I overheard it, actually, while some friends and I were walking on a street in Manhattan. I knew about other filthy epithets directed at other groups of people, words such as *nigger* and *spic* and *mick,* but these were just like any other dirty words to me. I did not know their special sting until I learned there was one that could be aimed at me as well. It was a terrible shock to realize that someone could hate me merely because of my religion, could totally ignore what I was like as an individual, what each Jew was like! I must have been reared in a very protected way. I had been able to maintain a belief that whatever I accomplished (or failed to accomplish) was purely the result of my own talents and efforts. I believed in **meritocracy,** the system of success based on merit alone. But on that memorable day I began to realize and think about the possibility that something other than individual merit powerfully influenced opportunities for success in my country.

It wasn't until I took a college sociology course on racism that I began to really understand how **prejudice** and **discrimination** work and why they exist. I had spent a good deal of time trying to figure out what I had done to make a bigot hate me. Then I decided that it was he (or she) who was flawed, that such a person was stupid, weak, or just mean. But as the sociological perspective taught me, neither of these explanations focused on the most important factor. Discrimination and prejudice, like other social facts, result primarily from the structure of the social order, not from flaws in our basic nature as humans.

To begin with, let's define some basic terms. Prejudice is an attitude, a system of negative beliefs and feelings about a group of people. Negative beliefs (the cognitive component of the attitude) are sometimes expressed as **stereotypes.** Such generalizations about the characteristics of a group of people depict them as inferior in a variety of ways. Jews are stereotyped as greedy, clannish, and shrewd; blacks as lazy, childish, and impulsive; Irish as alcoholic; Italians as criminal; women as emotional and scatterbrained; old people as childlike and senile; and the list goes on. Even generalizations about people that appear at first to be positive typically turn out to have negative meanings and consequences. For example, it might seem a compliment to be characterized as gentle and caring, as American women have been, but not if the belief is used to patronize women and deny them jobs, on the ground that they are too gentle and caring for the tough world of work outside the home.

The affective component of the attitude consists of the negative feelings aroused in the prejudiced person by members of the target group. The bigot may feel pity, fear, disgust, contempt, envy, or other unpleasant emotions in the presence of members of the group or even when just thinking of them. I can recall discussing race relations at a dinner party with a woman who revealed that she felt compelled to boil the lunch dishes used by the black cleaning woman working in her home. This emotional reaction of prejudiced people to the thought of contact with the people against whom they hold their prejudices is extremely tenacious. Long after we have changed our minds about some belief, we may retain the "gut" feelings that accompany the beliefs. For example, a man who has decided to eliminate his prejudiced stereotypes about women, and who largely succeeds in doing so, may have a much more difficult time ridding himself of the condescending or paternalistic feelings that he was raised to have toward women.

Sources of Prejudice

Like any other attitude, prejudice can be learned. Sometimes it is passed on by parents to their children. When this happens, prejudice is taught as is any other norm in which the parents believe. In a few places in the United States it is possible to see this process in its most blatant form. Filmed interviews of neo-Nazi communities have shown parents proudly parading their very young children carrying toy guns and urging them to chant memorized anti-black and anti-Jewish hate slogans. This view of how prejudice is acquired can be called the *normative theory* (Bogardus, 1959; Westie, 1964). Prejudice can be taught.

A major psychological theory about how prejudice is acquired focuses on a process called *scapegoating*. According to this view, most of the actions in which we humans engage are initiated to satisfy some need or desire. When we fail to satisfy such a need or desire, we feel the unpleasant psychological state of frustration. A great deal of research has suggested that we commonly reduce frustration by engaging in aggressive behavior (Dollard et al., 1939). Think, for example, of the schoolchild who is reprimanded publicly by the teacher and turns to "take it out" on the innocent kid next to him (or her). Aggression can be either physical (kicking the cat after you have hit your thumb with a hammer) or attitudinal. Experimental studies have illustrated that people who are subjected to frustrating experiences express greater hostility toward "foreigners" and other out-groups than do people not subjected to such frustrations (Holmes, 1972; Miller & Bugelski, 1948). The scapegoat theory of prejudice suggests that when our frustration is caused by a powerful source, such as a grade school teacher, it is too dangerous for us to vent our aggression against that source. We therefore choose a target that is in no position to strike back, one that is in some way distant or relatively powerless. The reason such targets for aggression are called scapegoats is that they may have nothing to do with causing the frustration but are chosen as targets nonetheless.

Prejudice is an attitude, a predisposition toward a behavior. Our highest national ideals teach that prejudice is wrong, yet bigotry persists. The Swedish economist Gunnar Myrdal, a Nobel Prize winner, concluded in his study of American race relations (1944) that we do have high-level norms against prejudice. We maintain our belief in the Constitution and the Bill of Rights. Yet he noted that we often have bigoted views that operate in everyday interaction with others. Myrdal pointed out that, if we are to fulfill the American promise of equality of opportunity, we must deal with what he called the American dilemma, the gap between our highest ideals and our everyday behavior. But in the United States we never legislate against or for attitudes, only behaviors. We may hold whatever *attitudes* we like, as long as we do not *act* to break laws. The relationship between the attitude (prejudice) and the behavior to which it is related (discrimination) must be clearly understood if we are to make any progress in reducing either.

Discrimination

Discrimination is the unequal treatment of an individual on the basis of his or her group membership. When we discriminate, we ignore the individual qualities of a person. Instead, we take a shortcut in our evaluation. By categorizing others in terms of group membership, we assume that they have all the characteristics that we attribute (often falsely) to their group. Suppose a company wants to fill an executive position and is looking for a cool-headed, aggressive person for the job. A woman who has applied is rejected by an interviewer who believes that all women are emotional and passive. That is discrimination. The female candidate was turned down on the basis of her group membership rather than on the basis of her individual characteristics. Notice that the interviewer might have found reasons to reject her application without discriminating against her. Let's assume that the interviewer considered the woman and determined that this specific candidate *actually was* emotional and passive. Denial of opportunity on the merits of a case is not discrimination. The problem is that so often evaluations are based on group, rather than individual, grounds. Also, when we evaluate others in terms of group membership, we never get the opportunity to discover that *not all* group members share the same characteristics. The bigot's assumptions remain unchallenged.

To discriminate, a person must be in a position to make choices. He or she must be able to distribute the rewards of the society, such as jobs, promotions, and status. Discrimination is literally choosing. So discrimination deals with the distribution of opportunity in the society. When individuals distribute opportunities on the basis of their private preferences for one group of people over another, that behavior is generally considered an instance of individual discrimination. But when the advantages for one group of people over another are built into the structure of society and into its rules, it is called **institutional discrimination** (Knowles & Prewitt, 1969). One example is the literacy requirement that was used in the South to keep blacks from voting. Literacy laws appear on the surface to be group-neutral. But they were created to penalize certain groups of people on the basis of disadvantages they had already suffered—lack of or vastly inferior education. As blacks gained literacy, bigoted whites institutionalized the custom of demanding a higher level of reading ability from them than from whites (giving the blacks more difficult passages to read and allowing fewer or no errors). Any group who has received inferior education in the past will necessarily have trouble with certain kinds of tests, whether they be the old, unevenly applied voter literacy tests or the scholastic aptitude tests of today.

When apparently group-neutral but actually discriminatory standards are built into the structure of society, it is difficult to point at specific discrimination. After all, if the entire society believes that skills are demonstrated by scores on scholastic aptitude tests, how can it be argued that such tests discriminate unfairly? Another example is the illegal practice called *redlining.* Banks outline in red on city maps the areas in which homeowners are considered bad risks for mortgage or home-improvement loans. An overwhelming percentage of the people inside such outlines may be black, but how can they prove that the practice is discriminatory? Banks merely point out that they are following good credit practices, without regard to color. Institutional discrimination is very difficult to combat.

The Relationship Between Prejudice and Discrimination

Having examined both prejudice (an attitude) and discrimination (a behavior), we can now consider the relationship between them. When an attitude and a behavior are in agreement, we consider them consonant; examples are a bigot who discriminates and an unprejudiced person who does not discriminate. When an attitude and behavior are not in agreement, they are considered dissonant. But how can a bigot refrain from discriminating, or why should an unprejudiced person discriminate? These are the questions raised by Robert Merton's (1976) model for the relationship between prejudice and discrimination (see Table 18-1). For both the active bigot and the all-weather liberal, the attitude and the behavior are consonant. But the timid bigot and the fair-weather liberal are examples of people in dissonant situations. Merton suggests that these occur because of social situational pressures. For example, the timid bigot may wish to act on his or her prejudice, but such behavior would not be tolerated in the neighborhood or by the law. Similarly, the fair-weather liberal would like to treat all people equally, but living in a highly bigoted neighborhood, he or she feels forced to discriminate in order to get along with neighbors.

Prejudice Leading to Discrimination

Clearly, situations of dissonance between an attitude and behavior are unpleasant to endure, and psychologists have suggested that they are generally avoided or resolved as soon as possible. There are two general models for how such dissonance is avoided in the first place or reduced when it does occur. The first assumes that we engage in behaviors in

TABLE 18-1
Merton's model of the relationship between prejudice and discrimination

		Is the person prejudiced?	
		Yes	No
Does the person discriminate?	Yes	Active bigot (prejudiced discriminator)	Fair-weather liberal (nonprejudiced discriminator)
	No	Timid bigot (prejudiced nondiscriminator)	All-weather liberal (nonprejudiced nondiscriminator)

order to express our underlying attitudes: "I like a certain food, so I eat it all the time." Because the behavior follows the attitude, the two are necessarily going to be consonant. Bigots discriminate, and unprejudiced people don't discriminate. Generally it seems safe to assume that this occurs. Prejudice can be passed on by simply teaching it as a norm (the normative theory) or as a consequence of psychological mechanisms (the scapegoat theory), and we can expect discrimination to follow. Thus, prejudice leads to discrimination in this model.

Discrimination Leading to Prejudice

A second model proposes that discrimination can come first and lead to the development of prejudice. Daryl Bem (1970) argues that it is possible for us to act without having attitudes first. According to Bem, when someone asks what we think of something (what our attitude is), we may not know, not having thought about it. To find out what we think, we look at behaviors and infer consonant attitudes from them, as we would when trying to guess about someone else's attitudes. "He must like hamburgers, because he eats them all the time." Or about ourselves, "I must not like Judy, because I avoid her all the time." Bem says that we may think we are reading our own attitudes directly when we answer such questions but that we actually "peek" at our behaviors to get an idea about what attitude we do or should hold. When an attitude is adopted to make sense of a behavior, the two will necessarily be consonant with each other. According to this view, then, attitudes (such as prejudice) can sometimes be developed to make sense of what we do (such as discriminate).

Why would we discriminate against others in the absence of prejudice against them? Well, look back at Merton's idea that sometimes social situational pressures lead us to do things with which we don't necessarily agree. Having discriminated against someone we did not hate, merely because others urged us to do so or because we felt we might profit by discriminating (gain friends, feel superior, get favors in return, save on labor costs), we need to make sense of such discriminatory behavior. According to this model, we justify and excuse discrimination by developing the prejudice to support it. Once the prejudice is developed, it tends to reinforce further discrimination, just the way any attitude tends to induce consonant behavior. Eventually, then, prejudice and discrimination are mutually reinforcing (Levin & Levin, 1982).

Minority Groups

Whatever the relationship between prejudice and discrimination, it is the latter that actually creates suffering among specific groups of people in society. Such groups have come to be called **minorities,** and the groups that dominate them are called **majorities.** Very simply, majority groups are in a position to distribute rewards in society, and minority groups are not. So the distinction between majorities and minorities is one of power, not numbers. A majority group can be much less than 50 percent of a population. For example, in 1991 51.3% of the population of the United States was female (U.S. Bureau of the Census, 1992) yet men continue to control the great bulk of the wealth and power in America. To the extent that men determine the distribution of resources such as employment, pay, and promotions, they must be considered the majority group, though they are in the numerical minority. So majority status really refers to who controls the majority of the social and economic power. Minority status is the result of discrimination. Minorities are created by the treatment they receive, as a result of which minority-group members cannot control their own destinies. Unlike majority groups, they certainly do not control or even influence the destinies of others.

One of the best known and clearest sociological definitions of a minority group comes from the work of Louis Wirth (1945:347): "We may define a minority as a group of people who, because of their physical or cultural characteristics, are singled out from the others in the society in which they live for differential and unequal treatment, and who therefore regard themselves as objects of collective discrimination." It is important to realize that the surface characteristics that differentiate minorities from majorities are completely unimportant in the process that creates minority status. They matter only to the extent that they make discrimination easier by making the target group more readily recognizable. The insignificance of the characteristics is shown by the wide variety of minority groups in a range of categories.

Racial minority groups are those singled out on the basis of genetically inherited physiological characteristics, such as skin color, eye shape, hair texture, and hair color. Generally we identify racial groups such as Negro, Mongol, and Caucasian as if clear distinctions could be made among them with biological precision. But actually such distinctions are determined socially, not scientifically. How narrow a nose or how straight and light-colored hair must a person have to qualify as caucasoid? These characteristics are all continuous, not discrete; so decisions clearly cannot be based on distinct physiological boundaries. Skin tone, for example, ranges from very light to very dark along a continuous scale. There are no clear divisions.

Ethnic minority groups are those singled out for discrimination on the basis of learned characteristics, such as language, clothing, patterns of leisure, diet, worship, and so on. In the United States we talk of Irish, Italian, Polish, Japanese, and Puerto Rican ethnicity. Ethnic groupings often carry national names (such as Italian) but not always. Ethnic characteristics are also shared among racial groups (American blacks) and among religious groups (American Jews).

Religious, racial, and ethnic minority groups are the most extensively studied categories in American sociology. But recently we have begun to pay increasing attention to groups such as women, the elderly, homosexuals, and the disabled. Their emergence as minorities is relatively new, but the discrimination directed against them is not. Among the elderly, a liberation group called the Gray Panthers has emerged. Women have been organizing in recent decades, notably in the National Organization for Women (NOW) and the National Women's Political Caucus (NWPC) and around passage of the Equal Rights Amendment to the Constitution. Nevertheless, the elderly are still being forced to retire on fixed incomes, and women still earn slightly under seventy cents for every dollar earned by men. Minorities in the United States have a long way to go toward equality, and they are increasingly organizing and demanding attention to their grievances.

Summary

Discrimination is the negative treatment of a person or persons on the basis of their group membership rather than their individual qualities. Prejudice is a set of negative beliefs and feelings about a group of people that acts to justify and reinforce discrimination against them. Discrimination and prejudice are mutually reinforcing. Each can lead to the development of the other.

Negative beliefs are sometimes called stereotypes (the cognitive component of prejudice). They are generalizations about the characteristics of a group that depict them as inferior. Negative feelings aroused in the prejudiced person by the target group (the affective component of prejudice) include fear, pity, envy, hate, and disgust.

Two major theories about the origins of prejudice are the normative and the psychological views. The normative theory proposes that prejudice is learned, just as any belief system is learned. The psychological theory (sometimes called the scapegoat theory) argues that people whose needs or desires are frustrated sometimes vent their frustration on a relatively unthreatening "scapegoat" in the form of prejudice.

Anyone can be prejudiced. But to discriminate, a person must be in a position to distribute some valued good. Sociologists distinguish between individual and institutional discrimination. Individual discrimination involves the denial of opportunities to a group of people by one person. Institutional discrimination is built into the laws of the society and the regulations of its institutions.

A minority is a group of people that has been singled out from others in society for discriminatory treatment on the basis of physical or cultural differences. Minorities lack the ability to control their own destiny, whereas majority groups control their own destinies and those of others. Minority groups may be identified by race, ethnicity, sex, religion, or even age. The specific traits of minority groups are unimportant in the process of discrimination. It is the identification and manipulation of them by majority groups that matters.

The Illustration for this concept applies the definition of minority group to the elderly in the United States. It argues that prejudice and discrimination are directed against older Americans; that is, they are the victims of "ageism." Like sexism or racism, ageism identifies a target group as different and inferior and singles it out for inferior treatment.

The Application for this concept examines the depiction of some minorities in the mass media. It has long been understood that the images of a group on television and in magazines and newspapers greatly influence the opportunities made available to them. Using the research method called *content analysis,* you will discover the extent to which blacks, women, and the elderly are represented in selected mass media. By extending your analysis, you may also discover whether depiction of these groups is more negative than depiction of their majority group counterparts.

Illustration

William C. Levin, "Age Stereotyping: College Student Evaluations," Research on Aging *10 (1988): 134–48*

One of the minority groups mentioned in the Definition section of this concept, the elderly, has only recently become of interest to sociologists. The term *ageism,* meaning prejudice and discrimination against people on the basis of their age, wasn't widely known until 1975, when Robert Butler used it in his powerful study of aging in America, *Why Survive?* He documented the multitude of stereotypes of old age, including slowness, diminished creativity and thinking ability, inflexibility, chronic illness, inability to learn, being excessively habitual and traditional, conservativism, childishness, egocentricity, irritability, cantankerousness, tendency to live in the past, talkativeness, aimlessness, feebleness, and being uninteresting. (This is the short list.)

These simplifications about the elderly, like other stereotypes, make the world seem understandable to the bigot who employs them. If all older people are assumed to be incompetent and frail, it is not necessary to make the effort to learn about the actual characteristics of any individual over a certain age. Stereotypes are also used to justify negative treatment of people on the basis of their group membership. The bigot who assumes that the stereotypes of the elderly are true feels justified in not hiring them. So stereotypes are not merely harmless words. They profoundly influence the fates of many minority-group members and may alert us to the fact that a group of people is vulnerable as a target of discrimination.

During the decade after publication of Butler's book, a good deal of attention was focused on the elderly in the United States. This was partly because the proportion of the population over the age of sixty-five was growing rapidly (from about 5 percent of the population in 1920 to over 12 percent in 1989), with many consequences for the society. At the same time, the rates of poverty for the elderly were significantly higher than those for other age groups. During the 1970s and early 1980s the elderly received increased government spending, especially for medical and income entitlements in retirement, and their poverty rate gradually declined to the point at which it was roughly equal to that of the country at large. It even seemed as though stereotyping of the elderly had declined, and surveys of attitudes toward them reported as much. As with black Americans after the civil rights movement of the 1950s and 1960s, when telling racist jokes generally declined in acceptance, ageist stereotypes appeared less acceptable. Was this a case of a minority group that Americans had lifted or released from its low status?

A Study of Ageism Among Students

In the course I teach called Aging and Society, students had, by the mid-1980s, very positive public attitudes about older Americans. They certainly did not agree with the ageist stereotypes Butler had listed in 1975. However, when they conducted surveys of elementary schoolchildren asking "What is an old person like?" they reported long lists of nasty stereotypes from these children. What was going on? I suspected that age stereotyping had not actually disappeared but had merely gone "underground." That is, the negative beliefs persisted, but there was such widespread disapproval of expressing most stereotypes, and especially against the elderly, that direct questions to adults about characteristics of the elderly would reveal only socially acceptable responses. The problem was how to determine whether ageist stereotypes were still around without revealing to the subjects of my study that I was trying to find this out. Once they knew what I was studying, they would not reveal their ageist attitudes, if any.

The answer was to conduct an experiment. I found photographs of a specific man when he was twenty-five years old (taken in 1939, actually) and then when he was fifty-two. These photos were closely matched for head position, clothing, lighting, hair style, degree of smile, and so on. I was then able to photograph the same man when he was seventy-three years old, again matching the qualities in the previous two pictures. As you can see from the photographs in Figure 18-1, he had not lost his hair or gained weight. He didn't even wear glasses.

These photographs were the key to the study. If I could get people to evaluate the man on the basis of his pictures, any differences in the

FIGURE 18-1

Photographs of the man used in the study of attitudes toward age. Differing evaluations of him can be due only to differences in his age.

evaluations could be attributed only to his age. And the people doing the evaluating would be unaware that their attitudes toward age were being measured. Here is how it worked.

College students from three areas of the country, at San Francisco State University, East Tennessee State University, and Salem State College in Massachusetts, served as subjects in the study. They were told that they were taking part in a study of the ability of some people to make "surprisingly accurate evaluations of a stranger on the basis of very little information." The photographs, along with attached questionnaires, were distributed at random to the students. The respondents were given a few moments to look at the photograph of the person and to read a few sentences about him. (Keep in mind that each subject, at random, got only one of the three photographs, either young, middle-aged, or old, and that no one knew that other students were evaluating different pictures.) The sentences they read were the same for all three age conditions. They told things such as where the man had gone to school, what his job and interests were, his marital status, and so on. The information was included to keep the subjects from thinking that they were making guesses solely on the basis of the photograph. All subjects were then asked to make guesses about other characteristics of the person in their photograph. That is, how smart would you say he is, how intelligent, how reliable, and so on?

Results

Following are Tables 18-2, 18-3, and 18-4, which give the average evaluations of the young, middle-aged, and old conditions. To help you read them, you should know that each of the characteristics was measured on

TABLE 18-2

Average evaluations of a man, when young, middle-aged, and old: Salem State College (Massachusetts) sample

Characteristics	Mean evaluations (lower scores indicate more negative evaluations)			Significance
	Young ($N = 53$)	Middle ($N = 55$)	Old ($N = 55$)	
Activity	5.17	4.16	3.80	*
Competence	5.42	5.40	4.58	*
Intelligence	5.11	5.35	4.38	*
Power	5.15	4.38	3.95	*
Health	5.89	5.00	3.95	*
Security	5.02	5.13	4.58	*
Creativity	4.15	4.00	3.86	—
Speed	4.66	4.20	3.51	*
Attractiveness	4.93	3.53	3.33	*
Pleasantness	4.93	4.51	4.06	*
Reliability	5.26	5.75	4.80	*
Energy	5.00	4.49	3.93	*
Calm	5.00	4.89	4.35	*
Flexibility	4.49	4.15	3.91	—
Education	5.49	5.87	5.20	*
Generosity	4.64	4.49	4.51	—
Wealth	5.09	4.86	4.93	—
Memory	5.38	5.13	4.04	*
Social involvement	5.38	4.46	3.51	*

a 7-point scale (called a semantic differential measure), with the extremes for each characteristic at the ends of the scale. For example:

Competent ___:___:___:___:___:_√_:___ Incompetent
7 6 5 4 3 2 1

In the above example, a respondent has evaluated the person he or she saw as quite incompetent. Also keep in mind when looking at the data in the tables that more positive evaluations for every characteristic are represented by higher numbers. In this example greater competence gets a score toward 7 and incompetence a score toward 1. Lastly, in the column labeled Significance, the characteristics marked with an asterisk are those in which the evaluation of one of the ages is different enough from the others that the difference is very unlikely to have occurred by chance. In other words, take special note of the characteristics that are marked, because something worthy of note is going on there.

TABLE 18-3

Average evaluations of a man, when young, middle-aged, and old: East Tennessee State University sample

Characteristics	Mean evaluations (lower scores indicate more negative evaluations)			Significance
	Young ($N = 42$)	Middle ($N = 52$)	Old ($N = 57$)	
Activity	5.07	3.90	3.57	*
Competence	5.17	5.29	4.70	*
Intelligence	5.21	5.31	4.56	*
Power	5.05	4.52	3.83	*
Health	5.93	5.19	4.02	*
Security	5.28	4.98	4.44	*
Creativity	4.55	3.98	3.56	*
Speed	4.76	4.15	3.29	*
Attractiveness	5.02	3.65	2.93	*
Pleasantness	4.98	4.67	4.19	*
Reliability	5.41	5.62	4.88	*
Energy	5.10	4.39	3.61	*
Calm	5.31	5.29	4.68	—
Flexibility	4.86	4.08	3.72	*
Education	5.91	5.81	5.00	*
Generosity	4.45	4.54	4.26	—
Wealth	5.33	4.81	4.68	*
Memory	5.43	5.29	4.35	*
Social involvement	5.83	4.37	4.12	*

Conclusions

What did you notice in the data? Was it clear that students had evaluated the man differently when he was old than when he was young or middle-aged? For all three regions of the country, this person was almost without exception more negatively evaluated as an older man. For example, in San Francisco his "competence" evaluation as a young man was 5.26, and as a middle-aged man it was still almost as high, at 5.19. But as an older man he scored only 4.29, a drop of almost a full point on the scale. Somehow, in the eyes of the students, he had become significantly less intelligent, active, competent, attractive, healthy, powerful, and so on. He was also evaluated as having a poorer memory and as being socially less involved. In only one characteristic, generosity, was he evaluated in all three samples as *not* having gotten worse with age. Apparently, ageist stereotypes and negative beliefs were still in evidence as of the late 1980s among these college students.

TABLE 18-4

Average evaluations of a man, when young, middle-aged, and old: San Francisco State University sample

	Mean evaluations (lower scores indicate more negative evaluations)			
Characteristics	Young ($N = 57$)	Middle ($N = 64$)	Old ($N = 52$)	Significance
Activity	4.74	4.09	3.31	*
Competence	5.26	5.19	4.29	*
Intelligence	5.18	5.27	4.50	*
Power	4.72	4.56	4.08	*
Health	5.61	4.75	3.71	*
Security	4.75	4.86	4.37	—
Creativity	4.47	4.06	3.69	*
Speed	4.65	4.39	3.17	*
Attractiveness	5.04	3.44	2.77	*
Pleasantness	4.75	4.55	4.48	—
Reliability	5.02	5.36	4.75	*
Energy	4.67	4.36	3.54	*
Calm	5.02	5.30	4.85	—
Flexibility	4.46	4.20	3.87	*
Education	5.54	5.83	5.27	*
Generosity	4.47	4.34	4.77	—
Wealth	4.91	4.89	4.75	—
Memory	4.70	5.16	3.98	*
Social involvement	5.07	4.47	4.19	*

The Perils of Stereotyping

As a final note on this study I want to point out, as I did to all the subjects after they had finished taking part in the study, how dangerous it was that they were willing to make guesses about what someone was like merely on the basis of a photograph and a bit of résumé data. True, I had asked for their cooperation, and to make it seem reasonable that they try to make the evaluations, I had (falsely) cited a series of studies (from a fictitious university) in which the ability to make such evaluations had been "documented among experienced personnel officers of corporations." Nevertheless, the willingness to make such guesses about strangers is a requisite for stereotyping. I merely made it easy for them to think that they were doing something harmless, even "scientific." More than once we have seen prejudices, and the discrimination they justify, flourish under such conditions.

Application

A great deal of social research has been conducted to discover the extent of prejudice in American culture. One approach has focused on the way prejudice is reflected in and transmitted by the mass media. Sociologists have long been aware of the fact that television, films, newspapers, magazines, and advertisements teach specific ways of thinking and believing. Because of their huge audiences, mass media can have mass impact. If the media contain stereotypes about various groups, such as blacks, women, and the elderly, they make it extremely difficult to eradicate prejudice from the culture. If certain groups are underrepresented in the mass media, they tend to become "nonpersons" for whom no public image, positive or negative, is evident.

The purpose of this application is to discover how certain minorities are presented in selected mass media. The method normally used to study written or pictorial data is called *content analysis.* Using content analysis you will conduct a study of the characters in television programs, advertisements on television or in magazines, or newspaper comic strips. The idea is to count the frequency with which blacks, women, and the elderly are represented. This will enable you to determine whether members of these minorities are represented in the mass media in the proportions in which they exist in the population.

It is somewhat more difficult to conduct a content analysis to discover whether members of these minorities are stereotyped or presented in a negative way. But if you would like to extend your study, methods for conducting this further content analysis are included at the end of the regular instructions.

Conducting a Content Analysis of Media Portrayal of Minorities

The first step is to choose whether you want to analyze the content of television programs, advertisements on television, magazine advertisements, or newspaper comic strips. The methods described in this assignment will work equally well for any of these. Usually, your choice will be simply a matter of access.

The second step is to learn how to use a Content Coding Sheet (one is provided in this application). For each television program, advertisement, or comic strip, a coding sheet must be completed. On the coding sheet, the name of the program, product advertised, or comic strip is recorded. Then the characters are counted and categorized. Characters are coded one by one. It will take some practice to get familiar with use of the coding sheets, especially when coding television programs or advertisements, because characters are not on the screen for long. (Printed materials, such as magazine ads or comic strips, are somewhat easier, for this reason.)

Coding Sex, Race, and Age

You will code each character for sex (M = male, F = female), race (B = black, W = white), and age (Y = young, M = middle-aged, O = old). For example, the first character (usually the most central, or leading character) might be male, white, and middle-aged. The next might be female, white, and also middle-aged. The third character might be young, male, and black. To give you the idea of how these would be coded, I have filled out a Sample Content Coding Sheet for a current television advertisement for office computers.

Sample Content Coding Sheet

Name of product, program, or comic strip: Office computers

Character	Sex	Race	Age
1	M	W	M
2	F	W	M
3	M	B	Y

Racial identification is limited to white and black, for the sake of simplicity, so you will have to eliminate programs, advertisements, or comics featuring characters of Asian, Hispanic, or other origin.

It is not as easy to categorize the age as the sex or race of a character (those are normally quite easy to code). Decisions about age will necessarily be less accurate. But my students have found that, with practice, their coding decisions for age become easier. For this assignment, consider characters young who are up to the middle twenties, (being sure to include children), middle-aged from the middle twenties to the early sixties, and old from the middle sixties on. If you question whether your age-group decisions are accurate, simply have a friend code the same programs, advertisements, or comics independently. If the two of you make the same coding decisions for a high percentage of the characters (over 70 percent) this "intercoder reliability" suggests that your coding decisions are acceptable.

Sampling

It is impossible to study all the images presented in even one form of mass media. Just as individual people are selected as samples from populations to be studied, so programs, advertisements, or comic strips are sampled. If you wanted to see how the representation of various groups had changed over the years, you could sample from among comic strips or television programs in archives covering the last twenty years. That would be well beyond the scope of this application. However, you will need to select about twenty programs, advertisements, or comic strips during the week in which you do this assignment. It is best not to choose them all from the same day (or time of day for television programs and advertisements), so that a broader spectrum will be represented.

Data Collection

Using copies of the Content Coding Sheet provided, code the characters for approximately twenty television programs, comic strips, or advertisements on television or in magazines. (Avoid ads with hundreds of people or with crowds. Sometimes there are too many characters to code.) Remember to practice a bit with the coding sheet, so that you become familiar with the coding categories.

Analysis

Once you have done the coding, the analysis is simply a matter of calculating the percentages of the characters depicted in each category. What percentage of the characters were male and what percentage female? What percentage were white and what percentage black? What percentages were young, middle-aged, and old? To get each figure, divide the number in each subgroup by the total number of characters. For example, if you coded a total of 78 characters in magazine advertisements and 58 were female, the female percentage would be 74 percent (58 divided by 78). Enter these percentages on the Data Summary Sheet provided.

Interpreting the Results

Studies using content analysis have found that minority groups are generally underrepresented in mass media programming (O'Kelly & Bloomquist, 1979). One clear exception is that women are overrepresented in advertisements, largely because they are targeted as important purchasers of household products and other consumer goods. But in television programming, women, blacks, and the elderly are not shown in proportion to their numbers in the general population. (Women constitute about 51 percent of the population, and blacks and the elderly slightly more than 11 percent each.)

Extending the Analysis

It is one thing to find that minority-group members are underrepresented in the mass media and quite another (and more serious) to discover that they are negatively depicted. Much recent research has found that when members of these minorities are shown, they tend to be put in positions of low power and high dependence. Women are overwhelmingly cast in stereotypical jobs, such as clerical worker, nurse, or homemaker. Older people are almost never shown actively involved in the work world (at least not in paying jobs). And although the stereotyping of blacks is not as blatant as it was some years ago, they are still much more likely to be shown in low-level or stereotypical jobs such as athlete or musician than in powerful positions such as corporate executive.

Coding Character Roles

To discover whether the minority-group members in your analysis are stereotyped, you need to make coding decisions about the role of each character. The role played by a character can generally be identified with an occupational name, such as business executive, physician, laborer, homemaker, or bus driver. (In the comic strips, the "occupations" will sometimes be quite different. Spiderman would have to be coded as a superhero, and his mother appears as a homemaker.) Coding the role played by each character will enable you to evaluate the extent of stereotyping in the media you examined.

To illustrate how roles might be coded, I have filled out another Sample Content Coding Sheet for the computer equipment advertisement, this time including roles of the characters. You can add a column for roles to the coding sheet forms you are using.

Sample Content Coding Sheet

Name of product, program, or comic strip: Office computers

Character	Sex	Race	Age	Role
1	M	W	M	Executive-employer
2	F	W	M	Secretary-employee
3	M	B	Y	Employee-onlooker

In some cases, the role of the character cannot easily be given an occupational name. For example, many advertisements show people as consumers of products (such as medications) or of services (such as health or homeowner's insurance). With practice, however, you will find it easier to characterize a wide variety of people with just a few words in the role column.

Extended Analysis: Roles

You need to evaluate the kinds of roles that are typically attributed to the various groups. Using the extended coding sheets, look at each subgroup separately. What percentage of the women were cast in stereotypical roles such as housewife, nurse, or secretary? What percentage were executives, construction workers, or some other nonstereotypical role? What percentage of the men were in roles stereotypically reserved for men? Make the same kinds of comparisons of roles for whites versus blacks and for the three age groups. What percentage of the blacks were shown in positions of power and control? What percentage of the older people were shown in jobs of any kind, especially those of higher status?

The results can best be discussed one category at a time. Begin with the depiction of male versus female roles, focusing on whether they tended to follow cultural stereotypes. Then discuss the relative power of the men and women in the content analysis results you found. Are men generally depicted as holding more powerful roles? Next discuss the roles of whites and blacks. Are whites more commonly cast in more powerful roles? Finally, compare the roles associated with the various age groups. To what extent are older characters cast in passive or weak positions, such as consumers of medical care or likely victims of crime?

What do your results indicate? Are members of these groups cast in stereotypical roles? Are they shown as needing more help, as more dependent or passive, than majority-group members? By looking at the roles given to minority-group members, you may be able to answer questions such as these and to see the implications of such mass media images for the perpetuation of negative stereotypes in the culture generally.

Content Coding Sheet

Name of product, program, or comic strip: ______________________

Character	Sex	Race	Age
1	______	______	______
2	______	______	______
3	______	______	______
4	______	______	______
5	______	______	______
6	______	______	______
7	______	______	______
8	______	______	______

Data Summary Sheet

Category	Number of characters		Percentage	
	Total	______		
Sex	Male	______	Male	______
	Female	______	Female	______
Race	White	______	White	______
	Black	______	Black	______
Age	Young	______	Young	______
	Middle-aged	______	Middle-aged	______
	Older	______	Older	______

CONCEPT 19

Social Change

Definition

Social Change Change in the beliefs and values that the people in a society hold and in the way they agree to act toward one another.

In my class one day we were discussing the factors related to marital success. I had read in the newspaper an estimate that *half* the marriages begun in the United States during the 1980s would end in divorce, and I was just about to shock everyone with that news when one of the students stole my thunder. "Well," she said, "the qualities I want in my first husband are ambition, dependability, and a desire to have children right away." I would certainly not startle *her* with news of high divorce rates. She already had plans for more than one marriage. She told us she figured to make lots of money and raise a family with mate number one and, having got that out of the way, find a more adventurous, fun-loving partner for phase two. Her kids would be on their own by then, and she wanted a chance to fulfill her own career and travel ambitions. And what of phase three? Although she had not yet thought that far, she speculated that if mate number two did not age gracefully (you know how those fun-loving types tend to burn out), it might be good to find another dependable, healthy partner for the golden years.

If it sounds as if I'm poking a bit of fun at her ideas, I suppose it's because I am still carrying around many of the beliefs about marriage that were commonly taught in the 1950s: marriage is for life, and you get a real feeling of failure if you don't manage it. But American culture has obviously changed, and my student's ideas reflect the changes dramatically. In fact, she was at the leading edge of an important **social change** of which we are all becoming increasingly aware, the change in the meaning of marriage.

We live in a time of extremely rapid social change. More has changed in the Western world in the last two hundred years than in all the preceding thousands of years of recorded history. Since the Industrial Revolution and all the social revolutions that accompanied it (such as the overthrow of feudal aristocracy), we have experienced an impressive range and number of changes in our values, beliefs, knowledge, and ways of living. To list just a few: The nature of work has changed—we rely much less on physical labor and more on machine power. This has increased both productivity and leisure time. Dependence on paternalistic systems of government has been replaced by a belief in the capability of individuals to govern themselves. People now expect to be judged on the basis of their productivity rather than their circumstances of birth. We have become increasingly secular rather than sacred in our systems of belief, urban rather than rural in our settlements, specialized rather than broadly skilled, and concerned with the future rather than with continuity with the past. Our life span has been dramatically increased, with the result that we are forced to reevaluate the meaning of later life. Male and female roles are currently being redefined (making some people very uneasy). Recently there has been a dramatic increase in our attention to the manipulation of information and the computer technology used for that purpose. Our attitude toward the desirability of change has itself changed.

Notice how different these changes are from one another. Some have taken many decades (or more) to develop, and others seem to have exploded upon us. Some are technological in origin, and others stem from new beliefs, knowledge, or values. Some are geographically limited, and others have spread throughout society and even across cultures. Some seem to repeat as patterns in history, some seem to be elements of a broader development, and some seem to be unprecedented. To make sense of such varied kinds of social change I'd like to divide this discussion into two parts. First I'll review four theories of social change. The first two, the linear and cyclical theories, deal with the direction, or "shape," of social change. The second two, the structural functional and conflict theories, deal with the underlying causes of social change. Then I'll discuss some specific sources of social change.

Theories of Social Change

Theories of social change have been developed to explain the wide variety of changes that have occurred throughout history. By proposing what all changes have in common, the theorists seek to enable us to predict future changes.

Theories of the "Shape" of Change

Linear Theory A **linear theory of social change** sees change as proceeding in a uniform, recognizable direction. For example, societies have been thought to develop from simpler to more complex forms and from a concern with collective well-being to a concern with individual well-being. What all linear theories have in common is the belief that social change occurs along a straight line, as if the change were being directed by some underlying force (Moore, 1963).

An early example of a linear theory is the evolutionary approach. **Evolutionary theories of social change** are based on the belief that societies develop in the same way that biological systems do in the natural world. Taking Charles Darwin's ideas about natural selection as a starting point, social evolutionary theories propose that human social orders differ in their ability to survive, just as animals and plants do. According to Herbert Spencer (1820–1903), a principal exponent of the view that has come to be called social Darwinism, those societies whose forms or organization best enable them to dominate others are examples of more highly evolved, more complex social orders. By the end of the nineteenth century, the social Darwinist Lewis Morgan (1877) had arranged human societies in order of advancement, with savages at the bottom, barbarians next, and, at the top of the hierarchy, civilized, or cultured, peoples. It should come as no surprise that the West Europeans who devised the theory included their own societies in the most advanced category. Thus, this theory of social change served to justify the domination of other cultures by the European empires of the time. By the logic of the theory, if the English, for example, were *capable* of dominating others, that in itself was adequate proof of the advanced, superior development of their social order. Thus, those who dominated *deserved* to do so. As a theory of social change, social Darwinism was more a political tool than an analytical device.

Other versions of linear theory have focused on the changes in the character of the societies of the Western world in the last two centuries. These changes, which have come to be referred to simply as the Industrial Revolution, actually included a broad range of revolutionary political, economic, and social structural changes. In one linear theory, Max Weber ([1925] 1946) described the nature of these changes in terms of what he called increasing *rationalization.* By rationalization he meant the application of objective, impersonal judgments in the construction of human relationships rather than the subjective, commonsense judgments that had dominated in prerevolutionary social orders. So, for example, justice in low-rationalization societies was carried out by a paternalistic aristocracy making subjective, highly personal decisions. By contrast, in highly rationalized societies, justice is carried out according to an impersonal system of written laws applied equally to all and based on an established set of principles.

A number of other linear theories of social change were developed by social analysts in their attempts to understand these changes in Western societies. Émile Durkheim ([1893] 1960) proposed that the nature of social cohesion was shifting from what he called a *mechanical* form (based on similarity among all the societal members) to an *organic* form (based on interdependence among a highly differentiated population). This theory was discussed in Concept 10. Ferdinand Toennies ([1887] 1957) saw the changes as proceeding from simpler, rural, traditional communities (which he termed *Gemeinschaft*) to more complex, urban, large-scale societies (*Gesellschaft*). And Karl Marx's ([1867] 1967) linear theory attributed the process of social change to economic forces, which would inevitably lead societies from feudal forms through capitalist, socialist, and, finally, communist forms of organization.

Linear Theories and Modernization The theories of Weber, Durkheim, Toennies, and Marx all focus on changes in the social order that arose with the end of feudalism. Feudalism was a form of political, economic, and social order that dominated Europe during the Middle Ages, and was ended by the social and technological revolutions of the eighteenth and nineteenth centuries. The general term **modernization** has been used to describe all the changes that occur as societies move from a more traditional, small-scale, rural character (such as feudal orders) to a more industrial, large-scale, urban, complex character (such as current Western orders) (Black, 1976). Theories of how modernization takes place have been linear in their assumption that the process follows a uniform, recognizable direction. Accordingly, societies that are modernizing become increasingly urban as opposed to rural, populations increase in both size and density, agriculture is increasingly replaced by industry, work becomes increasingly specialized, the base of technical knowledge grows increasingly large and complex, and physical work is increasingly replaced by machine power.

Most studies of modernization have focused on changes in the Western societies to which I have been referring, but the process is still taking place all over the globe. Third World nations, often called *developing nations,* are modernizing right now. Because there are models for modernization outside the West, such as Japan, it is no longer accurate to think of modernization as Westernization.

Cyclical Theory of Social Change **Cyclical theories of social change** deny the assumption of linear theories that change takes place in a uniform, recognizable direction. Rather, they propose that societies change first in one direction, then in another. Some cyclical theories suggest that the pattern of social change is like that of biological growth—a period of development, followed by maturity, eventual decline, and death. In his application of this model to Western societies, Oswald Spengler (1880–1936) called the mature stage of societies their "golden age" (1932). He argued that Western societies, such as the United States, were already in decline and that the cyclical process was inevitable. Cyclical theories are sometimes called *rise-and-fall theories*.

By contrast, Pitirim Sorokin's (1889–1969) cyclical theory of social change (1937) does not predict inevitable decline in societies. He proposed that societies moved back and forth between two, opposite cultural points of view. What he called *ideational* cultures stress spiritual or religious values and evaluate experiences in terms of them. Iran under the rule of the ayatollahs is an example of an ideational culture. By contrast, *sensate* cultures stress the value of immediate physical (sensory) needs and evaluate experiences in terms of them. American culture is sensate in its concern with material comforts and gratification of physical needs. According to Sorokin, as cultures move between ideational and sensate forms, they pass through (and sometimes settle at) a compromise form he called *idealistic*. Idealistic cultures combine the ideational and sensate forms, so that spiritual values and sensory needs are integrated and balanced. One example was Greek culture during the fifth and fourth centuries B.C., with its tradition of art, literature, and other aesthetic (sensory) pursuits matched by its belief in reason and the value of truth (which are ideational).

Theories of the Underlying Causes of Change

Structural Functionalist Theory The **structural functionalist theory of social change** was developed primarily by the American sociologist Talcott Parsons. He saw society as a huge system of interrelated parts held together by the strongly shared norms and values of its members. Because we are taught by preceding generations to *want* to play our parts in the structure of society, Parsons believed, we act so as to contribute to the stability of the overall system. In Parsons's terms, the ideas and actions of members are functional for a society because they contribute to the stability, or equilibrium, of the entire system (Parsons, 1937, 1954, 1966, 1971).

Parsons's view of the nature of the social system puts great emphasis on its stability. He proposed that society naturally tends toward a state of equilibrium. Thus, when changes occur in knowledge or technology, society adapts by changing its structure somewhat and returning to a new, stable order. In Parsons's scheme, new knowledge and technologies are examples of approved pursuits in a society like ours. Structural functionalism sees them as sources of controlled, even desirable social change. Computer technology has changed the structure of work in American society, and the process of our adaptation is an example of such social change.

But Parsons attributed the equilibrium of society to the fact that people are taught to conform to the rules of *accepted* social behavior. He had no way to allow for the existence of apparently disruptive behaviors (such as conflict) in the normal operation of society. So he was forced to assume that conflict came from outside a social system (for example, a threat or attack from another society) or from the behavior of improperly socialized members who failed to function in society as they were intended (for example, criminals or poor people who broke the law in protest of their condition). Structural functionalism does not see conflict as a positive force for social change but, rather, as an undesirable cause of tension, which must be controlled by the system.

So this theory actually has two distinct views of social change. When change results from socially approved behavior, structural functionalism describes the system response as adjustment or adaptation. When conflict intrudes to change the system, the response is called an effort to manage (suppress) the strain or tension.

Conflict Theory of Social Change In contrast to structural functionalism, conflict theory proposes that conflict is at the very heart of the social system. Conflict theory sees the social order, and the strains and changes in it, as resulting from struggle between society's members as they compete for valued resources. When individuals or groups conflict with one another, only two outcomes are possible. Either one side defeats the other and gains the *dominant* position, or neither side can gain a lasting advantage, and a *balance of forces* develops. In either case, a form of social stability results from the conflict. In the dominance situation, the stability is due to the fact that one side is able to control the other. Slavery is an example of social domination of one group by another. In the balance-of-forces situation, the stability results from the fact that each side keeps the other from gaining a further advantage. Between World War II and the dissolution of the Soviet Union in the late 1980s there was a balance of power between the United States and the Soviet Union, given force by frequent calculations of military equality. Either was capable of annihilating the other with missiles tipped with nuclear weapons. Conflict theorists propose that much of the stability of social order is maintained by either dominance or balance-of-forces outcomes of social conflict.

Conflict theorists have an easier task than do structural functionalists in explaining social change, because they can attribute it to the same forces that create social stability. According to the **conflict theory of social change,** the struggle between contending groups of individuals does not stop just because a situation of dominance or balance of forces is reached. Dominated groups continue to struggle against those who control them, and contending groups or individuals often upset their delicate balances as conflict between them continues. Societies are basically battlefields on which the conflicts between contending forces are played out, and social change is merely the visible evidence that a set of relationships has been altered by conflict.

Perhaps the best known of the conflict theorists was Marx ([1844] 1964, [1867] 1967), who saw social conflict as focused on the struggle for ownership of the means of production in a society. For Marx, the contending groups were the owners of the means of production (entrepreneurs) and the people who could make a living only by selling their labor (workers). According to this theory, workers would inevitably become aware that they were being exploited by owners, and the economic structure of capitalist societies would be overthrown by armed worker revolution.

Marx was an active proponent of such change, both predicting and hoping for the eventual replacement of capitalism by socialist and then communist socioeconomic systems. In a way this is the exact opposite of the conservatism of Parsons's structural functionalism. Whereas Parsons proposed a theory in which social orders are conserved by self-balancing mechanisms, Marx proposed a theory predicting and advocating massive, inevitable changes in social order.

Some Sources of Social Change

So far this concept has focused on conflict as a source of social change. But a number of other sources can be identified. I will briefly discuss some of the more important ones.

Technological Innovation and New Knowledge

Since the beginning of the Industrial Revolution the world has been flooded by technological innovations and new knowledge. Each has resulted in social change. It would be impossible to trace all the consequences of any single change in technology or knowledge, but I will present a few simplified examples to illustrate the principle. Steam power has changed the nature of work and the shape of the economy. Increasingly efficient means of travel have changed the distribution of the population and the way we conduct commerce. The telephone, radio, and television have changed the distribution of information and the way we spend our leisure time. Computers have had influences in almost all those areas—work, the economy, information management, even leisure.

Technological innovation and the development of new knowledge are closely related. Knowledge can lead to the development of new technologies, which can, in turn, generate more knowledge. For example, an understanding of optics (largely arising from the desire to see the planets more clearly) led to the development of the microscope, which made possible the development of the germ theory of disease, which changed the way we treat illness (with vaccines and antibiotics rather than bloodsucking leeches), which led to the creation of a vast medical industry.

The most recent technological tool that is likely to result in widescale social change in modern societies is the computer. Computers are capable of storage and rapid manipulation of tremendous amounts of data, both numerical and textual. Most people saw early computers as curious toys, much as the initial reactions to automobiles, television, airplanes, and the telephone failed to predict the impact these would have on daily life. However, in the less than fifty years since the first large computers were developed, they seem to have become part of the everyday life of Americans.

Merely fifteen years ago computers were so large, complex, and expensive that only big businesses, universities, and government agencies had the necessary skills and money to take advantage of their capacities. With recent developments in miniaturized printed circuitry, computers have become available in a wide range of sizes, powers, and prices. Now engineers design buildings on computers; banks do their accounting on computers; supermarkets use computers at checkout counters to price items and keep inventory; businesses of every size plan marketing strategy and keep accounts on computers; home computers are used to keep track of dairy herd production, play video games, and write documents (as I am currently doing). A complete listing of the uses of computers in the United States today would take days. This list just scratches the surface.

The futurist Alvin Toffler (1970) forecast that Americans would live in the "electronic cottage." It seems as though this prediction is coming true. Increasingly, the "bookkeeping" details of daily life are being handled automatically by computers. But what actual social changes has the technology of computers brought about, and what further changes can be expected?

True social change refers to change in the beliefs and values that people hold and the ways they deal with one another. The mere existence of computers does not constitute social change. But computers have obviously changed the way we deal with one another. For example, the computer has spawned a number of interrelated industries—from those to manufacture them and write instructions for their operation (called *programs,* or *software*) to those for sales, service, repair, operator training, and publication of books for users and owners. In any city's Sunday newspaper job advertisements, the list of programmer openings may outnumber any other single category.

In addition, people have come to expect, and trust the accuracy of, computer-generated letters, bank statements, billings, and so on. Development of a faith in the reliability of computers has spread across our culture. That is a major change from the early days of computers (and machines generally), when machine-made products were considered inferior. So our assumptions about how work can be accomplished have been changed by computers.

By their very nature, computers have made possible the centralization of vast amounts of information. A large proportion of the crime data in the country have been computerized cooperatively among criminal justice agencies and are rapidly available to them. The same is true of financial information shared by banks and other credit-checking agencies. Data that existed in two different places and could not easily be compared can now be matched by computer. For example, in some states (such as Massachusetts) computerized records of welfare recipients have been cross-checked with information about their personal savings, to remove ineligible individuals from the rolls.

Some of the social changes that computers might bring about can only be the subject of speculation. For example, the wide availability of home computers is likely to greatly increase the number of people who will write books. It is so much easier to produce written material on a word processor than on a typewriter that one of the important barriers to writing will be overcome by computers (Walters, 1983). But the existence of computers will have no effect on the quality of writing, so the increase in manuscripts submitted to publishers may not be a blessing.

An increase in the number of small, home-based businesses is another prediction. Home computers make correspondence, shipping, and accounting easy to accomplish, even for those with little training. Computers have for some time been used to develop specialized lists of potential customers, which small businesses can purchase from central vendors. It is possible that computers will eliminate (or severely decrease) the reading of books, magazines, and newspapers, just as television seems to have done. As home computers are hooked into specialized news and entertainment services, subscribers will be able to read on their screens or printouts (or listen to) whatever they choose at home. In fact, contact between people in public places may actually be reduced, as shopping, banking, and other tasks formerly requiring trips are done from the home. Even more exotic changes brought by computers are being discussed (such as direct medical diagnosis and care via computer-linked sensors). Only imagination limits the possible predictions.

New Beliefs

Some beliefs can never be confirmed, so they cannot be called *knowledge.* But our beliefs about what is true do not have to be proved or verified in order to cause social change. For example, on Halloween eve, 1938, Orson Welles narrated a radio broadcast of H. G. Wells's "War of the Worlds." Hundreds of thousands of listeners believed that the United States had been invaded by Martians; they fled their homes in panic, warning neighbors to get away or protect themselves. In spite of the fact that their beliefs could not be verified (although some people did "see" ordinary objects in their environment as monsters), the social changes were frighteningly real. It's true, of course, that this was only a temporary social change. An enduring social change resulted from the development of what Weber ([1905] 1958) called the *Protestant ethic.* He showed that, with the rise of Protestant beliefs about the relationship between humans and God, the growth of capitalism became possible. Calvinists (the denomination of Protestantism on which Weber focused) taught that human fate was predetermined by God and that people were on earth only to glorify God. Humans could do nothing to influence whether they would go to heaven or hell, but they might get some clue about their standing with God by measuring the extent of their good works on earth. Good works on earth came to be measured by success in business and the accumulation of capital. Making money was not undertaken to enjoy life, which Calvinists believed was evil, but rather to reinvest in the growth of business for the glorification of God. When large numbers of people work hard and deny themselves personal comforts in order to reinvest in business, the growth of capitalism becomes ensured. Thus, a major social change was made possible by a change in beliefs. To this day, the value for hard work is referred to in the United States as the Protestant work ethic.

Environmental Changes

Events in nature can also cause social change. Changes in weather patterns, for example, influence the food supply, increasing or decreasing the wealth and physical well-being of people and changing the time they have available for pursuits other than food production. In turn, economic and governmental structures become altered. A drought can bring down the government that fails to deal with its effects, as surely as an armed revolution can. On a much smaller scale, an environmental event in the United States recently brought about a clear social change. In the winter of 1979, Chicago had a huge snowstorm that brought activities to a standstill. Mayor Michael Bilandic was blamed for the inability of the city to clear the streets, and his challenger, Jane Byrne, was elected, defeating a Democratic machine that had been in power for years.

Many environmentalists and scientists have for years been warning of much more serious, longer-term environmental changes, such as global warming and damage to the earth's ozone layer. Should these occur to the degree that some have predicted, the social changes (not to mention the economic and political changes) that accompany them will be extensive. Environmental issues were taken seriously in the 1992 presidential race to a greater extent than ever before, and environmental safety has become a real factor in the American marketplace. For example, manufacturers have begun advertising many products' safety in the environment as a selling point.

Population Changes

Population growth, decline, or migration has also caused social change. For example, the increase in Americans over sixty-five years of age has caused a boom in housing and related industries in areas of the country that are popular for retirement, placed increasing demands on the medical resources and Social Security funds of the nation, and increased the power of the elderly as a political and economic force. The age makeup of the American population is predicted to continue to change, with increases in longevity (the average lifespan is still increasing) and decreases in fertility (the average number of children born to women in their childbearing years) has dropped since World War II). When the baby-boom cohort (those Americans born between 1946 and 1964) approaches retirement age, a much smaller percentage of the population will be working to support a disproportionately large group of retiring elderly Americans. It is likely that something will change in our social structure to accommodate these demographic trends. For example, productivity rates may rise; the age at which people retire may be pushed back; the percentage of people allowed to retire may decline; and salary structures may change so that in order to qualify for full retirement benefits workers will be forced to stay on the job full-time for a number of years but with salaries reduced from their mid-life peaks. Should these changes come about, it is clear that the relationships between various age generations would also change.

During the late nineteenth and early twentieth centuries, the United States experienced a mass internal migration from rural to urban areas, with the effect that services in cities such as Chicago, Washington, Philadelphia, and Detroit were tremendously strained. The character of life in those places was permanently changed. The same sort of strain happened in Boston during the middle to late 1800s, as tens of thousands of Irish immigrants fled the potato famine in their homeland. The Boston run by an aristocracy with roots in England and the Revolutionary War eventually became the Boston run by the Irish political machine.

The Interdependence of Sources of Social Change

Although the sources of social change can be listed and discussed separately, it is important to realize how much they overlap. A single example will illustrate. In the early 1970s the major oil-producing nations came to see themselves as having a common purpose. They united in an organization and began to raise the price of oil. Accustomed to cheap petroleum, Western countries were saddled with two technologies (large cars and inefficient heating systems) that increased the impact of these price increases. A particularly cold winter made things even worse. Research into more efficient cars and heating systems and the increasing belief that conservation was a good idea led Americans to change lifestyles and develop new industries. The reduced consumption of oil decreased the power of the oil cartel, with the result that prices dropped.

In this chain of events, greatly simplified here for the purpose of the illustration, the influences of beliefs, knowledge, technology, and the environment are all intertwined. I haven't even mentioned the impact of conflicts among competing forces, such as domestic oil companies, the oil cartel, individual consumers, small businesses that tried to develop the new technologies, and investors who wanted to make profits from the tremendous shifts in capital. It should be clear that social change, like social order itself, is complex.

Summary

Social change refers to changes in the beliefs and values that people in a society hold and the way they agree to act toward one another. We live in a time of extremely rapid social change, the most recent evidence of which is the increasing use of and interest in computers. But the bulk of the changes that have influenced our lives in modern societies have derived from the Industrial Revolution and the social changes that accompanied it. In comparison with prerevolutionary societies, we make great use of machine power, rapid systems of transportation, division of labor into very specialized tasks, and centralization of authority to a great extent. We have even come to accept rapid change as a part of life.

A number of theories have been proposed to understand the process of social change. Linear theories see change as proceeding in a uniform, recognizable direction. One example is the evolutionary model, which is based on the belief that societies develop by a system of natural selection, in the same way that biological systems evolve in the natural world. Another linear theory, proposed by Max Weber, focuses on the changes brought by the Industrial Revolution (a process often called *modernization*). Weber described social change in the West as a process of increasing rationalization, in which the effort to apply objective, impersonal judgments in the construction of human relationships replaced the subjective, paternalistic judgments of the aristocratic world.

In contrast with linear theories, cyclical theories of social change propose that change can proceed in more than one direction, repeating a back-and-forth or cyclical pattern. One example is the view of Pitirim Sorokin that societies move back and forth between an ideational form (stressing spiritual or religious values) and a sensate form (stressing the value of meeting immediate physical needs). Sorokin further suggested that, between these forms, societies can experience a compromise idealistic form, in which spiritual and sensory needs are balanced.

Linear and cyclical theories of social change are largely descriptive. Other theories focus on the causes and nature of social change. The structural functional theory argues that society tends naturally toward equilibrium and that social change is evidence of the social system's natural adjustment (or adaptation) to new elements introduced into the society. But structural functionalism does not treat all sources of change the same. Some, such as conflict within the society, are seen as undesirable, causing tension that has to be managed.

By contrast, the conflict theory of social change proposes that conflict is at the very heart of the social system and is actually a source of both social order and social change. When conflicts between contending groups result in the domination of one side by the other or in a balance of forces between them, no change is likely to occur. When one side overthrows a dominating group or manages to change the balance of forces between them, social change occurs.

Social change can result from changes in knowledge and technology, beliefs, environmental conditions, or the size and composition of the population. Several of these sources of social change can occur simultaneously, and they can influence one another.

The Illustration for this concept summarizes a study of social change in attitudes toward the family. As the structure of the American family has changed, so have our ideas about family issues such as sex roles, divorce, and having children. A study by Arland Thornton reveals differing rates of attitude change among men and women.

The Application for this concept focuses on evidence of social change in the area of racial equality in the United States. Using data readily available in the *Statistical Abstract of the United States,* you will determine whether the figures for income, unemployment, and poverty rates show that black and Spanish-speaking Americans have improved relative to white Americans over the last few decades.

Illustration

Arland Thornton, "Changing Attitudes Toward Family Issues in the United States," Journal of Marriage and the Family *51 (1989): 873–93*

At the beginning of this concept I told the story of one of my students who fully expected to be married more than once during her life. Not only did she expect to be divorced at least once, she had actually shaped her multiple-marriage plans around the demands of her career and her desire to have a family. It seems to me that her remarks reflect a social change in American culture, a change characterized by new attitudes toward marriage. That is, she apparently had attitudes toward marriage different from the ones I learned growing up and certainly different from the ones my parents were taught. Though I know many people who believe that these changes are undesirable, I know of no one who is unaware of them. But where did her changed attitudes come from? Are they like the attitudes of other Americans? Some of the answers are available in recent research conducted by Thornton on attitudes toward the family. First, I'll summarize some of what he found. (A complete summary of his findings would take too much space. If you are interested in seeing the rest, try reading the original article. It should be in your library, and this summary should give you help in reading and understanding it.) After this summary, I want to point out some of the problems of studying social change, problems that Thornton had to confront in his study.

Thornton examined the responses of Americans to a range of questions asked in surveys that were conducted between the early 1960s and mid-1980s. He had access to the data from attitude surveys that others had done. For his study he reanalyzed those responses (a technique called secondary analysis) to see how they had changed over time. The data he wrote about were actually drawn from three different studies: the "General Social Survey," "Monitoring the Future," and the "Study of American Families." In order to present the results as clearly as possible, I will identify only the questions asked of the respondents, ignoring, for the time being, which of the studies the questions were from. But the differences in the three studies will be an important part of the later discussion of research methods.

The Results of Thornton's Study

Attitudes About Gender Roles in the Family

A number of questions were asked of respondents about their attitudes toward the family roles of men and women. One of the questions involved this statement: "Most of the important decisions in the life of the family should be made by the man of the house." Respondents were asked whether they agreed or disagreed with the statement. Those disagreeing with it were considered to have the more egalitarian attitude toward family gender roles. That is, they did not think that the important family decisions should necessarily be made by "the man of the house." Table 19-1 shows the responses of a sample of women and of their children to this question.

The table summarizes data from a 1962 sample of women who were chosen for the study because they had just given birth. They were interviewed that year and then again in 1977, 1980, and 1985. By 1980

TABLE 19-1

Attitudes toward family sex roles (percentage of respondents with egalitarian sex roles, 1962–1985)

Mothers				Their daughters		Their sons	
1962	1977	1980	1985	1980	1985	1980	1985
32.3%	67.6%	71.2%	77.7%	66.1%	85.3%	44.9%	64.9%

their firstborn children were eighteen years old and could also be interviewed about their attitudes. In 1980 and 1985 these children were asked the same questions that their mothers had answered. The responses show a pattern of increasingly egalitarian attitudes. The mothers began with clearly nonegalitarian attitudes in 1962, when only about a third (32.3 percent) disagreed with the statement that men should make the "important family decisions." By 1977, however, two-thirds (67.6 percent) of these women disagreed with the same statement. The increase in egalitarian attitudes continued, with 77.7 percent of these women holding egalitarian gender-role attitudes by 1985.

Their children, both daughters and sons, also showed increases in egalitarian gender-role attitudes between 1980 (when they were eighteen years old) and 1985 (at age twenty-three). In 1980 the daughters were slightly less egalitarian than their mothers (66.1 percent of the daughters and 71.2 percent of the mothers disagreed with the statement that year). But five years later the daughters had become *more* egalitarian than their mothers (85.3 percent of the daughters disagreed with the statement and 77.7 percent of their mothers disagreed). By contrast, although the sons became increasingly egalitarian between 1980 and 1985, they remained less egalitarian than their mothers. In 1980 fewer than half (44.9 percent) of the sons disagreed with the statement, but five years later the sons had become more egalitarian on the subject, though still not as egalitarian as their mothers (rejecting the statement at the rate of 64.9 percent).

Attitudes Toward Divorce

In the same study the mothers and children were asked about their attitudes toward divorce in families with children. Respondents were asked whether they agreed with this statement: "When there are children in the family, parents should stay together even if they don't get along." Their responses are summarized in Table 19-2.

Here we see another change in attitudes about an important family issue. In 1962, 51 percent of the mothers in the sample disagreed with

TABLE 19-2

Attitudes toward divorce (percentage disagreeing with the statement that parents should stay together, 1962–1985)

Mothers				Their daughters		Their sons	
1962	1977	1980	1985	1980	1985	1980	1985
51.0%	80.4%	82.1%	82.3%	82.8%	89.9%	65.2%	70.7%

the statement. By 1977 the rate of disagreement had jumped to 80.4 percent, and then it increased very slightly over the next eight years, to 82.3 percent. These women had developed a much less negative attitude toward divorce. At age eighteen their daughters were already as accepting of divorce as their mothers (82.8 percent disagreeing with the statement in 1980), and three years later they were even more accepting (89.9 percent acceptance). Again the sons' attitudes changed from 1980 to 1985 (going from 65.2 percent disagreement to 70.7 percent), but the men were less accepting of divorce than their mothers.

Attitudes Toward Childlessness

The mothers and children were also asked this question: "Do you feel almost all married couples who can *ought* to have children?" Their responses are summarized in Table 19-3.

The responses are a bit different from those in the first two tables. But the pattern of change in attitudes is still there. Almost 85 percent of the mothers agreed in 1962 that couples who could have children should. By 1980 the rate of agreement was only half the 1962 rate, at 42.8 percent. (There was no report of 1977 responses for this question.) Again, the attitude change was most dramatic earlier in the study, and the trend slowed in the 1980s. In this case, both male and female children retained their attitudes between 1980 and 1985, with sons agreeing somewhat more than daughters with the idea that almost all couples should have children if they could.

Attitudes Toward Premarital Sex

In a final question, samples of American women and men were asked about their attitudes toward premarital sex. (Note that unlike the data cited in the previous two tables, these data were not assembled from a sample of one group of people who were interviewed several times over a period of years. Rather, they are from samples of different people in the various years. So a sample was taken in 1965, another in 1972, and so on to 1985.) The question was worded as follows: "If a man and woman have sex relations before marriage, do you think it is always

TABLE 19-3

Attitudes toward having children (percentage agreeing that couples should have children, 1962–1985)

Mothers			Their daughters		Their sons	
1962	1980	1985	1980	1985	1980	1985
84.8%	42.8%	42.6%	35.8%	32.6%	41.9%	40.6%

TABLE 19-4

Attitudes toward premarital sex (percentage stating that premarital sex is always wrong, 1962–1985)

Respondent category	1965 sample	1972 sample	1977 sample	1982 sample	1985 sample
Women under 30	49.7%	16.9%	15.5%	13.6%	13.9%
Women over 30	70.5	48.7	42.3	38.9	37.5
Men under 30	47.5	13.0	9.1	13.8	7.6
Men over 30	47.2	34.9	28.0	25.2	23.7

wrong, almost always wrong, wrong only sometimes, or not wrong at all?" Table 19-4 summarizes the responses.

The data in this table again show consistent changes in family-related attitudes for both men and women. Between 1965 and 1985 both male and female respondents, whether they were over or under the age of thirty, disapproved of premarital sex less and less over time. Again, the bulk of the attitude changes appeared in the decade between the mid-1960s and mid-1970s. Change tended to slow down through the 1980s, though it did continue in the same direction. In 1965 about half the women under the age of thirty (49.7 percent) responded that premarital sex was always wrong. By 1972 only 16.9 percent of these women felt that it was always wrong. By comparison, in 1965 more than 70 percent of the women over the age of thirty felt premarital sex was always wrong, and though the percentage feeling this way had dropped below 50 percent by 1972, older women remained more opposed to premarital sex than younger women during each year of the study.

The men in the study also declined in the degree of their disapproval of premarital sex, though they started off disapproving less than women in the first place. Men under the age of thirty were particularly accepting of premarital sex, dropping from 47.5 percent disapproval in 1965 to only 13 percent disapproval in 1972. As in other cases, their attitudes remained relatively stable through the 1980s. Men over the age of thirty started off in 1965 with disapproval rates about the same as those of younger men (47.2 percent disapproval). Their rate then declined only slightly, to 34.9 percent disapproval in 1972 and about 25 percent in the remaining years.

Interpreting the Data

What do the data from these four tables show? It is clear that attitudes about family issues have changed, with increasing acceptance among both sexes of (1) egalitarian sex roles in the family, (2) divorce, (3) childless marriages, and (4) premarital sex. A clear pattern appears in the responses to these questions. The greatest percentage of these attitude changes occurred in the first decade of the research, with the direction of the attitude changes continuing into the 1980s, though at a slower rate of change.

According to Thornton, these results show that our norms and values about family life, like our attitudes in a number of other areas of social life, have been changing. Citing a number of other surveys, Thornton suggests that Americans have come to accept a greater diversity of behavior generally. For example, the idea of increasing freedom of choice shows up in areas such as how we socialize our children, our declining allegiance to political parties and churches, and our support for civil liberties. The overall trend toward tolerance and independence is simply reflected in our family attitudes as well.

A Methodological Note

Recall that in order to study social change in the family, Thornton used data from three studies. Why did he do this? Wouldn't it have been easier to use just one? The fact is that it is rare for researchers to find that all the data they want about a given topic are available in a usable form. So measuring social change requires some inventiveness. For example, Thornton wanted to know if there had been changes in the attitudes of Americans toward a range of family issues since the 1950s. He was able to find surveys of attitudes done forty years ago, but no single study asked all the questions he was interested in. So he combined them. But, as you may have guessed, this created problems. Let's look at just one.

Take a look at the data presented in Table 19-4. They show a clear pattern of decreasing disapproval of premarital sex from 1965 to 1985. What I did not tell you is that the data in that table were from two different studies. The data for the years 1972 to 1985 were taken from a question in the "General Social Survey," whereas the data for 1965 were from a different study, conducted by the National Opinion Research Center. Both studies employed national samples, but the questions used in 1965 and the remaining years were worded differently. In 1965 respondents were asked whether it was "always wrong" if "a man had intimate relations with a woman he is engaged to and intends to marry." By contrast, the question used from 1972 through 1985 asked, "If a man and woman have sex relations before marriage, do you think it is always wrong, almost always wrong, wrong only sometimes, or not wrong at all?" Do you think the differences in the wording of the questions could cause the responses to differ? Looking back at Table 19-4, do you think it is possible that the sharp decline in disapproval between 1965 and 1972 was caused by the fact that the questions were worded differently? Do you find one statement easier to disapprove of than the other?

Thornton did not try to hide this problem in his article. In fact, like other students of social change, he openly admits the compromises he is forced to accept in using data collected years before by other people. But having revealed the weaknesses in the data, he leaves it to the reader to decide whether the results are still valid. It is often the best we can do in research on social change.

Application

I can recall hearing my parents say, "Things aren't what they used to be." And they remember their parents saying the same thing. Just to confirm that every generation feels this sense of "things" changing (generally for the worse), I have consulted my students. They say it also. But a bit of further examination reveals that people often have little or no evidence for their beliefs beyond a sense of what "everyone knows to be true." For example, you may be confident that black Americans have made significant economic progress since the civil rights demonstrations and legislation of the 1960s, but *how* do you know, and how much progress has there been? This application will show you one way to gather information on such questions about social change.

A Vital Source of Information

One of the most useful sources of information available to social researchers is a publication of the U.S. Bureau of the Census titled *Statistical Abstract of the United States*. Every year an updated version is published that includes the most recent information on a wide range of characteristics of the population of the United States. Most of the data are generated by the national census, which is conducted every ten years. But smaller scale analyses are being conducted all the time, the results of which are used to update and refine the core statistics. In addition to summarizing the most recent data, each new edition of *Statistical Abstract* includes data from past years. This allows us to make comparisons across time for specific data (such as average incomes of black Americans and white Americans) and thus test our notions about social change. That is what this application is designed to do.

Statistical Abstract is available in the reference section of virtually every library—certainly in every college and university library. This application focuses on data about changes in the economic well-being of black Americans, white Americans, and Americans of Hispanic origin. You will look up data for specific categories of information in given years. The data you find should provide you with the evidence you need to better gauge the degree and types of social change that have occurred in these areas.

This application was prepared using the 1990 (112th) edition. If you use a more recent edition, you will probably have to use data from different comparison years, but the procedure will be the same. Using the most recent edition available, look up the appropriate charts for the following issues and copy the data into the spaces provided. For the purpose of this application I have selected only a portion of the data available in this source.

Measuring Economic Changes in Racial Groups

Have black Americans made economic progress since the 1960s? If so, how has their progress compared with the progress made by white Americans? To compare the economic well-being of blacks, whites, and Americans of Hispanic origin over the past few decades, I have chosen three separate statistics: (1) median household income for each group, (2) percentage of unemployment for blacks and whites, and (3) percentage of families in each group who are below the official poverty level. The data for each of these categories are available in *Statistical Abstract.* Each table named below lists the figures for the most recent year and for previous years. As you copy out the data, ask yourself if the comparison of the most recent data with the data from the early 1970s supports your assumptions about the degree and type of social change that you believed had occurred.

Income of Households

The median income for households over time is listed on page 445 of the 112th edition in Table 696 entitled "Money Income of Households—Median Household Income in Current and Constant (1990) Dollars, by Race and Hispanic Origin of Householder: 1970 to 1990." You can find this table in another edition of *Statistical Abstract* by looking in the table of contents for the section called "Income, Expenditures and Wealth" and the subsection on "Money Income of Families." Using this table, copy the data into the spaces provided. (Use the figures on the left of the table under "Median Income in Current Dollars," rounding to the nearest thousand.)*

---

* It might help you in reading these tables to know the difference between incomes reported in "current dollars" versus those reported in "constant dollars." Current dollar figures are reported for the value of the dollar in the year in which the data were collected. For example, the median income of all households in 1970 was $8,734 at the purchasing value of the dollar in that year. By comparison, if the value of the dollar in 1970 were adjusted to its purchasing power as of 1990, by which time the dollar bought much less, that $8,734 would have been equivalent to a yearly income of $29,421. Look to the right side of the table to find this equivalent figure. This is an adjustment in which yearly incomes are reported in "Constant (1990) Dollars." That is, incomes for previous years are adjusted to the value of the 1990 dollar. For this application either of the two styles of reporting would do; however, the differences are more dramatic when the effects of the cost of living are not removed from the data.

Median income for households

1. Most recent income level:

   White ________ Black ________ Hispanic ________

2. Income level in 1970:

   White ________ Black ________ Hispanic ________

How do these data square with your beliefs about the relative well-being of these groups? Have blacks and people of Hispanic origin made progress since the 1960s? To calculate the percentage of improvement for each group, divide the earlier income figure into the more recent one. (For example, $7,000 in 1967 divided into $22,000 in 1984 is an improvement of 3.1 times, or 310 percent.)

3. Percentage of improvement in median income:

   White ________ Black ________ Hispanic ________

Have blacks or Americans of Hispanic origin improved relative to whites over these years? Have they kept pace?

Unemployment

The rates of unemployment for white, black, and Hispanic Americans are listed in the 1992 edition of *Statistical Abstract* in Table 635 on page 399, entitled "Unemployed Workers—Summary: 1980–1991." This table allows you to see a number of changes in unemployment rates over time. (Once again, you can find this table in another edition of the *Statistical Abstract* by looking for the section called "Labor Force, Employment and Earnings" and the subsection on "Unemployment.") Let's start with racial differences over time. Using this table, copy the unemployment rates, expressed as percentages (in the bottom half of the table), into the spaces provided.

Rates of unemployment

1. Most recent unemployment rate: Total (1991) 6.7%

   White ________ Black ________ Hispanic ________

2. Unemployment rate in 1986: Total (1986) 7.0%

   White ________ Black ________ Hispanic ________

3. Unemployment rate in 1980: Total (1980) 7.1%

   White ________ Black ________ Hispanic ________

What happened to unemployment rates between 1980 and 1991? For many years the black unemployment rate was approximately twice as high as the white unemployment rate. Did this ratio change? To find out, divide the rate of black unemployment into that for white unemployment for a given year. For example, in 1972 white unemployment was 5.1 percent, and black unemployment was 10.4 percent, yielding an unemployment ratio of 2.04. In that year black unemployment was slightly more than twice white unemployment. What were the ratios for the years listed below?

Ratio of white to black unemployment

4. 1980 ________ 1986 ________ Most recent year ________

Poverty Level

Another measure of well-being is the percentage of a population that falls below the official poverty level. In the section on "Income, Expenditures and Wealth" is a table entitled "Families Below Poverty Level and Below 125 Percent of Poverty Level: 1959 to 1990." In the 1992 edition this is Table 724 on page 459. It reports that in 1973 (when data on Hispanic Americans was first tabulated) 8.8 percent of all American families were below the official poverty level. In the same year (1973) 6.6 percent of white families were below the poverty level, compared with 28.1 percent of blacks and 19.8 percent of Hispanics. What has happened since? For each of the following categories record the appropriate percentage of families that were below the poverty line. (The figures for 1973 are already filled in.)

Families below the poverty level

1. Whites: 1973 6.6% 1978 ______ 1983 ______ Most recent ______

2. Blacks: 1973 28.1% 1978 ______ 1983 ______ Most recent ______

3. Hispanics: 1973 19.8% 1978 ______ 1983 ______ Most recent ______

What changes have there been in the percentage of families in each group that are below the poverty line? In less than twenty years the percentage of all American families below the poverty line increased by almost 22 percent, from 8.8 percent in 1973 to 10.7 percent in 1990. Who accounted for this increase? What happened to the poverty rate for white, black, and Hispanic families between 1973 and the most recent year recorded in your edition of *Statistical Abstract?*

Overall, what do these data tell you about the degree of social change in the relative economic well-being of whites and others in the United States? It might be useful just to browse through some of the other tables in the *Statistical Abstract,* especially those dealing with changes in the makeup of the work force in the 1970s and 1980s. The statistical evidence of changes in gender roles is everywhere in them.

CONCEPT 20

Deviance

Definition

Deviance Behavior that does not conform to society's norms for expected behavior.

Every winter the main branch of the Boston Public Library serves as home base for a number of special people. One January morning, as I walked up the broad stone steps to the main entrance, one of the denizens stepped up to me. He was wearing a long, green, army coat, black-and-white checked slacks, orange running shoes, and a navy watch cap. His silvery, shoulder-length hair seemed streaked with tobacco stain, and he had not shaved for several days. "Do you take photographs?" he asked. For some reason I did not react in the normal, urban way—which is to not acknowledge a bum and just keep walking. Instead, I told him I had no camera. Fifteen minutes later I continued up the library steps, but in the interval I had learned something of his life. Carl Sisten had once been married, had two children, had repaired air conditioners for a living, and really couldn't say how he had arrived at his present way of life.

I suppose I had expected something more exciting—that he had been a famous photographer with a tragic story or perhaps that he was crazy and would tell of trips to Saturn from those very steps. But just as he was no fallen star, neither was he a raving lunatic. Though he was in a sad condition, after our conversation he had become more ordinary to me, certainly much less strange and shocking than at first. But he was undeniably a deviant, a person who does not conform to society's norms for expected behavior.

We tend to shrink from contact with such people, from mentally ill or handicapped people, alcoholics, criminals, or even people who are merely eccentric. We commonly wish that they would be isolated from us or punished for their behavior. This concept is intended to explain the character of **deviance** and why society reacts to deviance as it does.

Normative Order and Deviance

In its simplest sense, social order is nothing more than the predictability of social behavior. When we deal with other people, we wish to know how they are going to act and how we are expected to act in return. As we saw in Concept 8, every society develops and teaches norms, widely accepted rules for behavior that apply to specific social situations. To the extent that such norms are accepted and obeyed, social life is predictable and orderly. Therefore, violation of norms—deviance—is generally seen as a threat to social order. Yet people violate norms all the time. Why?

There are two major views about the origins of deviance. One argues that people break norms because something in their individual makeup influences their behavior. According to this view, the process by which social order is maintained is basically sound, and most people fit into the system quite well. But some individuals are flawed in some way. They are society's deviants.

The other view is more sociological, arguing that deviance is created by social forces. Society determines what is considered socially "right" and "wrong" when it creates norms. A number of **social forces theories of deviance** have suggested how deviance is created. One (the learned deviance theory) holds that some people are taught norms that conflict with those of the larger society and so are automatically defined as deviant. Another (the anomie theory) argues that, when the means to obey norms are not available, some people are forced to deviate from those norms. Two others (the conflict and labeling theories) suggest that powerful social groups often have the ability to advance their own interests by strategically manipulating the definition of deviance. According to each of these "social forces" approaches, deviance is due not to abnormalities in individuals but to the way social forces operate. Let's look first at some of the theories of deviance that focus on individual characteristics and then at those concerned with social forces.

Individual Theories of Deviance

Biological Theories **Biological theories of deviance** have proposed that some people are born rule breakers. Most of these theories have focused on criminal behavior. In the late nineteenth century, the Italian criminologist Cesare Lombroso (1911) described the physical appearance of the "criminal type": a person with a big jaw, high cheekbones, sparse beard growth, and, in general, the looks of an evolutionary throwback. Although these ideas are strongly rejected by today's criminologists, it is still common to hear people say, "He looked like a criminal."

Taking these ideas a step further, the American psychologist William Sheldon (1949) identified what he considered three major body forms: the *endomorph,* with a round, puffy body; the *mesomorph,* with a muscular, tough, angular form; and the *ectomorph,* with a slim, fragile form. Sheldon also attributed to each body type a set of personality traits. Endomorphs were supposed to be easygoing and relaxed; ectomorphs restrained, private, and sensitive; and mesomorphs energetic, insensitive, and assertive. From observation of delinquent boys, Sheldon concluded that mesomorphs were most likely to be criminals.

More recently, biological theories of deviance have taken a new turn. Instead of looking at a criminal's external appearance for clues about his or her deviant tendencies, some scientists have begun to look directly at the person's sexual genetic makeup (Reid, 1979). Females typically have two X chromosomes (XX) and males one X and one Y (XY). In comparing convicted criminals with the rest of society, researchers found that criminal populations had a higher percentage of males with an extra Y chromosome (XYY). It was proposed that the abnormal genetic makeup of these men caused them to be more violent and, therefore, more likely to be criminals.

The problem with all the biological theories is that they have been unable to separate the influence of social expectations from the influence of physiological makeup. Lombroso and Sheldon, for example, quite probably applied the prevailing cultural picture of what criminals *should* look like. They had no way of showing that the physical characteristics they described actually caused deviance. An even more likely causal sequence is that people who "look like criminals" in a society are treated so badly that they are forced into lives of poverty and rule breaking to survive. Similarly, men with the XYY genetic makeup often appear quite strange in comparison with other males, and people who look different are often treated cruelly in our society. Theories of the effect of an extra Y chromosome have not been able to show that the treatment their subjects received at the hands of society did not cause their deviance (Witkin et al., 1976).

Psychological Theories **Psychological theories of deviance** focus on the mental and emotional capacities of individuals, which are developed primarily in childhood. Deviants, then, are believed to be people who, for a variety of reasons, are incapable of adequately learning or following the rules of society.

Freudian psychology maintains that all people are driven by unconscious, irrational impulses for sexuality and aggressiveness (**id**). Normally, by identification with his or her parents, a child learns to control the impulses of the id, internalizing the society's rules for social behavior (**superego**). Mediating between the impulsive demands of the id and the control demands of the superego is the **ego.** In Freudian terms, a person whose early socialization proceeds properly is capable of balancing the deviant impulses of the id against the restrictive rules of society. But psychologically flawed individuals are considered prone to deviance. For example, a person with an inadequately developed superego will readily give in to the id, committing sexual or aggressive acts of deviance. Alternatively, a person with an overdeveloped superego feels extreme guilt for even having private feelings of sexuality or aggressiveness. Freudian theory suggests that such people also commit acts of deviance to bring about the punishment they feel they deserve.

By comparison, behavioral psychologists place much less emphasis on early learning. Instead, they propose that adults act to satisfy their needs and desires in ways that are suited to the situation in which they find themselves. Thus, a behavior that is rewarded in a social setting will be repeated, and one that is punished will be extinguished. The psychologists Albert Bandura and Richard Walters (1959) compared the environments of delinquent and nondelinquent boys and found that delinquents had been raised in homes in which violent behavior was rewarded. Either the boys had been given support for anything they did (which allowed them to believe that the same would be true outside the home), or they had been beaten by parents (allowing the boys to conclude that violent behavior was an acceptable way of achieving an end).

Social Forces Theories of Deviance

Learned Deviance Theory The norms by which one culture operates can be quite different from those of another culture. For example, in the United States the rule is one husband for each wife (and vice versa), whereas in some Arab cultures one man is permitted to have several wives. This cultural variation is possible because humans create their own rules for social order—and can change the rules. Larger, more complex cultures even tolerate within their boundaries **subcultures,** groups whose beliefs and styles of life differ in some respects from those of the larger culture.

Several **learned deviance theories** have concluded that subcultures may teach norms for behavior defined as deviant in the larger culture. People who grow up in a subculture that teaches theft and violence as norms can learn to steal in just the same way that those in the larger culture learn to go to work. Edwin Sutherland (1924) used the term *differential association* to describe how some people, having closer contact with a subculture than with the dominant culture, come to reflect the deviant norms by which they were socialized.

Notice how much this approach has in common with that of the behavioral psychologists discussed earlier. The views are, however, somewhat different. Whereas Bandura and Walters focus on the *psychological process* by which deviant behavior is reinforced in individuals, social learning theories focus on the *existence of subcultural norms* as a source of deviance.

Walter Miller (1958), for example, described what he saw as a delinquent subculture, which taught its members the following norms: (1) toughness—being physically strong; (2) excitement—looking for "kicks" in an otherwise boring and stifling world; (3) smartness—having "street wisdom," including the ability to avoid getting caught; (4) autonomy—wanting to be free of the restraints of authority; and (5) fate—believing that what happens is not the result of responsibility for one's actions but, rather, of luck. Young people who are more loyal to a delinquent subculture than to the larger culture learn to fit in with their friends. The price they pay is that the larger culture defines them as deviant.

Another theory arguing that deviance is learned points to what Oscar Lewis (1968) called the *culture of poverty*. Lewis proposed that poor people live in a subculture that prevents them from taking advantage of whatever opportunities may become available. They are deviants in their learned inability to hold jobs or plan for the future. Edward Banfield (1968) specified the supposed norms of the subculture of poverty in his description of slum dwellers. These "lower-class individuals" lack a concern for the future, a sense of control over their fate, and self-discipline; they prefer "action" to work, are insensitive to the value of property and the filth around them, and do not value education (Banfield, 1968:61–72). Clearly, anyone who learns such subcultural norms is destined to be seen as deviant by the rest of the culture. It is important to realize that many other sociologists contend that poor people do *not* actually hold norms like these. Rather, these researchers explain deviance as the result of social processes such as anomie, conflict, and labeling.

Anomie Theory Among the norms that a society teaches its members are norms for what is worth having or achieving (goals) and norms for accepted ways of achieving goals (means). Robert Merton ([1949] 1968) focused on the problems that occur when there is a disparity between the norms for goals and those for means. This condition is called **anomie,** a state of social ambiguity in which rules for behavior are unclear or in some way unsatisfactory. Merton described four categories of deviant behavior that he saw as adaptations to a means-goals disparity and one category of nondeviant behavior for the condition in which both means and goals are accepted and available (see Table 20-1). According to this model, the *conformist* is not a deviant. He or she accepts the society's goals and means and, having access to means to achieve goals, acts in ways approved by society. The *innovator,* having accepted the goals of society, achieves them by disapproved methods and is considered deviant. Perhaps the innovator rejects the approved methods as too difficult, time-consuming, or clumsy—for example, when a manufacturer illegally

TABLE 20-1

Merton's model of adaptation to goals and means

Type of adaptation	Are goals accepted?	Are means accepted and available?	How does society see the adaptation?
Conformist	Yes	Yes	Not deviant
Innovator	Yes	No	Deviant
Ritualist	No	Yes	Deviant
Retreatist	No	No	Deviant
Rebel	No (creates new goals)	No (creates new means)	Deviant

dumps toxic wastes in a stream to save money. Or perhaps the innovator is a person who wants to earn money (the goal) but is unable to get a job (the approved means) and is forced to steal. The difference between these two examples of innovators is that one has *rejected* legitimate means and the other has *been denied* legitimate means. The *ritualist* has rejected the goals but continues to follow the approved means of the society. For example, a person can continue to go to school or work even though she or he has no expectation of benefiting from the behavior. The *retreatist,* having rejected both goals and means, withdraws from society, either in drug or alcohol addiction or in some other isolating style of life. Carl Sisten, the man I met on the Boston library steps, is an example of a retreatist. The *rebel* not only rejects society's goals and means but also replaces each with his or her own versions. Political rebels (often called *revolutionaries*) or cultural rebels (such as *punks*) are examples.

Conflict Theory Some sociologists, such as Karl Marx, have suggested that conflict is central to the everyday operation of society. They have proposed that society is made up of large numbers of individuals and groups who compete with one another for the resources valued by them. So, for example, business and labor are constantly in conflict for control over wages, working conditions, and influence in the way business is conducted. The stability of society is merely the result of the conflict between such forces. When one side defeats the other, the stability of domination prevails (as in slavery). When neither side is able to gain a clear advantage over the other, a balance of forces results (as in the relationship between business and labor unions).

The **conflict theory of deviance** sees deviance both as a direct consequence of such struggles and as a tool used by contending sides in the pursuit of their goals. In the first instance, dominated groups of people are likely to belong to the category of innovators identified by

Merton in his anomie theory of deviance. Having lost a struggle for desired resources, members of dominated groups often lack the means to achieve goals and so must resort to disapproved means. For example, American blacks are often denied the education or training needed to compete with other members of society for jobs. Thus, some deviance is a direct consequence of the cruel treatment people experience at the hands of their competitors or oppressors.

The second insight into deviance from the conflict approach is more subtle. When individuals or groups conflict with one another, they use whatever tools are available to aid them in their struggles. Brute, physical force is common in street fights or in war, but when the conflict is social in character, the tools are also more social. Although it is possible for a dominant group to physically enslave or jail a subordinate group, this is a very expensive and inefficient mechanism for control. A more efficient and cheaper method is for the dominant group to cause the society to define the target group as deviant (by a process called *labeling*), thus solidifying the target group's position of inferiority. Ever since slavery, for example, blacks have been "kept in their place" in the United States by a set of beliefs that defined blacks as inferior and even dangerous. To the extent that this definition of blacks as deviant has been accepted, their inferior education, wages, and social status have been "justified," and their domination ensured. So the definition of a target group as deviant is used as a tool in social conflicts. And the groups subjected to the will of those more powerful lack any such control over society's norms. So in pursuing their aims in the struggle over scarce resources, less powerful people are limited to "weapons of the weak," such as protest and physical confrontations.

Labeling Theory The process by which more powerful individuals and groups create social definitions of deviance has been studied in great detail. The **labeling theory of deviance** begins with the recognition that, because we create norms for social order, we can also change them or apply them differentially—to some people or situations and not to others. From this point of view, deviance is merely the type of behavior that society defines as deviant. The people who shape definitions do so in their own interests, and they must be powerful enough to influence the process by which norms are created. As the sociologists Peter Berger and Thomas Luckmann say it (1967:19), "He who has the bigger stick has the better chance of imposing his definitions of reality."

Another sociologist, Howard Becker (1963), has illustrated how interested parties successfully campaigned in the early part of this century to label marijuana use (and, later, alcohol use) as deviant. The Bureau of Narcotics pushed for passage of the 1937 Marijuana Tax Act as an expression of its administrative influence. Urging official agencies and legislators to act were people whom Becker called "moral entrepreneurs." They claimed to want to help those beneath them achieve a better status by stopping their drug use. But Becker noted that these moral crusaders, people who wished to have their values expressed in social norms (including laws), had other interests as well. Because moral entrepreneurs are typically members of the upper levels of the social system, they enhance their power and reaffirm it simply by successfully influencing the normative structure.

Becker also pointed out how other powerful groups benefit from the labeling process. For example, "Some industrialists supported Prohibition because they felt it would provide them with a more manageable labor force" (Becker, 1963:149). At the level of conflict between groups, labeling is a tool for maintaining or enhancing control of the opposition.

Labeling theory has also focused on how norms are applied in a case of individual deviance. Edwin Lemert (1951) recognized that not all norm breaking results in application of the deviant label. He uses the term **primary deviance** to refer to all norm breaking, whether or not it is ever labeled deviant. If norm breaking comes to be recognized as deviance, he terms it **secondary deviance.** People break norms all the time. But some people are never caught (such as the person who cheats on taxes and isn't audited), some are excused (such as children and teenagers who are testing the boundaries to see what they are allowed to get away with), some break norms considered to be trivial (such as parking illegally), and some break norms that have been suspended for special circumstances (such as killing during war).

Social Functions of Labeling Deviance

From the point of view of the secondary deviant, being labeled and subjected to the punishments of society is unpleasant at best. Typically, deviance is considered extremely disruptive to social order. But the identification of deviance may also have some important positive consequences.

The French sociologist Émile Durkheim pointed out that identifying deviance highlights the meaning of a society's norms by showing the consequences of rule breaking ([1893] 1960). Deviance, then, provides standards of normality for rule followers. The stability of society is also enhanced by providing a focus for community unity. Communities in which deviant individuals or subgroups exist often unite to defend their way of life against the perceived threat. In fact, Durkheim argued that, without deviance, a society would stagnate, taking norms for granted and losing respect for their importance in maintaining social order.

Dealing with Deviance: Social Control and Motivation

Social order is maintained by two main mechanisms, motivation and social control. **Motivation** is the giving of rewards (often called **positive sanctions**) to people for engaging in socially approved behavior. **Social control** is the application of punishments (**negative sanc-**

TABLE 20-2
Relationship between type of sanction and level of sanction experience

Level of experience		Type of sanction: Positive (reward)	Type of sanction: Negative (punishment)
Level of experience	External	Praise, raises in salary, promotions for good work Medals for outstanding acts Diplomas and titles for completing a degree	Ridicule of laughter for socially inappropriate behavior Dismissal for poor work Fines or jail sentences for lawbreaking
	Internal	Feelings of satisfaction, pride, or accomplishment for socially approved behavior	Guilt or shame for socially disapproved behavior

tions) for violating norms. Such positive and negative sanctions can be experienced internally or externally. Table 20-2 illustrates the relationship between the two kinds of sanctions and the level at which they are experienced. It gives examples of the sanctions in each of four sets of conditions: positive external, positive internal, negative external, and negative internal.

Internal motivation and social control work because members of a society usually adopt social norms as their own (a process called **internalization**) and come to believe that they ought to be obeyed. When we feel pride in a job well done or guilt for wrongdoing, we are expressing internalized norms. External rewards (such as money) and punishments (such as jail sentences) operate in two ways. First, they reinforce internalized norms, providing clear illustrations of what good things happen if rules are followed and what bad things happen if they are not. Second, external sanctions have functions on their own, beyond the reinforcement of internalized norms. Making money, for example, is considered worthwhile as an end in itself in the United States. And societies maintain systems of external social control (including police, courts, lawyers, and jails) to punish wrongdoing when internalized social controls have proved inadequate to deter deviance.

Social Control and the Causes of Deviance

A great deal of money is spent in the United States to maintain external systems of social control. Court calendars and jails are overloaded, police forces have been increasing steadily, and the Supreme Court has asked for relief from an overwhelming caseload. With limited resources for social control, it is important to ask how the money is being spent. We can reduce the costs of dealing with deviance only if our programs target the true causes of deviance. Increasingly, sociologists have pointed out that we have overemphasized social *control* of deviance and ignored some of its important social *causes.*

Blaming the Victim Versus Blaming Social Forces

We tend to assume that the causes of deviance (and, therefore, the problems that result) are to be found in the characteristics of deviants. This is a view of social problems that William Ryan (1971) has called **blaming the victim.** Most people believe that, if *they* can live decent, successful lives, those who do not must be flawed in some way. The victim-blaming approach is used to justify social control programs (such as jail) and efforts to "correct the flaws" of deviants by resocializing them. Ryan has identified two types of programs that can deal with social problems such as deviance. These programs, which are developed out of the victim-blaming perspective, are called *exceptionalist programs.* In the area of criminal deviance these are designed to correct the flaws of prisoners by resocializing them, teaching them new trades, or otherwise improving their motivation to obey the law on the outside. It is efforts such as these that lead many to refer to our prisons as correctional institutions.

Certainly, efforts at rehabilitating deviants are preferable to pure punishment and isolation in institutions. But there are two important problems in this approach to deviance. First, resocialization rarely works, especially in the extremely abnormal world of prisons or mental institutions. Second, if the causes of deviance are not in the individual deviant but rather in the operation of social forces, then no amount of rehabilitation or correction aimed at the individual will reduce the incidence of deviance. Programs based on blaming the victim may be limited to dealing with symptoms rather than causes of deviance.

A social forces approach to deviance provides an alternative. By contrast, *universalistic programs* are aimed at changing the forces that influence deviance from *outside* the deviant. For example, instead of looking for evidence that some people *reject* legitimate means for achieving goals (the victim-blaming view), we can focus on how access to means *is denied* some people (the social forces approach). Of course, those who were responsible for the original denial of access (and who benefited as a result) will not be happy to have the balance of forces changed. Perhaps this is why the victim-blaming approach still dominates.

Summary

Deviance is behavior that does not conform to society's norms. There are two major views about the origins of deviance. One argues that people break norms because something in their biological or psychological makeup influences their behavior. An example of this view is the belief that certain individuals are born with a biological tendency to break rules. Another example is the theory that, due to problems in their upbringing, some individuals are psychologically flawed and become incapable of adequately learning or following social rules. The tendency to focus on the characteristics of deviants to discover the sources of rule breaking has been called *blaming the victim.*

The other major approach argues that deviance is created by social forces. When society determines what is socially "right" and "wrong," it identifies people who will be labeled deviant. Among these social forces theories of deviance are (1) the learned deviance theory, (2) the anomie theory, (3) the conflict theory, and (4) the labeling theory. According to learning theory, children raised in a subculture whose beliefs, values, and ways of life are different from those of the larger culture may learn to behave in ways that others label deviant. Robert Merton's anomie theory suggests that deviance is a response to disparities between social norms for goals and the means for achieving them. Conflict theory argues that the definition of deviance (and, therefore, of who is deviant) is a tool used by conflicting social forces (either individuals or groups) as they contend for the resources they value. According to this view, the people who have the power to decide who is deviant (a process called *labeling*) are in a position to protect their advantage against threats from others. As Edwin Lemert recognized, not all rule breaking results in application of the deviant label. He uses the term *primary deviance* to refer to norm breaking whether or not it is labeled deviant. He applies the term *secondary deviance* to norm breaking that is labeled deviant.

Societies maintain systems of motivation to encourage their members to engage in approved behavior. Such rewards (or positive sanctions) may be external (promotions, high salaries, and so on) or internalized in individuals (feelings of pride in a job well done, for example). Social controls are maintained to punish deviant behavior. Such punishments (or negative sanctions) may be external (such as jail sentences or fines) or internalized in individuals (feelings of guilt or shame at having broken a rule).

The Definition section for this concept focused on how specific behaviors are defined by society as deviant, a process called *labeling.* The Illustration for this concept deals with labeling from the point of view of the deviant. That is, how does a person who has been labeled as *deviant* deal with it? The Illustration presents three examples: a classic study by Erving Goffman called *Stigma,* and recent applications of some ideas from that study to the labeling of working-class students at elite law schools and to the labeling of people who work with animals used for experiments.

The Application for this concept employs crime data collected by the U.S. Bureau of Justice and the Federal Bureau of Investigation and printed in the *Statistical Abstract of the United States.* You will be able to find data to answer questions about where in the United States it is most dangerous to live (with respect to the crime rate) and whether the national crime rate has been increasing.

Illustration

Erving Goffman, Stigma: Notes on the Management of Spoiled Identity *(Englewood Cliffs, N.J.: Prentice-Hall, 1963)*

Arnold Arluke, "Going into the Closet with Science: Information Control Among Animal Experimenters," Journal of Contemporary Ethnography *20 (1991): 306–30*

Robert Granfield, "Making It by Faking It: Working-Class Students in an Elite Academic Environment," Journal of Contemporary Ethnography *20 (1991): 331–51*

Deviance is behavior that does not conform to society's norms for expected behavior. Decisions about who will be labeled as deviant are made by those who create and apply those norms. So deciding who is deviant is a matter of social definition and context. For example, killing another human is deviant behavior in most situations, but not when society sends its citizens into war. Also, you can wear a chicken suit to a costume party, but outside such a context it would be considered deviant. The discussion of the labeling theory of deviance in the Definition section for this concept focuses on how specific behaviors are defined by society as deviant. However, a great deal of sociological research has also examined how the person labeled as deviant deals with that. The Illustration for this concept presents three examples of such work.

Erving Goffman's Study of Labeling

Erving Goffman's classic study of deviance, *Stigma: Notes on the Management of Spoiled Identity* (1963), examines labeling from the point of view of the deviant. Goffman begins by describing how in our interactions we anticipate what a person's characteristics will be. "We lean on these anticipations that we have, transforming them into normative expectations, into righteously presented demands" (p. 2). When we encounter a person who is not as we expect and demand, someone who is different in some way outside the normal range of variation, that person is diminished in our minds from a whole and normal person, to a spoiled, incomplete one. Goffman calls the tainting characteristic a *stigma*. The range of stigmatizing characteristics is huge. Goffman names three categories of stigma: (1) physical deformities, such as being crippled or blind;

(2) flaws of individual character, such as being alcoholic, mentally ill, or a convict; and (3) flaws of group membership, such as being a member of a racial minority, an outcast religion, or an offensive occupation such as a drug dealer. Goffman's focus in this work is on how a person copes with the stigma of deviance. In Goffman's terms, how does the deviant "manage his or her spoiled identity?"

Goffman distinguishes between the *discredited* person, who has been stigmatized by others on the basis of any of the three kinds of stigmatized attributes, and the *discreditable* person, who has some characteristic that, if discovered, would lead the individual to be stigmatized. For the discredited person, the challenge is tension management. That is, in social situations the individual who has already been labeled as a deviant must develop ways of limiting the embarrassment or ridicule that flows from being stigmatized. For example, blind people may act as "sighted" as possible, moving about with confidence in familiar surroundings or picking up helpful cues about the environment with senses heightened in compensation for their loss of sight. Goffman also found that some stigmatized persons respond with defensive cowering (such as the girl I knew in seventh grade who at any social gathering would stare at the floor or shrink into a corner) or with hostile bravado (for instance, the vagrant walking up to strangers at a sidewalk cafe and announcing, "I'm a citizen of the street, and that's my country!").

For the discreditable person, the task is information control. That is, if I have a characteristic that might stigmatize me, I must find ways to control what others learn about me. Goffman found that discreditable people regularly hide from others (such as the facially deformed individual who simply never goes out) or "pass" for normal (for example, the prostitute who lies about her occupation, the homosexual who stays "in the closet," or the ex–mental patient who conceals his or her hospitalization).

Later Applications of Goffman's Ideas

Earlier I referred to Goffman's work as a classic. What I meant is that it has remained useful long after being published. People cite it in their current work, and its ideas are tested in settings never even imagined by Goffman. Two studies, one by Arnold Arluke and another by Robert Granfield, illustrate the continuing usefulness of Goffman's ideas on stigma management. In his work on the management of spoiled identities, Goffman tended to pick people with stigmas that were extreme. Those with physical deformities included cripples, blind people, and a girl without a nose, and those with discreditable occupations included hangmen, prostitutes, and vagrants. Arluke and Granfield both contributed to the body of information about stigma management by studying people who were not such obviously "untouchable" Americans.

Arnold Arluke's Study of the Animal Research Setting

Over a period of two years, Arnold Arluke conducted a study of twelve biomedical laboratories and animal facilities in which animals were being used for medical research. As a participant observer (see Concept 4 for a discussion of some of the qualitative methods used in social research) he took part in much of the work being done at these places, and conducted extensive interviews with 135 research chiefs, veterinarians, technicians, graduate students, and animal caretakers who were involved in animal studies. He was interested in showing that the techniques of identity management described by Goffman are used by people whose work is not so crassly deviant as the hangman's or the prostitute's. In fact, he reports that when many of these people started their careers in research they were not subjected to being labeled as occupational deviants. One man whose research involved the infection, and inevitable death, of dogs was quoted as saying, "Twenty years ago people thought of me as a hero because I was trying to find a cure for a deadly disease, but now people think of me as a criminal because I kill puppies." How do these people manage their spoiled, or potentially spoiled, identities?

According to Arluke, the higher the job in the hierarchy of an animal lab, the less likely that the job holder would be confronted by the discrediting evaluations of others. For example, those who headed the research projects (called principal investigators) typically moved in social and professional circles in which people would be either sympathetic with the aims of animal research or connected somehow with such research themselves. However, those at the bottom of the labs' occupational ladder, the technicians and animal care workers, were much more likely to be accused of being "animal murderers." In addition, those working on "higher" animals, such as dogs, cats, and chimps, ran a greater risk of being stigmatized.

The techniques of identity management identified by Goffman were used by many of the animal researchers. For example, it was common for workers to conceal what they did for a living. Some would say merely that they studied a disease, leaving out that the work involved animals. Arluke saw lab technicians drape cloths over the cages of lab animals before wheeling them on a cart through hospital corridors. He also found that access to animal labs was carefully controlled. One worker said that getting into the animal room was like "getting into Fort Knox."

Arluke discovered that animal research personnel had dealt with four ways of being stigmatized by others: The *reproaching other* tended to ask such questions as "How could you do that?"; the workers' identity management techniques included walking away from the situation and a number of long debates on the relative merits of the research versus the animals' well-being. A second type of stigmatizing person was the *confrontive other*. This person did not ask questions, but made direct statements about the immorality of animal research. The debates with confrontive others, when they occurred, tended to become heated, and the efforts to avoid exchanges with them more strenuous. The *dangerous other,* seen by animal researchers as capable of harassment and personal assault, led to a "bunker mentality" among some. The fear of attacks by animal rights activists, for example, made concealment and avoidance critical in certain situations. Lastly, the *distorting other* was seen as capable of misrepresenting the nature and value of animal research. The fear of such distortion led workers to attempt to present themselves in as positive a light as possible.

Robert Granfield's Study of the Law School Setting

Robert Granfield's study of stigma and identity management led him to a very different setting: an elite national law school in the eastern United States. He was interested in how students from working-class backgrounds manage to fit into the distinctly upper- and upper-middle-class environment of such a place. What happens when new law students discover that the speech patterns, the clothing styles, and even the deeply held beliefs to which they had been socialized might damage their social and career chances? Granfield found that Goffman's ideas helped clarify this form of stigmatizing as well. Granfield, like Arluke, conducted a participant observation study, in this case to investigate how law students manage their spoiled, and potentially spoiled, identities. Between 1985 and 1988 he joined a number of student groups at the law school, later conducting in-depth interviews with 103 law students, and finally administering a survey to half the 1,540 students in attendance. He found that the working-class students did, in fact, feel out of place at this elite law school. Many reported noticing that the values expressed by faculty and wealthier students were different from their own. It was assumed, for example, that all the students in one business-oriented class had "fathers that worked in business and that we all understood about family investments" (p. 337).

Granfield reported that although some students aggressively maintained their working-class styles and beliefs, this was not the most common strategy for coping with the possibility of being labeled as deviant. Instead, Granfield found that under the pressures to fit in, many of these students' working-class identities soon began to fade. Some began "faking" (Goffman used the term *passing*) to conceal their class background. For example, one student told the story of buying new clothes at the conservative Brooks Brothers store in order to adopt the style of dress he saw on the more "typical" law student. Several working-class students reported that on job interviews they learned to avoid talking about their backgrounds, instead leading the discussion toward the type of work in which they were interested. Those who decided that it was in their best interests not to "stick out" at law school employed the techniques of identity management that Goffman had identified almost thirty years earlier, just as Goffman had seen them used by cripples, blind persons, and prostitutes.

Summary

This Illustration has focused on some of the ways in which individuals attempt to deal with the fact that they have been, or may be, labeled as deviant. According to Erving Goffman, people who differ in some way from the normal range of characteristics or behaviors develop a range of techniques for managing the information about their differences that might lead them to be stigmatized, or might worsen the experience of already having been labeled deviant. Arnold Arluke discovered that the techniques of identity management Goffman described in 1963 are used by people who work in various jobs within animal experiment labs; and Robert Granfield found that they are also used by working-class students at an elite law school. To the extent that each of us has characteristics that might lead us to be labeled as deviant, however mild the stigma might be, it is likely that we use these techniques as well.

Application

Most of the applications in this book involve collecting and analyzing your own data. It is this process of getting new information, whether by participant observation, survey, or experimental methods, that attracted me to the behavioral sciences in the first place. There are a drama and an immediacy to it all. In every study I do, I am eager to find out what the answers turn out to be this time. Perhaps you have felt some of that in doing your own research, even if it has had to be limited in scale. But you should be aware that a large proportion of the research published in sociology is what we call "secondary analysis." That is, we often turn to data that have been collected by others and reanalyze it to shed light on issues that may not have been of concern to the original researchers. After all, collecting data is expensive, and there is no need to reinvent the wheel by doing large-scale surveys when they have already been done.

Data like these exist in many places. For example, the U.S. Department of Commerce's Bureau of the Census conducts a national census every ten years. It provides policymakers and researchers (including sociologists) with a wealth of information about rates of birth, death, marriage, divorce, employment, income, education, occupation, ethnicity, and much more. All these data are further broken down by region, state, age, gender, and other important variables. Just by using this single source of data from among the many others we have available, we can discover much of interest about social behavior without having to leave the library.

This application takes advantage of existing data about deviance that are available in the most recent edition of the Census Bureau's *Statistical Abstract of the United States.* It is available in any college or university library in the reference section, and it can be bought at a government bookstore in larger cities or ordered from the U.S. Government Printing Office in Washington, D.C. I am using the 1992 (112th) edition for this application, but the format and the categories of data presented change very little from one edition to the next, so last year's edition or next year's will do fine. It's just that some of the exact figures I cite will change yearly.

Most of the data presented in Section 5 of *Statistical Abstract* (the section on law enforcement, courts, and prisons) comes from two sources, the Bureau of Justice Statistics (BJS) and the Federal Bureau of Investigation (FBI). The BJS publishes yearly reports on a range of topics, including capital punishment, prison populations, and victims of crime. Its major publication is the *Sourcebook of Criminal Justice Statistics,* which is available in many college and university libraries but not all. If the issues raised in this application interest you, it is the single intensive source of data that I would suggest you buy. The FBI publishes a yearly report called *Crime in the United States,* which compiles data reported by state and local law enforcement agencies. Through its Uniform Crime Reporting Program, the FBI receives monthly and yearly data on crimes for local, county, and state law enforcement personnel. The crimes included in these reports have specific definitions, which may not be the same as your normal usage of the terms. They are as follows:

1. *Murder and nonnegligent manslaughter* include only intentional (or willful) homicide and exclude attempted murder, suicide, accidental deaths, and deaths caused by negligence.
2. *Forcible rape* includes in its definition attempted rape. (By the way, I can't figure out how "nonforcible rape" might be defined. It seems to me that force is included automatically in the word *rape.*)
3. *Robbery* means taking something of value by force or threat and also includes attempted robbery.
4. *Aggravated assault* includes assault with intent to kill the victim.
5. *Burglary* means a crime of theft accomplished by illegal entry, including attempted burglary.
6. *Larceny* means theft without the use of violence but excludes crimes like embezzlement and confidence games.

7. *Motor vehicle theft* is what it sounds like.
8. *Arson* means intentional burning of property such as buildings or cars.

Statistics on Crime in the United States

For this application you need to sit down with the *Statistical Abstract* and look up the answers to the questions posed here about certain forms of deviance (usually crime) in the United States. As you find the information in the tables, record the information in the appropriate spaces on the Data Recording Sheet. Remember that the figures will vary from year to year, depending on which edition of this source you use. I hope you will see, after going through this application, that a great deal of information is readily available but also that special skills in reading and analyzing these data are worth learning. For example, the various forms of deviance are not always measured as you might expect, and the data are sometimes presented for areas called SMSA's, which you will have to read about to understand.

Question 1: Where are the most dangerous areas of the country to live? Where are the least dangerous?

To begin with, answering these questions depends on what we define as dangerous. Does "dangerous" mean life-threatening? What if it includes the likelihood of being a victim of assault, rape, or aggravated assault? Let's look at a few kinds of data on the subject.

In the 1992 edition of *Statistical Abstract,* Table 289 on page 181 summarizes "Crime Rates by State, 1985 to 1990, and by Type, 1990." These are FBI data. (If you are not using the 1992 edition, find the equivalent table in your edition, and use it to fill in the actual numbers.) First look at the top of the table, where the information for regions is listed. The four regions are Northeast, Midwest, South, and West. (The specific states included in each of these regions are shown in a map printed on the inside front cover of the book. Ohio, for example, is included in the Midwest region rather than the Northeast.) Now look at the column listing the reported murder rates for each region. These are rates rather than numbers of cases. That is, if a region's murder rate is 9.1 (the figure for the West region in 1990), that means 9.1 murders were reported for each 100,000 people who live in that region. Using this one indicator of the "danger" of living in a region, which was the most dangerous? In 1990 it turned out to be the South. (Record the murder rate for each of the four regions in item 1 of the Data Recording Sheet, and circle the highest and lowest murder rates.) Were you at all surprised (as I was) that the Northeast had so low a rate in comparison with the South?

A closer look at the data for murder rates reveals more detailed, and equally surprising, information. The data are further broken down into smaller areas called "divisions" (also shown on the map on the inside cover), revealing that in 1990 New England had the lowest murder rate for any division in the country, with 3.9 murders per 100,000 population. Which region had the highest? (Record these murder rates in item 2 of the Data Recording Sheet.) Can you come up with some ideas about why the West South Central division (which includes Arkansas, Louisiana, Oklahoma, and Texas) should have earned this questionable honor? Speculation, especially when informed by some reading of the studies that might have been conducted on the subject, generates much of our further research. In the spaces provided in item 3 of the Data Recording Sheet, try listing two or three possible reasons for New England's low murder rate and the West South Central's high rate. For example, do you think it could be the weather?

Let's move on to other types of personal danger. What does the same table show about the likelihood of robbery? This time, the Northeast does not compare so well. In 1990 it had the highest robbery rate in the nation, and the Midwest had the lowest. (Record the regional rates of robbery per 100,000 population in item 4 of the Data Recording Sheet, and circle the highest and lowest rates.) What about the rates for forcible rape? In this case the Northeast is back at the bottom of the list with the lowest rate, and the West, South, and Midwest are about tied for the highest rate. (Record the regional rates of forcible rape per 100,000 population in item 5 of the Data Recording Sheet, and circle the highest and lowest rates.)

Well, things must be fairly confusing already. The data show that regions differ in terms of crime rates for murder, forcible rape, and robbery, but not consistently. The Northeast, for example, has a relatively high rate of robbery but relatively low rates for murder and rape. How can we make sense of these differences? One way is to speculate about them. Another is to group the data into larger categories that essentially ignore smaller differences. For example, the table you have been looking at lumps together the data for rates of murder, rape, robbery, and aggravated assault in a column called "Violent Crime." (These grouped data are found in the column headed "Total" just to the left of the data for murder rates.) If we assume that these four crimes constitute the major threat to an individual's safety, perhaps we can answer the question about which are the most dangerous areas of the country from this column. So, which were the most dangerous and least dangerous places to live in the country in 1990? Looking at the data in the column labeled "Total" under "Violent Crime," you can see that, overall, the likelihood of being a victim of *some* sort of violent crime was highest in the West and lowest in the Midwest. (Record the regional rates of violent crime per 100,000 population in item 6 of the Data Recording Sheet, and circle the highest and lowest rates.) If you look at divisions within the country, you can see a dramatic difference in violent crime in comparing the Pacific and West North Central divisions. In 1990 the West North Central division had less than half the rate of violent crimes per 100,000 population (432) than was reported for the Pacific division (910 per 100,000). It looks as if that was the least dangerous place to live. Once again, try listing two or three possible reasons for the West North Central division's relatively low rate of violent crime in the spaces provided in item 7 on the Data Recording Sheet.

Question 2: What has happened to the rates of crime over time in the United States? Has it been getting more or less dangerous in the country?

In the 1992 edition of *Statistical Abstract,* Table 293 on page 183 gives FBI data for rates of forcible rape per 100,000 people for the years from 1970 to 1990. As the table shows, reported rates of rape have increased dramatically. (Record the rates for forcible rape per 100,000 people for the years 1970, 1975, 1980, and 1990 in item 8 of the Data Recording Sheet.) Note that the rate in 1990 was more than double the 1970 rate. What has been happening? To this point your speculations have probably focused on explaining the differences in the data. (These are differences over time rather than differences over regions, but the process is the same.) Now consider the method by which the data are collected for these tables. It has been suggested that the apparent dramatic increase in rates of rape is really a result of an increased willingness to report rapes, largely as a consequence of the growth of feminism during those reporting years. In other words, merely because the FBI data seem to provide national rates of rape, what they really give us is rates of reports of crime. Secondary analysis is limited to the accuracy of the data that are being reanalyzed.

What about changes over time in the rates of murder? Table 292 on page 183 in the 1992 edition of *Statistical Abstract* reports data from the National Center for Health Statistics on "Homicide Victims, by Race and Sex: 1970–1989." On the right half of the table are figures for the homicide rate per 100,000 citizens. Look first at the column headed "Total," which lists the homicide rates for all citizens without regard to sex or race. The homicide rate increased by 29 percent between 1970 and 1980, from 8.3 per 100,000 population in 1970 to 10.7 in 1980. (This table in previous editions of *Statistical Abstract* also reported the homicide rate in 1960, which was only 4.7 per 100,000 population.) However, between 1980 and 1989 the homicide rate actually went down by about 14 percent, from 10.7 per 100,000 population in 1980 to 9.2 in 1989. (Record the rate for homicide per 100,000 people in the population for the years 1960, 1970, 1975, 1980, and 1986 in item 9 of the Data Recording Sheet.) This change is interesting given the great concern expressed in the 1980s in the United States about the rates of violent crime, especially homicide. Apparently our fears do not precisely parallel the real dangers we face.

It is vital to keep in mind that the data you reexamine in secondary analysis were collected by someone other than yourself, and in all likelihood they were collected for purposes other than yours. Many of the data can be bent to your purposes quite successfully, but the word *bent* is intentional. In this kind of research the data are essentially always adapted to the researchers' needs. Sometimes, in the interest of making things work out, we are tempted to assume that data called "rape rate," for example, are exactly what they say and that it is better not to look too closely at how the variable was defined and measured originally, lest we be compelled to toss the data out as not at all what we would have collected.

Data Recording Sheet

1. The 1990 murder rate per 100,000 population for the four major regions of the United States:

Northeast _____ Midwest _____ South _____ West _____

2. The divisions of the United States with the lowest and highest 1990 murder rates, along with their rates:

Division with the highest murder rate ____________________ Rate ____________________

Division with the lowest murder rate ____________________ Rate ____________________

3. Some possible reasons for New England's low murder rate and the West South Central's high rate: ____________________

____

____

____

4. The 1990 robbery rate per 100,000 population for each of the four major regions of the United States:

Northeast _____ Midwest _____ South _____ West _____

5. The 1990 rate of forcible rape per 100,000 population for each of the four major regions of the United States:

Northeast _____ Midwest _____ South _____ West _____

6. The 1990 rate of violent crime per 100,000 population for each of the four major regions of the United States:

Northeast _____ Midwest _____ South _____ West _____

7. Some possible reasons for the West North Central division's relatively low rate of violent crime:

____

____

____

8. The rate of forcible rape per 100,000 people in the population for the years 1970, 1975, 1980, 1985, and 1990:

1970 _____ 1975 _____ 1980 _____ 1985 _____ 1990 _____

9. The rate of homicide per 100,000 people in the population for the years 1970, 1975, 1980, 1985, and 1989:

1970 _____ 1975 _____ 1980 _____ 1985 _____ 1990 _____

CONCEPT 21

Anomie and Alienation

Definition

Anomie A condition of social ambiguity in which an individual does not know how to act because the rules for behavior (norms) are either unclear, entirely absent, or in some way unsatisfactory.

Alienation A condition of social ambiguity in which an individual has lost the meaning of his or her participation in social roles.

And he was rich—yes, richer than a king—
And admirably schooled in every grace:
In fine, we thought that he was everything
To make us wish that we were in his place.

So on we worked, and waited for the light,
And went without the meat, and cursed the bread;
And Richard Cory, one calm summer night,
Went home and put a bullet through his head.
[Robinson, (1897) 1978: 104]

These are lines from an 1897 poem by Edwin Arlington Robinson. You may recall a 1960s song by Simon and Garfunkel that was based on the Robinson poem. Stories like this bring into sharp focus all our concerns about how a person can lose enthusiasm for taking part in everyday life or can lose the very will to live. It can happen even to a person like Richard Cory. People were concerned about this issue more than ninety years ago, just as they are today. This concept, on anomie and alienation, deals with the kinds of problems we sometimes experience in making sense of our work, our participation in society, even our sense of who we are. Let's start by examining how someone like Richard Cory could commit suicide.

Anomie

Every day, we take part in the routines of social life. We go to work or school, deal with friends, merchants, strangers. In every interaction we feel comfortable as long as we know what to do and why we are doing it. The rules for how to behave in social situations are the norms we were taught when we were still young. They make sense to us in a quiet sort of way. We don't usually examine them consciously. When these rules satisfy the expectations that others have of our behavior, they seem to be working well. Social situations are comfortable for us when everyone involved agrees about how to act, so everyone can predict what will happen.

Sometimes, there is confusion about how to act. Norms seem to have been suspended, are contradictory, or seem nonexistent. This situation can be caused by a sudden disruption in patterns of social life, such as the death of someone very close, a sudden change in our financial well-being, or a severe societal upheaval, such as war. Suddenly, the rules that regulated daily life are thrown into turmoil, and the resulting confusion about how to act is called **anomie,** or the state of normlessness.

Durkheim's Theory

I find it fascinating that Émile Durkheim explained the concept of anomie in 1897, the same year that Robinson's poem "Richard Cory" was published. Could they both have been reacting to the social upheavals accompanying the Industrial Revolution, which had been shaking the normative foundations of Western society since the middle of the nineteenth century? Durkheim's study, entitled *Suicide* ([1897] 1951), suggested that large-scale changes in the way society was organized could bring about anomie. So could disruptions in a person's private social world, such as sudden poverty or wealth. Durkheim then showed that, in conditions of extreme normlessness, a person can feel so lost about how to behave that he or she commits suicide. That is a measure of how powerful a stabilizing force norms are. We have all heard about how the stock market crash of 1929 caused scores of ruined investors to kill themselves. Everything they had worked for had been wiped out, in spite of the fact that they had operated according to the rules of the market. Their anomie resulted from the fact that the rules by which they had run their lives had not worked. It was not the poverty itself that caused these people to kill themselves, but the sudden loss of faith in the rules that had been the foundation of their lives.

The fact that poverty was not the cause is made clear by the suicides of people like Richard Cory. Imagine working all your life to become wealthy. You put in long hours, deny yourself luxuries, take gambles, and wait long years for the payoff. It becomes *the* way of life for you—the set of rules by which you operate in the world. Then, one day, success comes to you; you are rich. The problem is that all the rules by which you lived, those deeply held norms for behavior that had guided you, are now unnecessary or even inappropriate. What will get you going in the morning? What will tell you how to act in the new situation? Richard Cory killed himself because he had no rules to guide his life—no way to know what to do amid all his wealth and fame.

Suicide is, of course, the most extreme reaction to feelings of normlessness. We all experience mild or momentary insecurity about how to act in given situations. But even if no clear resolution develops, we don't kill ourselves over these ambiguities. Depending on the extent of the anomie experienced, we have various reactions, ranging from mild unease or anxiety to severe withdrawal or deviant behavior, such as committing criminal acts.

Merton's Theory

Robert Merton's ([1949] 1968) approach to anomie helps explain this process. According to Merton, norms give members of a society the means to attain valued goals of the society, such as money, power, and status. Anomie may result when there is a gap between those goals and the approved means (norms) available to reach the goals. For example, a person may be taught that the way to get rich is to get an education, get a job, work hard, and climb the corporate ladder. But that person may then be denied access to the education, job, or promotions. The norms for achieving the goals are unavailable to her or him, so the goals are also.

Notice the link between Merton's view of anomie and Durkheim's. Merton's idea that means (norms) are inadequate or unavailable is much like Durkheim's idea that a disruption in social order suspends the usefulness of norms. Merton claimed that the typical reaction to this form of anomie was what he called *innovation*. (Other reactions are discussed in Concept 20.) Basically, innovation is a kind of deviant behavior, that is, behavior not approved of by society, employed to reach the goal when approved means are unavailable. Such deviance can range from minor cheating (such as enhancing how one looks by means of wigs, makeup, or padding) to more serious cheating (such as on income taxes) to very serious crimes (such as extortion, theft, and murder). The illegal means that people choose are determined by the extent to which they are denied approved means and the situation in which they live. A poor person will not choose cheating on taxes (because he or she may have no income to tax in the first place). The subtler forms of deviance practiced by wealthy people are not available to the poor, so they must resort to more dangerous, less profitable crimes.

Anomie is more likely to occur in huge, complex societies like our own than in smaller, simpler ones, where it is easier to keep close track of the rules of everyday life and the extent to which people follow them. We have so many sets of rules for behavior and these rules change so rapidly that it is easy for us to become confused about what is proper to do in a given situation. For example, have changes in the role of women made it wrong for a man to try to be protective? What is marriage supposed to be like once the older notions of male dominance have been shed? This ambiguity about norms may be at the root of many people's search for systems of belief in which the rules for behavior are extremely stable and clear. Since the 1960s, young people, especially, have joined religious and semireligious groups in which daily life is regulated in almost military fashion. Diet, sexual behavior, travel, and work are closely controlled. This may be a reaction to anomie that reduces dissatisfaction with unclear guidelines for behavior by replacing them with new, clear norms. It remains to be seen whether such groups will integrate their aims with the values of the larger American culture.

Alienation

We feel comfortable as long as we know *what* to do and *why* we are doing it. Anomie was defined as confusion about what to do. **Alienation** focuses on why: the meaning of our social behavior, the reasons that help us make sense of what we do. As we saw in Concept 7, any culture has **values,** shared ideas about the most abstract goals the group believes are worth achieving. In American culture they include freedom, achievement, equality, justice, work, progress, representative government, and so on. From such values flow a variety of rules and guidelines (norms) that regulate our daily lives, telling us how to behave in given situations. The values give meaning (the why) to the norms. Alienation occurs when we fail to understand or agree with the meaning of social participation. We may continue to participate in routines, but we are just going through the motions.

The Sociological Meaning of Alienation

Probably the best-known discussion of alienation was presented by Karl Marx (1964). He argued that industrial assembly lines would rob workers of their feelings of control over the creative process. There is a big difference between building an entire engine and merely being responsible for turning one bolt in the assembly process. Marx defined meaningful work as work in which the worker maintains control of the creative process.

Because this is not possible in assembly lines, such work is inevitably alienating (meaningless) to the worker. A worker may need to keep working at an alienating job because it is the only way to earn a living, but it becomes a matter of stultifying routine.

Marx thought of social behavior in primarily economic terms. For him, alienation was alienation from one's work. But it is possible to think of other aspects of social existence that might lose their meaning for people. A family member might wonder, after years of unhappy marriage, why he or she got married in the first place. A student, not certain about what to study or whether more education would lead to a satisfying career, can come to question the purpose of continuing in school. A citizen who disagrees with much of what is done in the country at the various levels of government might question the purposes of government and the usefulness of voting or participating in other ways. Just as it is possible to experience alienation from work, it is possible to experience alienation from family, school, government, or other forms of social participation.

The General Meaning of Alienation

In this discussion of alienation I have focused on the special, sociological meaning of the term; but alienation has also developed a more general, everyday meaning. It is often used to refer to the feelings of powerlessness, isolation, meaninglessness, and self-estrangement brought on by mass, urban life. Robert Blauner (1964) identifies these four dimensions of alienation in his discussion of industrial workers:

1. *Powerlessness* is the feeling that one cannot do anything to influence one's destiny.
2. *Isolation* is the feeling that one is separated from others by the scale and complexity of modern life.
3. *Meaninglessness* is a feeling of lack of agreement with or understanding of the meaning of one's social participation.
4. *Self-estrangement,* a consequence of meaninglessness, is the empty process of social participation without any feeling of fulfillment.

The "what" and "why" of social behavior are intricately interconnected. We need to know both to feel comfortable. Disruption in one is likely to be accompanied by disruption in the other. Anomie and alienation are two sides of the same coin.

Summary

Two important sources of disruption in social order are anomie and alienation. Anomie is a condition of social ambiguity in which an individual does not know how to act because the rules for behavior (norms) are either unclear, entirely absent, or unsatisfactory in some way. The term *anomie* was used by the French sociologist Émile Durkheim to describe what happened during the overthrow of traditional social order that accompanied the Industrial Revolution. He concluded that large-scale changes in the way societies are organized can destroy belief in social norms, bringing about widespread anomie. Depending on the extent of the anomie experienced, individual reactions may range from feelings of unease or anxiety to withdrawal, deviant acts, or even, Durkheim suggested, suicide.

Robert Merton's use of the term *anomie* focuses on the process that leads to deviance. He proposed that, when individuals adopt the goals of a society but are prevented from having the means to attain them, they experience a form of anomie. Their response can be what Merton called *innovation,* the use of nonapproved methods of attaining approved goals.

Whereas anomie refers to confusion about *what* to do as a member of society, alienation refers to confusion about *why*—the meaning of social behavior. Alienation occurs when an individual fails to understand or agree with the meaning of social participation, even though he or she may continue to go through the motions of participation. Perhaps the most famous use of the term comes from the work of Karl Marx. He believed that workers who have no say in the work they do (such as workers on an assembly line) cannot experience work as a meaningful activity. They are, in Marx's terms, alienated from their work. More generally, alienation has come to refer to feelings of powerlessness, isolation, meaninglessness, and self-estrangement brought on by the complexity and scale of life in modern societies.

The Illustration for this concept summarizes Durkheim's classic 1897 study, *Suicide.* Though modern Americans tend to think of suicide as the ultimate act of desperation by sick individuals, Durkheim clearly showed that it is a social phenomenon. There are consistent differences in the rates of suicide in different societies and among different groups in a single society.

The Application for this concept employs a well-known measure of anomie for a brief survey of people in your community or on your campus. Beyond discovering anomie in your sample, you are asked to choose an independent variable or two that you think might distinguish between respondents with high levels of anomie and those with low levels.

Illustration

Émile Durkheim, Suicide: A Study in Sociology, *trans. J.A. Spaulding and G. Simpson (New York: Free Press, [1897] 1951)*

On first view, it seems that there can be no more personal and private act than suicide: killing oneself is normally seen as the ultimate statement of individual failure to cope. This view is so dominant in American culture that our efforts to understand and prevent suicide are limited almost exclusively to consulting psychiatrists and psychologists. They tend to focus on failure of will in a person, a tragic unhappiness, or personal inability to cope with reality. What is wrong with the suicidal person?

This focus on the individual probably dominates because American society has always emphasized individual achievement. The New World was an adventure from its beginnings. Newcomers fought against the land for a foothold, against Native Americans and one another for domination, then against England for independence. Capitalism has also been a battlefield for independent action and risk taking. Because success was credited to individual initiative in American culture, failure has been blamed on flaws in *individual* character. If those who succeed are thought to have superior will, talent, and energy, those who do *not* must lack those qualities.

Given this set of cultural assumptions, it is understandable that theories of human behavior that do not focus on individual characteristics have not been broadly accepted in American society. This has been especially true of sociological explanations of acts such as suicide that appear to be so intensely individual. Durkheim's classic work provides us with an alternative social explanation of this behavior and teaches some important lessons about the need to look beyond individual characteristics in explaining human behavior.

Durkheim began his study with the discovery of some fascinating patterns in the data about suicide. He recognized that there were (and we know there still are) consistent differences in the *rates* of suicide among various categories of people: (1) Protestants are more likely to commit suicide than Catholics; (2) people living in cities are more likely to commit suicide than those living in smaller communities; (3) single and divorced people are more likely to commit suicide than those living in families; (4) people living during times of rapid economic change (boom or bust) are more likely to commit suicide than those living during times of economic stability; (5) people living in times of rapid social change are more likely to commit suicide than those living in times of social stability; (6) soldiers are more likely to commit suicide than civilians; and (7) officers are more likely to commit suicide than enlisted men.

If suicide were the result of flaws in individuals, in their private ability to cope with the world, how could these stable differences in rates of suicide be explained? Durkheim compared the statistics for some psychological causes of suicide (such as insanity) and concluded that individual characteristics could not explain the differing suicide rates he found. What was the common explanation for suicide rates as varied as these? Why, for example, should the suicide rate be high during times of great economic well-being? Durkheim's answer lay in the nature of social membership.

He proposed that suicide is directly linked to a person's feelings of social integration. When our relationship with the social world is operating normally, we know what the rules of social behavior (the norms) are, and we believe they should be obeyed. Given adequate opportunity to act according to widely shared norms, societal members feel socially integrated, bound into the social network. But Durkheim recognized that the normal system of social integration can be disrupted in a variety of ways. Each can greatly reduce people's feelings of social bonding, to the extent at which they become more likely to commit suicide. He identified three types of suicide, based on different types of disruption in social integration: *anomic suicide,* which occurs when rules for social integration are unclear or absent, and the individual lacks meaningful rules for behavior; *egoistic suicide,* which occurs when social integration is excessively weak and cannot support the individual; and *altruistic suicide,* which occurs when social integration is excessively strong and overwhelms the individual.

Anomic Suicide

Durkheim proposed that during times of rapid social or economic change, the norms by which people operate become unclear or confused. The condition of anomie is a state in which the rules for behavior are so unclear that people do not know how they are supposed to behave. Thus, when social revolutions (such as the Industrial Revolution) occur, former rules become obsolete, and before new ones become clear, the suicide rate rises. During times of economic change, the same thing occurs. The interesting thing is that increases in the rate of suicide occur whether the change is economic disaster (such as the stock market crash of 1929) or sudden prosperity. The important common denominator in these seemingly opposite events is rapid change that disrupts norms.

Take, for example, the case of a person who suddenly becomes famous. An actress has been putting in long hours, studying, auditioning for parts, calling casting directors, focusing intensely on her single goal. Suddenly, she achieves the goal; she gets the part, is a hit, is famous. Normally you would think that she would be happy and satisfied. But, as Durkheim's theory points out, her sudden success has made obsolete all the rules by which she lived up to that point. All the rules for "making it" no longer apply to a person who is already "there." Unless she is capable of quickly replacing the lost norms with appropriate new ones, she is likely to feel a special kind of anomie intensely enough to commit suicide. Think of the film, television, and rock stars Elvis Presley, Jimi Hendrix, Janis Joplin, or Freddie Prinz, who killed themselves (either quickly or slowly by drugs) after having achieved great success. Like Richard Cory in the poem at the beginning of this concept, they had "everything" but rules by which to make any of it meaningful.

A more common kind of anomic suicide occurs when a person is left suddenly alone by the death of a spouse or loved one or by divorce or separation. In these cases, the state of anomie is brought about by the sudden loss of a person whose presence has been crucial in everyday patterns of interaction. With a loss like this, all the reasons for following social rules seem lost or unclear, and suicide, again, becomes more likely.

Egoistic Suicide

By contrast, egoistic suicide occurs when rules for social integration are clear enough but the amount of social integration is too low. According to Durkheim, a person who is excessively committed to personal beliefs and aims, rather than to those of the group, will be inadequately socially integrated. Because strong social integration is needed to provide a person with a stable sense of self and with support during difficult times, anyone with few social bonds and with privately held aims is a candidate for egoistic suicide. This theory explains the relatively high rates of suicide among Protestants (Christian sects that not only tolerate but in some cases also encourage highly individualistic moral decisions) and among single, divorced, and urban people (whose lack of contact with others often forces them to develop extreme self-reliance).

Altruistic Suicide

Altruistic suicide occurs when a person is excessively socially bound, Durkheim believed. In this case, the person's sense of responsibility to the group is so strong that it overwhelms the individual's sense of self, and the person sacrifices his or her will and life to the group. An example is the soldier who throws himself on a grenade to save his buddies. A more specific (and tragic) example was the mass suicide at Jonestown, Guyana, in 1978. In the mid-1950s a preacher named Jim Jones founded a group called the People's Temple, and by 1974 he had built a devoted congregation with whom he established a religious commune named Jonestown in the South American country of Guyana. Jones exercised total control over the lives of his disciples, determining what they ate, where they went, what information they received, and all the rules of life in Jonestown. By 1978 he had become everything to them and had convinced the believers that their entire lives were dependent on the survival of the group. When news came that an American congressional investigation team was soon to arrive, Jones became convinced that the commune was under attack by the outside world. It is hard to imagine how isolated the residents of Jonestown were and how dependent on the group for their sense of themselves. But their immersion in the group must have been total, for, when Jones announced that their proper course was to commit suicide and ordered a large vat of poisoned Kool-Aid prepared, hundreds and hundreds of his disciples lined up to drink and to die. Although some were later found to have been shot, most of the over nine hundred members of Jonestown committed mass, altruistic suicide. They had come to believe that without Jonestown, they were nothing.

Application

Anomie, as we have seen, is a condition of social ambiguity in which an individual has lost, or no longer believes in, the meaning of his or her participation in social roles. It has been suggested that anomie is most likely to occur in times of social change or when norms become weakened or vague for a variety of reasons. Doesn't that seem to describe our culture, or at least the American culture we see analyzed and criticized daily through dozens of sources? We are often described as a people whose styles change rapidly and whose rules for behavior have become increasingly lax and vague. If these statements are even partly true, shouldn't we find evidence of anomie easily?

Conducting a Survey on Anomie

Anomie is essentially a condition of damage to one's feeling of social belonging. Do you ever see evidence of this damage among friends or in yourself? It need not be so severe that is leads to suicide, though Durkheim made a persuasive case that it does. Assuming that it can be experienced at less intense levels and that people around you could be walking around and operating in everyday life despite feelings of normlessness, who might these people be? This application is designed to enable you to use a well-known measure of anomie in brief questionnaire form to try to discover evidence of this condition in your community or campus.

This application is different from the previous ones in one important way. Unlike the other applications, in which I have chosen the independent variable, here you are asked to choose your own independent variable. For example, in the Concept 14 survey application I suggested that the size of the work group might be a good independent variable that influences feelings of solidarity. Similarly, in the Concept 16 survey of dependence on religious institutions, I suggested age of the respondent as the independent variable, reasoning that it should influence the dependent variable. In this application, however, you are asked to speculate about what would be a good independent variable to use and to collect the information that will allow you to see if the independent variable you choose does make a difference in the scores for the dependent variable (anomie). In fact, if you are interested enough, you could collect data for two or three independent variables and reanalyze the anomie scores for each one.

Collecting the Data

Leo Srole's Anomia Scale (Srole, 1956) consists of five statements with which a respondent may agree or disagree. By adding up the number of statements to which the respondent agrees, a score for anomie ranging from 0 (low anomie) to 5 (high anomie) is calculated. The questions from the anomie measure take just a few moments to complete, and your approach to respondents will be introduced as with the other survey Applications. You should interview about twenty people to yield decent results. However, whom you choose to interview depends on the independent variable or variables you select.

The Survey Sheet contains just the five statements for the measure of anomie and some blank spaces at the bottom for you to fill in the independent variable(s) you decide to use. The same is the case on the Data Analysis Sheet. It will be up to you to decide how to measure the independent variable and to analyze the data to see the effect of that independent variable. To help you get the idea, consider an example.

Selecting an Independent Variable

Let's say you decide that living at home creates anomie for a student. The reasoning might be that two sets of rules for behavior exist for commuting students, whereas resident students have only one. Thus, you expect levels of anomie to be higher for commuter students than for residents. It is an easy matter to ask respondents whether they are commuters or resident students and to record their answers in the space provided for the "Independent variable" at the bottom of the Survey Sheet. Then, you would list the responses to the anomie measure on the Data Analysis Sheet in groups of ten, with commuter-student responses in one block and resident-student responses in the other. The comparison of the average anomie scores for the two groups will give you the result of your survey.

Of course, there are many possible independent variables to use. A little thought and discussion with others will easily yield many good possibilities for students on your campus and even more if you interview beyond the campus. The key to choosing a good independent variable is to have some logic for why you think it should yield different scores on the anomie measure. I have no idea whether the commuter-resident variable I used above would work. I just made it up to illustrate the point here. But I do have a logic (about dual sets of rules for commuters) that seems to justify trying that independent variable. At the bottom of the Data Analysis Sheet is a series of lines for you to write in the reasoning you used to select your independent variable.

Lastly, some independent variables are very easy to measure (such as resident status, gender, or major in school), and a single, inoffensive question is often enough. However, others, such as age, belief in God, or social class are more sensitive. If you are in doubt about your ability to measure a given independent variable accurately and without offending a respondent, either try another independent variable or get some help in devising a measure of it. Your library should have a range of books on social research that will have a chapter on measurement or questionnaire design. You can also use them to get ideas about good independent variables. And the best help is to read about studies of anomie that have already been done.

Here are a brief introduction and some instructions to say to each respondent. Practice them until they sound fairly conversational; people are much more likely to respond if you don't read stiffly to them from a sheet of paper.

Introduction: I'm doing a survey for a class I'm taking. It's a study of some general attitudes that people have. I don't need your name, so your answers will be completely anonymous. The questions take less than two minutes to answer.

Instructions: I'll read you five statements. After each one, please say whether you generally agree with the statement or generally disagree with it.

After reading each statement to a respondent, indicate agreement with a statement in the space marked 1 and disagreement in the space marked 2.

Survey Sheet

1. In spite of what some people say, the lot of the average person is getting worse.
1. ______ 2. ______

2. It's hardly fair to bring children into the world with the way things look for the future.
1. ______ 2. ______

3. Nowadays, a person has to live pretty much for today and let tomorrow take care of itself.
1. ______ 2. ______

4. These days, we don't really know whom we can count on. 1. ______ 2. ______

5. There's little use in writing to public officials, because they often aren't really interested in the problems of the average person. 1. ______ 2. ______

Independent variable response: ____

____

Independent variable 2 response (optional): ____

____

Data Analysis Sheet

Name of independent variable used: ______________________

Group 1 Responses		Group 2 Responses	
______________		______________	
(Independent variable category name)		(Independent variable category name)	
Respondent number	Anomie scale scores	Respondent number	Anomie scale scores
1	________	11	________
2	________	12	________
3	________	13	________
4	________	14	________
5	________	15	________
6	________	16	________
7	________	17	________
8	________	18	________
9	________	19	________
10	________	20	________
Total for Group 1 anomie scores	= ________	Total for Group 2 anomie scores	= ________
Average Group 1 anomie score (Total ÷ 10)	= ________	Average Group 2 anomie score (Total ÷ 10)	= ________

Reasoning for the selection of the independent variable(s):

____

____

____

____

____

____

____

____

____

PART SIX

Sex and Gender in Society

Sociologists develop skill with the tools of their discipline so we all can understand human social interaction. Why do we behave toward one another as we do? What are the social forces that enable us to get along with one another, to be happy, or to predict the behavior of others; or that make us struggle, unequal, or hopeless and miserable. We hope that what we learn can be used to make our lives better.

All human inquiry depends upon our ability to recognize order and disorder in the world. Astronomy was made possible when humans recognized patterns in the movements of objects in the heavens. So the recognition of patterns in the interactions of humans made possible the development of sociology. Interestingly, in most cases it has not been the stable patterns of these interactions that has drawn the attention of sociologists, but disruptions in those patterns. When societies operate without serious disorder for long periods of time, they do not present their citizens with a reason to examine how they work. But if something happens to disrupt the normal order of social interaction, concern for the restoration of order leads people to examine the sources of the disorder. Change makes people nervous or scared, and they want things to get predictable again. For example, the great social and political revolutions that occurred in the seventeenth and eighteenth centuries in Western societies led to questions about the nature of social order, and, eventually, to the creation of sociology as a discipline. If the old ways of organizing societies were disappearing, what would take their place?

In much this way, the relationship between women and men in American society has increasingly become a matter of interest to sociology. Times are changing. The pattern of interaction between the sexes that was widely accepted thirty years ago is, in much of the culture, no longer accepted. This is apparent throughout society. Women are increasingly employed in jobs that once were all but closed to them. Pressure for equal compensation for equal work done by women and men is pursued in a number of ways such as court cases and organized campaigns by interest groups. In everyday interaction the norms for the behavior of men and women toward one another reflect the broader movement toward equality. As with any social change, there are those who are upset by new expectations and oppose them. And there are those who welcome such changes and work to accelerate them. But however individuals react personally to what has been called the sexual revolution, sociologists have been alerted to what may be a fundamental change in the way society will operate for a long time to come.

To this point the concepts which form the basic tools of sociology have been treated as relatively distinct from one another. It may be apparent, even without my mentioning it, that concepts like culture and socialization have something to do with one another. In some cases, such as the relationship between culture and norms, I have discussed the terms together. However, it was an unavoidable consequence of the concepts approach of this book that the chapter topics would be difficult to present as a coherent whole. However, in this final part of *Sociological Ideas* the concepts that have been discussed and illustrated in the first twenty-one concept chapters of the book are finally brought to bear on one topic: sex and gender in society. The idea is to demonstrate that each of the many concepts in the text has its use in the analysis of a contemporary issue. It is well beyond the scope of this book to show all the ways in which social concepts can be applied to this (or any other) topic. But I do hope Part Six will serve to illustrate that anyone well trained in the sociological vocabulary can use its many concepts to study human social behavior, even while it changes before our eyes.

CHAPTER 22

Sex and Gender in Society

Solve this riddle: Late one evening a night guard in the state capitol notices a light on in the office of one of the senators. The guard phones down to his friend at the reception desk to see if the senator has any appointments, and is told that the senator is, in fact, scheduled to meet with four members of a committee that is planning a new hospital. The receptionist does not know their names, just that one is an architect, one a physician, one a hospital administrator, and one a lawyer. The guard walks by the office to see how late the meeting will be going, and hears through the door the senator's familiar voice, only it doesn't sound normal. He is yelling: "No, no, Fred! Put down the gun! Don't do it!" The guard tries the door, but it is locked. And as he finds the right key he hears two shots and then a heavy thud. He unlocks the door, enters the office, and sees the senator on the floor—dead—with the four committee members standing nearby. The guard then pulls out his gun and immediately arrests the hospital administrator. How did he know who did it? The answer is that the hospital administrator was the only one of the four who was a man; the other three—the accountant, the architect, and the physician—all were women. (The guard simply assumed that none of the women was named Fred.)

If you automatically assume (as I did) that architects, physicians, hospital administrators, and lawyers must be male, then it is impossible to solve the crime on the basis of the given information. After I heard this riddle I tried it on about twenty people to see how common these assumptions were (besides, I was embarrassed to have made these gender assumptions myself). Most people could not solve it, though the women I spoke to were more likely than the men to figure it out. I believe this is because women usually are more aware of the facts of gender inequality in America that make this into a riddle in the first place. For example, if you assumed that all four people in the senator's office were men, it is probably because in the United States as of 1991 83 percent of all architects, 80 percent of all physicians, and 81 percent of all lawyers were male (U.S. Bureau of the Census, 1992). We jump to conclusions about a wide range of things in order to save time and simplify our thinking. If our experiences teach us that 99 percent of firefighters are male and that 98 percent of dental assistants are female (both true as of 1991 [U.S. Bureau of the Census, 1992]), it is merely efficient to fill in the likely gender for each job if not told otherwise.

I have opened with this riddle to show how subtle and pervasive gender assumptions can be. But by presenting the issue in the form of a riddle, I do not mean to diminish the seriousness of the topic. It is clear that relationships between women and men in America have been changing in the last few decades. The topic is discussed everywhere, with sharp differences voiced in print, on television, in our classrooms, and over the dinner table. The stakes are great. It is an issue that can involve all humans, potentially dividing us into equal-sized, and opposed, camps. Nothing should engage the attention of sociologists more immediately than a social change with the potential scale of this one. But why present such a topic now, at the end of the text? It is, after all, a substantive topic rather than a core concept such as those already covered.

This book has been devoted to showing you how the core concepts of the sociological perspective are defined and used in our field. One of the pitfalls of this concepts approach is that the material can become fragmented. That is, each of the concepts can seem distinct from the others, just as the operation of a hammer is distinct from the operation of a saw. These concepts, the tools of the sociologist's work, are like the tools of the carpenter; they can be used in combination to work on any of a range of tasks. To employ them effectively, we need to understand that they all have their uses and, in various combinations, can work together. In this chapter I want to demonstrate that all the concepts described in this book can be applied to the discussion of topics of great concern to our time, in this case, sex and gender in society. We are living at a time of rapid and wide-reaching social change in America. It is, of course, a difficult time. I remember being told that it is a curse to live "in interesting times." But it is also true that such times of change present sociology with the opportunity to do important work, just as the great, early sociologists, such as Émile Durkheim and Max Weber, studied the social changes of their time.

It would take too much space to apply all of the ideas covered in this text to the topic of sex and gender in society. But the main concepts from each chapter will be included here. Let me begin by defining a few terms.

Then we can move on to see just how many of the concepts in this book apply to the topic at hand.

Sex and Gender

In normal conversation, people typically use the terms *sex* and *gender* interchangeably. Sociologists, however, distinguish between them. (The words *male* and *female,* however, apply to both *sex* and *gender.*) **Sex** refers to the set of biological characteristics that distinguish females from males. These include a range of physical characteristics. At the core of sex differences is the chromosomal information that determines the sex of a child. The mother provides an X chromosome and the father either an X chromosome (producing a female child) or a Y (producing a male). **Primary sex characteristics** are the male versus female genitalia, which are used in the reproductive process. **Secondary sex characteristics** are the physical traits other than those of the reproductive organs, many of which are used to identify the sex of an individual in normal, public interaction. On average, males are taller and more heavily muscled and have more body and facial hair than women. By contrast, women develop wider hips than men, narrower shoulders, breasts, and a layer of fatty tissue throughout the body. But secondary sex characteristics are only categorical averages. They vary so greatly among individuals that there is room for overlap in the physical appearance of the sexes. There are, of course, many women who are taller or more heavily muscled than many men. In the Olympic Games, secondary, and even primary, sex characteristics have been rejected as adequate sex identification of athletes. Now contestants are subjected to genetic tests as proof of sexual identity.

Gender refers to the set of learned behaviors and beliefs that distinguish males from females. For example, the lessons about how to "be like a girl" that parents teach their female child create her gender. To the extent that she is taught to play with dolls, to pay attention to her hair style and put on makeup, to be passive, domestic, and concerned with pleasing others, she is likely to adopt the characteristics of the female role so well documented in American experience. For the male, the gender role includes learning to be assertive, aggressive, and independent, to play competitive sports (for example, football) rather than cooperative games (such as skipping rope). As with any cultural characteristics, these are generalizations. There is great variation in the extent to which members of each gender adopt the stereotypical characteristics of their category. However, the overall patterns are clear and persistent enough to influence the opportunities available to people because of their gender, and clear and persistent enough to draw the attention and anger of people who wish to change things.

Applying Sociological Concepts to Sex and Gender in Society

Thousands of sociologists have conducted studies of sex and gender in society. Just as no two woodworkers should be expected to fashion the same table or use the same tools in the same order, each sociologist has employed some of the concepts presented here to conduct their analyses, and each has combined them in unique ways. Some focus on differences in gender **roles** and others on **socialization** to those roles. Some are concerned with the unequal distribution of **power** and **authority** between the sexes or on the problem of **group cohesion** among women. In some cases the perspective of **conflict theory** or of **symbolic interactionism** dominates the research. In short, there are as many ways of using the concepts as there are analysts. Here, for example, are the titles of a few studies that exemplify some of the different ways researchers have used concepts found in this book to study sex and gender. (The concept from the text is highlighted in boldface type.)

1. "The **Power** War: Male Response to Power Loss Under Equality" (Kahn, 1984)
2. "Sex **Role Stereotyping** in the Language of the Deaf" (Jolly & O'Kelly, 1980)
3. *Labeling Women **Deviant:** Gender, Stigma and Social Control* (Schur, 1984)
4. "Women as **Generalized Other** and **Self** Theory: A Strategy for **Empirical Research**" (Deseran & Falk, 1982)
5. ***Family, Socialization,** and **Interaction** Process* (Parsons & Bales, 1955)
6. "**Cultural** Myths and Supports for Rape" (Burt, 1980)

My approach in this chapter is artificial in comparison with a real sociological analysis: I will merely go through the text from the first concept to the last to show that each has application to the topic of sex and gender in society. In fact, now that you have studied these core concepts, you should be able to anticipate how they would be used by sociologists to study the topic of sex and gender.

Concept 1: The Sociological Perspective

The **sociological perspective** on the topic of sex and gender in society is different from other perspectives. You would expect biologists, psychologists, economists, political scientists, and researchers in any field to study male and female differences with questions appropriate to their disciplines. Sociologists, however, are uniquely trained to focus on the nature of the bonds between people *as they are influenced by their sex or gender.* In Concept 1 of this book I used a metaphor to illustrate how sociology is concerned with social forces the way physics is concerned with gravity, a force that cannot be directly observed but that influences the behavior of objects in nature. Social forces are created by the fact that we live in the world with others, and these forces influence our behavior in ways just as concrete as the way a rock is influenced by gravity. In the case of studies of sex and gender, sociologists pay particular attention to the way behaviors are influenced by the fact that people differ by sex and gender. Here are a few examples.

The nature of the bond between individuals can be revealed in the way they communicate with one another. Studies of conversations between females and males have revealed that females tend to use more tag questions than males, often ending a statement with a qualifier such as "Don't you think?" or "Isn't it?" By contrast, men on average tend to dominate their conversations with women by interrupting, changing the topic, or speaking more rather than listening (Lakoff, 1975, 1990). Another illustration of the sociological perspective comes from the study of marital relationships: The division of labor and income in families has changed significantly in the last three decades as women have increasingly entered the labor force. In the early 1960s, about half of American marriages depended solely on the male's income. By 1988 American marriages depending solely on the male's income were down to 17%. Almost two-thirds of the American marriages in 1988 involved dual-earner couples (Wilkie, 1991). However, women's increased contributions to household income have not been matched by a decrease in responsibility for domestic work. Among dual-earner couples, women consistently are found to be more responsible than their mates for housework and childcare (Hochschild, 1989). The focus in these studies is on the nature of the relationship between people as it is influenced by the fact that they are males and females.

The sociology of sex and gender is just one example of the many ways sociologists study social bonds. Our behavior toward one another is also influenced by our social class membership (which in sociological study is called social stratification), by our age (social gerontology), by our race (sociology of race), by the type of community in which we live (urban or rural sociology), by the amount and type of education we receive (sociology of education), and so on. Look through the listing of courses in any department of sociology and you will find many different substantive areas of interest. What they have in common is that they all focus on the nature of social bonds.

Concept 2: Ideal Type, Model, and Paradigm

Male and *female* are **ideal types,** abstract definitions focusing on the typical characteristics of some real-world phenomenon, in this case the typical behavior of men and women we observe in the real world. The ideal type for *female* has included expectations for nurturing, passivity, affective (emotional) skills, domesticity, and supportiveness; the ideal type for *male* has included expectations for aggressiveness, achievement, competitiveness, and cognitive (reasoning) skills. I'm sure I don't have to point out how stereotyped these ideal types seem today. In the 1950s, however, they were relatively unremarkable descriptions of the genders. So was the ideal type for the nuclear family: a mother who stayed at home, a father who went to work as the sole earner, and one or more children. However, things have changed, and they continue to change. We can gauge the degree of change when we assess the difficulty people have in agreeing on any ideal type. When elements of a given social order are stable (for example, when there is widespread agreement among people about what male and female roles should be), ideal types are easier to establish.

Since things seem to be changing so fast in the area of gender in America, it might be useful to apply the concept of the **model** at this point. Models allow sociologists to speculate what the world would be like under specific conditions. For example, if social changes in gender behavior and expectations in America continue, it is reasonable to assume that the ideal types for *female* and *male* eventually will be very much more like one another. Numerous gender models currently are being debated in America. For example, what would the military be like if women were given the opportunity to serve in combat? What would be the eventual character of the American family if the current trend toward full-time female careers continues? And what would be the consequences for religious institutions if males and females had equal access to positions of leadership?

Underlying all the gender models and ideal types we create are the normally unstated assumptions we carry around about the nature of *male* and *female,* our gender **paradigms.** Given the depth at which these are held, they can be expected to resist change until long after more superficial changes in gender relations occur, if ever. Americans typically have viewed men as providers and women as nurturers. However, if Americans increasingly consider women as people who should work at careers outside the home, and men as people who should nurture their children, then our gender paradigm inevitably will change. In turn, ideal types and models for gender will change also.

Concept 3: The Scientific Perspective and Quantitative Research

In Concept 3 on the **scientific perspective** I stated that scientists try to measure the real world as objectively as possible. That is, they try to remove their personal biases from the process of discovering how things (and people) work. But how successful are they? Is it possible that scientists have characteristics that bias their work and that they are unable to control or eliminate? Recently it has been suggested that gender is just such a characteristic. There is no debate about the fact that scientific research is dominated by males: as of 1991, 74 percent of all natural scientists in America were male; and the federal government employed almost 88,000 persons considered "scientific personnel," only 18.6 percent of whom were female (U.S. Bureau of the Census, 1992).

Is the objectivity, and the accuracy, of social research influenced by the gender of the researchers? Clearly, social research would be demonstrably biased if male researchers were sexist in the ways they design, collect, and interpret research data. But it is just such individual biases that are supposed to be filtered out of the research process by careful adherence to the principles of science. So if the scientific method is followed, how could gender bias enter the research process?

Consider the way scientific inquiry works: All research begins with the selection of questions the researcher (or the research funding source) chooses to ask. Posted on the door of my office is a cartoon that makes this rather serious point. It shows two lab-coated scientists, one of them only knee-high to the other, who is of average height. The short one is pointing up at the other one and arguing with great energy. Off to the side two other scientists are looking at the scene, and one remarks to the other, "There goes Wilson, again. Making the case for his 'little bang' theory of the origins of the universe." It should come as no surprise, then, that some members of minority groups, such as black Americans who have experienced racial discrimination and Jews who have strong memories of the Nazi Holocaust, think it is vital to study the sources of prejudice and discrimination. Some victims of cancer can be expected to devote their careers to discovering its causes. And who is more likely than a manufacturer to want to learn what motivates workers? That is, people bring to the research process the experiences of their lives, and these lead them to choose certain research topics, and perspectives on them. So, from this perspective, one should expect different research questions from males than from females.

The questions we choose to ask grow inevitably from the lives we live, and it is no challenge to demonstrate that the lives of men and women differ in America. Consequently, if men conduct research, they will, even in the absence of bias or malice, overlook the perspective females bring to the research. For example, Lawrence Kohlberg (1966) developed his now-famous theory of the stages of moral development, based largely on his experiences as a male and on data he collected almost exclusively from male subjects. All the subjects answered the same, apparently objective, questions, that Kohlberg had devised for the research. Years later, Carol Gilligan (1982), using these same questions, found that young girls did not fit his theory very well. In fact, her results suggested that females failed to attain the moral development found in males of the same age. Specifically, Kohlberg had concluded that when confronted with a moral dilemma, (such as whether it is moral for a person to steal expensive medicine to help a sick family member), higher moral development was indicated by a willingness to subject oneself and others to neutral sets of rules, even if it meant doing damage to a relationship between individuals. So the male subjects were much more likely than females to say something like "Stealing is wrong, no matter what."

Gilligan, on the other hand, found that females often wanted to find some way to avoid or tailor such strict, neutral rules in the interest of maintaining relationships between individuals. So the female subjects were much more likely than males to say something like "Couldn't the person convince the druggist that the medicine was really important and he would be paid eventually?" It was not until Gilligan brought her (female) experiences to the research process that it was possible to measure moral reasoning another way, and then to propose that females develop a *different,* rather than an inferior, style of moral reasoning.

Concept 4: Symbolic Interaction and Qualitative Research

Symbolic interactionists point out that human social interaction occurs at the level of symbols, and that we must interpret the meaning of every interaction in which we take part. If this were not true of human social interaction there would be no need to distinguish between sex (the set of biological characteristics that distinguish females from males) and gender (the set of learned behaviors and beliefs that distinguish males from females). But we do attach meaning to behaviors that are learned and symbolic. Behaviors must be interpreted. If two people share a definition of the meaning of their interaction, they can then fulfill one another's expectations. The very nature of our social order is rooted in the definitions we share as a culture about the behavior appropriate for every social circumstance. So, for example, we share expectations about how teachers and students are supposed to act in classes versus in a hallway, and of how parents and children are expected to act at home versus in public. All cultures also establish and teach expectations for the behavior of women versus men at the level of symbols. Interactionists have spent a great deal of energy describing not only how this defining process works but how it perpetuates inequalities between the sexes.

Some of the best-known work in this area, and the easiest to observe in everyday interaction, has focused on the way language use (both verbal and nonverbal) differs between men and women. When parents tell their three-year-old daughter to talk or act "like a little lady," they are teaching her to present herself in ways that will fulfill the gender expectations of society. As a consequence of such training, men are consistently found to "talk rough," to interrupt other speakers (especially if they are female), and to change the subject more than women do. In nonverbal communication patterns, a man is more likely than a woman to intimidate physically the person with whom he is talking by striking informal postures such as leaning back in a chair or clasping his hands behind his head, which spreads the shoulders and elbows. When men do this they are reinforcing their role in society as the more assertive, achievement-oriented persons they were taught to be. By contrast, females who interact this way are likely to be criticized for being "aggressive," "pushy," or "masculine." Females who use male communication styles suffer these negative evaluations because they are expected to be more passive, nurturing, and gentle. Gender patterns in language reflect and reinforce the gender roles taught in a culture (Lakoff, 1975, 1990).

Because social interaction occurs at the level of symbols, control over language has the power among humans that muscle and sharp teeth have in the world of dogs and other animals. Because interaction occurs at the level of symbols, language can be a battleground for the control of one's fate in society. Understanding this should explain why members of minority groups struggle to control the language used to describe them. For example, feminists have good reason to wish to rid the language of words that automatically gender-type occupations (such as *fireman, businessman, chairman*) and to replace them with words that do not (*firefighter, entrepreneur, chair*). They also understand that when a man addresses a woman with apparently affectionate terms like "honey" and "sweetheart," the effect can be to demean the woman. This is especially true if the woman does not know the man well, since his use of these terms defines the situation as one in which he is free to be familiar, without her permission. It is something like the feeling you get when the dentist introduces himself (or herself) as, say, Dr. Benton but calls you by your first name without asking your permission. The clear signal is that you are less powerful in that setting. I have chosen to focus on language. However, there are countless other symbolic expressions of gender-appropriate behavior.

The research method of participant observation that is so closely associated with the perspective symbolic interaction has been used extensively in the study of gender relations in society. One example is the study Barrie Thorne (1986) conducted on the way girls and boys interact in elementary school. For a period of eight months Thorne observed children in a California elementary school. She found that the sex segregation they practiced was not uniform or total, but had a rhythm in which boys and girls sometimes chose to play together and also they chose to separate in certain circumstances. That is, decisions about the kinds of people these children chose to be with were specific to the situations in which they found themselves. For example, some sports activities were clearly sex-segregated, with boys in control of the play areas for team sports. By contrast, girls grouped together nearer the building for female-typical activities like jumping rope and hopscotch. However, in some sports that were not so clearly identified with either sex, such as kickball, Thorne found much more integration of the sexes. In the lunchrooms sex segregation was minimal among first graders, but almost total among sixth graders.

There were also subtler patterns of relations between the sexes. Thorne found, for example, that boundaries between the sexes were drawn and maintained by a number of social behaviors including the following: (1) cross-sex chasing, often in response to a taunt; (2) contests that were sometimes created by teachers ("let's have the boys against the girls"); (3) rituals of pollution in which members of either sex accused members of the other of some form of defect (such as "cooties"); or (4) invasions in which the violation of the space controlled in a sex-specific activity, such as baseball for boys, was intentionally violated by a child of the "wrong" sex (often leading to a chase). Participant observation allowed Thorne to discover and describe the specific situations in which sex segregation and interaction are practiced among children. Questionnaires or interviews would never have worked.

Concept 5: Function and Dysfunction

At the root of **functionalism** is a view of society as a huge and complex system with many interrelated parts that operates with some stability. How much stability is a matter of debate. There are plenty of disruptions in the operation of American society, such as divorce, poverty, and crime. For the moment, just allow for the fact that the stability of American society is demonstrated by our ability to predict with some confidence that we will be able to go to work or school tomorrow, that the government will deliver the services for which our taxes pay, that contracts and business deals will be negotiated and put into effect, and that we will be able to deal with one another in everyday interaction without too much friction. In short, within bounds, American society works. Though I am a ready critic of the shortcomings of our society, I also recognize that there are many societies which, by comparison, demonstrate the degree to which a society can exist in turmoil. In 1992, for example, we have (sadly) Somalia. Its people cannot count on the predictable operation of the law, government, transportation, commerce, or even civil interaction in its streets.

Functionalists define **functions** as the actions, ideas, and objects that contribute to the stability of a society at any level (for example, the psychological well-being of its members, and the stable operation of social systems such as the economy and government). In his early statement of the functionalist model, Talcott Parsons emphasized the search for the functions served by the persistent characteristics of American society. So in the 1950s he saw the American family as a small social system in which the different roles of male as provider and female as nurturer had complementary functions for the stability of the unit (Parsons and Bales, 1955). However, the days when the division of labor in American families was so clearly delineated by gender are gone. Nowadays, a functionalist perspective on gender would be hopelessly out of date if it made such an argument.

Today you are likely to hear functionalists explain gender differences in America in terms more suited to Robert Merton's extension of Parsons' form of functionalism. Recall from Concept 5 on functionalism that

Merton recognized that actions, ideas, and objects can have either **functions** (contributing to the stability of some system) or **dysfunctions** (diminishing the stability of some system). In addition, he described how functions and dysfunctions can be either **manifest** (intended and/or recognized), or **latent** (not intended and/or recognized). So a functionalist armed with Robert Merton's ideas would be more likely to suggest that gender differences are functional for the maintenance of advantages enjoyed by males in America but dysfunctional for females. From this point of view, the traditional division of labor in American families may be complementary, but it is not equal. Men who go out to earn a living for the family get paid salaries and control the money they earn, while their wives raise the families and care for the homes without direct pay for their work. They often must negotiate an income from their husbands. Another example of a Mertonian-style functionalist statement about American gender differences is the argument that teaching women to value looks and fashion has the manifest function of making women feel attractive and more appealing to men (at the level of personality) but the latent dysfunction of making them less able to compete with men for work (at the level of social structure). This is because when men and women learn to evaluate female worth in terms of how women look, women's capabilities become less important. Another example of this sort of functionalism is the contention that women's skills at nurturing may be manifestly functional for the child they raise and the husband they soothe after a hard day of work, but that those same skills have the latent dysfunction of making women less able to compete with males for power (Graves, 1965). In each case, the female quality is more functional for men than for women.

Concept 6: Conflict: Individual, Group, Class, and Societal

It is a short step from Merton's version of functionalism, with its acknowledgment that gender roles can have functions and dysfunctions, to conflict theory views of sex and gender in society. **Conflict theory** takes the position that people struggle for control over scarce resources, such as money, power, and status. The possible outcomes of that struggle are the dominance of one group (minority group) by another (majority group), or some kind of balance of forces between them in which neither group dominates the other absolutely and neither can gain further influence over the other. When groups are actively engaged in ongoing struggle with one another, we can speak of social conflict or, in the extreme, revolution. According to conflict theory, then, the relationship of women and men is that of the minority to the majority group. Gender inequalities in any society reflect the extent to which men dominate women in the struggle for valued resources. In short, conflict theorists argue that women are a minority group and that the form of discrimination that enforces the superior position of men in society is sexism.

Sexism is expressed at the levels of attitude (prejudice against women) and behavior (discrimination). Sexist attitudes are easy to find in the stereotypes of women as weak, dependent, illogical, emotional, and so on (Basow, 1992). **Stereotypes** often are negative versions of characteristics that, under other circumstances, might be seen as positive. For example, what a sexist would call "assertiveness" and "confidence" in a male would be labeled "offensive aggressiveness" in a female. You can find sexist stereotypes without looking very hard—in jokes about women (What characteristics are women assumed to have, and how are they ridiculed?) and in television advertisements (What products are aimed at female audiences, and what qualities of women are portrayed?). Notice the way language attaches gender to certain occupations (for instance, business*man*) or assumes a male voice unless specified otherwise ("Everyone wants *his* life to have meaning").

By contrast, evidence of sex discrimination is a bit less obvious (though not too much less). There are two types of evidence of sexist behavior: evidence of *acts,* and evidence of *outcomes.* Discriminatory acts, such as sexual harassment and firing (or failing to hire or promote) otherwise-qualified women, are often documented in legal cases. It is illegal to act against a person on the basis of her or his group membership (such as, religion, age, race, gender). This does not mean an employer has to hire or promote an incompetent woman merely because she is a woman. But it does mean that an employer who makes a personnel decision against a woman merely because she is a woman can be sued for sex discrimination. Cases like this are won in American courts every day, though it is necessary for the woman who brings the complaint to demonstrate that the employer's action was not based on an evaluation of her job-related merit. So, for example, a woman college professor who is denied promotion and whose evaluations and publications are all comparable with (or superior to) those males in her department who were promoted has a good chance of winning a sex discrimination case against her boss. As for sex discrimination in everyday interaction, law professor Anita Hill's nationally televised accusations of sexual harassment against Supreme Court nominee Judge Clarence Thomas appear to have accelerated the rate of such complaints in the workplace and given them increased legitimacy. And, of course, sexual harassment pales in comparison with violent and deadly forms of sex discrimination such as rape and physical assault. These behaviors are to sexism what lynchings are to racism.

The second type of evidence that there is sex discrimination in America is evidence of outcomes. That is, there is such persistent inequality in the distribution of opportunity between the sexes in America that it is commonly inferred (especially by conflict theorists) that discrimination must have occurred for such inequalities to exist. For example, as of 1990 the median income of all American women who worked full-time was $20,586, approximately 70 percent of the $29,172 earned by American men who worked full-time (U.S. Bureau of the Census, 1992). In addition, women are concentrated in the lower-salaried occupations. For example, in 1991, women accounted for less than 21 percent of all physicians in America, 11 percent of dentists, 18 percent of architects, and 19 percent of lawyers. However, they also accounted for more than 74 percent of all precollege teachers, 83 percent of all librarians, 99 percent of dental hygienists, 97 percent of all receptionists, and 99 percent of all secretaries (U.S. Bureau of the Census, 1992).

Although these facts about the gender inequalities of income and work in America are easily documented, there isn't agreement on how to account for such differences. Conflict theorists argue that they result from acts of discrimination against women, beginning with early training of girls that they should, for example, want to become nurses rather than doctors (with doctors earning, on average, between four and five times what nurses earn in a given year in the 1980s). Even when women held the same jobs as men, their earnings levels were significantly lower than men's. For example, in 1991 women working in managerial and professional positions earned approximately 70 percent of what men earned for the same work, and women in technical sales earned only 60 percent of what men earned for that work. The pattern continues in school: males are encouraged to study more cognitive and career-oriented areas, such as math and the sciences, while females are directed more toward domestic and human service fields, such as teaching, nursing, and social work. Ultimately, gender inequality in society is attributed to acts of discrimination in the home and in the workplace, and only some of them ever come to light. Conflict theory assumes that gender inequality in America is the result of the struggle between men and women for control over society's resources, and that men are winning.

Concept 7: Culture

Culture is a way of life that is learned and shared by human beings and is taught by one generation to the next. Anthropologists have reported that in every culture they have studied, males and females are taught different sorts of roles. In the overwhelming percentage of cases, males learn to play the dominant, aggressive roles, and there is no well-documented case of a culture in which females dominate (Giele and Smock, 1977). In some cultures, the degree of gender difference (or, if you prefer, inequality) is dramatic in comparison with the United States. The domination of males is apparent in poor, agricultural societies, such as India, Bangladesh, and Mexico. However, even rich countries, such as Kuwait, and highly modernized ones, such as Japan, teach strict expectations for female deference to males, though there are pressures for change building in each. In some cultures women are prevented from talking to men they do not know or from showing their faces in public (Iran). Often women are expected to show deference to men by always walking a few steps behind them (Korea). However, there are a number of instances in which cultures teach little sex-role differentiation. One example is the Israeli kibbutz, a communal settlement in which men and women are taught that it is equally appropriate for a male or a female to run farm machinery, cook and serve food, care for children, and carry a rifle on guard duty. To the extent that cultures differ in their pattern of gender roles, there is evidence that sex differences are cultural rather than biological.

The content of a culture includes: (1) the knowledge people have about how things work (empirical knowledge) and about the meaning of life (existential knowledge), (2) the values people hold about their highest moral goals, and (3) the forms of symbolic expression that represent their feelings, experiences, and tastes. In each of these areas, sex and gender play important parts. For example, our empirical knowledge includes beliefs about the differences between men and women. As late as the 1960s in America it was taught, almost uncritically, that women are best suited to the tasks of homemaking and raising a family, and men to the harsher challenges of making a living outside the home. While this is still commonly taught in America, lessons of gender similarity and equality are increasing. You can see evidence of this in Arland Thornton's study of changing attitudes toward family issues in the United States (1989), summarized in Concept 19 on social change.

Our cultural values, including beliefs in equality of opportunity, privacy, progress, and competition, have all been critical in the debate about gender equality in America. Any feminist speech or legal case against sex discrimination I have heard about makes arguments based on these values. They make the case for women's rights to equality of opportunity, for their ability to compete on the basis of merit, and for their desire for progress for themselves and society. As for the value of privacy, the 1973 Supreme Court decision in *Roe v. Wade* ruled that state restrictions on access to abortion were unconstitutional. That decision was based on a woman's constitutional right to privacy.

And the forms of symbolic expression that represent feelings, experiences, and tastes have always been strongly influenced by women. The fact that American culture is affected by the debate about sex and gender in society is evident in, for example, recent clothing, art, music, and films. Just think about how seriously Madonna's highly publicized work and image have been taken by everyone, from preteen fans to intellectual critics. Her costume, song lyrics, dance, and irreverent sexuality (and our reactions to them) are both cultural productions and symbols of our battle over the meaning of gender in America. Madonna is far from the only example of the symbolic expressions of gender in our culture; she's just one of the most interesting. Tamer versions are seen everywhere in daily life, such as in male and female images in advertisements for jeans, beer, and cars.

Concept 8: Norms: Folkways and Mores

Norms are widely accepted rules for social behavior that specify how to act in given situations. American norms for social behavior clearly distinguish between what is acceptable for males and what is for females. The norms that provide rules for everyday interaction, our **folkways,** generally teach women to dress, speak, and even move differently than men. These norms are visible expressions of the more deeply held values of the culture, which, in turn, differentiate male from female. So, if we teach that males should be the ones to fulfill our values for aggressiveness, competitiveness, and mastery of nature, then the folkways for male behavior will reflect these expectations. American males are, in fact, taught to speak and carry themselves in ways that are more aggressive than females. However, as you are no doubt aware, the differences between female and male folkways are changing. We have come a long way from the time when women were ridiculed for wearing pants, drinking liquor, or smoking cigarettes in public but men were not. The last twenty years in America have seen a clear weakening of folkways that differentiate females from males. However, some normative differences still persist; as a few hours observing the depiction of females and males on television programs and advertisements will demonstrate.

Mores are norms that are thought by group members to be so vital to their welfare and survival that typically they are written in law and their violation formally punished. Examples of mores include prohibitions against theft, assault, and murder; rights of assembly and free speech; and enforcement of such responsibilities as the care of one's children and the fulfillment of contracts. With the exception of laws dealing with physiological differences, such as the right to obtain an abortion, American mores generally are written to apply equally to females and males. However, there are some areas in which laws are written that treat the sexes differently. In some cases laws have been created to protect women, as in laws prohibiting sex discrimination in employment, harassment, or dangerous working conditions. Laws on rape in America specifically treat the sexes differently. In fact, *rape* normally is defined as a "sex-specific" crime, that is, "unlawful sexual intercourse with a female person without her consent" (Berger, 1977). But not all laws protect women; some laws have been written to their clear disadvantage. As of 1986 it was still the law in twenty-three states that a husband could not be prosecuted for raping his wife, because marriage was assumed to be a bond of consensus in which the sexual access of the male to the female could not be restricted by the state.

From the point of view of the sociologist, the mores that operate in society exist at both the manifest level (what they say) and the latent level (how they are actually applied). Even when laws seem nonsexist, they often are applied unequally. In rape cases, for example, courts often allow defenses that require a woman to demonstrate that how she behaved or the way she dressed did not encourage the sexual advances of the man she has accused. Also, though laws about child custody are not written to favor either sex, females have overwhelmingly been awarded custody by American courts. Lastly, women have the same difficulty making cases of discrimination as do any minority-group members. The employer who fires a woman or denies her promotion or employment simply because she is a female cannot be successfully prosecuted if there is no concrete evidence of an act of sex discrimination. That is, if the employer never specifically states that the personnel decision at issue was made on the basis of the employee's sex, or if there is not a clear set of criteria for personnel actions against which the facts of the case can be measured, there is little a complainant can do.

Concept 9: Socialization and the Development of Adult Character

Socialization is the process by which individuals internalize the rules of the social order in which they are reared and by which the character of the society is maintained from one generation to the next. Countless studies have documented the fact that females and males are socialized differently in America. The typical pattern is that boys and girls are given different kinds of toys to play with (dolls and tea sets for girls, sports equipment and cars for boys), clothes to wear (party dresses and jewelry for girls, coveralls and sneakers for boys), chores to do (domestic and caretaking for girls, maintenance and mechanical for boys), and stories to hear (family- and friendship-oriented for girls, adventure for boys).

As with all the other issues discussed so far, these patterns are changing in America, but the data still show that girls and boys are socialized differently (Basow, 1992). It makes sense that this will continue as long as gender differences are part of the culture. After all, socialization functions to reproduce the culture in which it occurs. The beliefs, values, and ways of life of adults are deeply held, and are passed on to their children with pride. Parents teach their children their way of life because they believe it will help the child to live among others in society. So if the members of a culture believe that women should be domestic, social, dependent, and as decorative as possible, that is what girls will be socialized to become. And if men are expected to be aggressive, independent, competitive, and goal-directed, that is what parents, teachers, and even peers will teach boys to be like. To the extent that we diminish the distinction between the roles males and females should play in America, gender socialization will reflect that fact.

The most commonly recognized agencies of socialization are the family, teachers, and peers. However, there are other, less apparent socializers, including the mass media and language. The images we see in our films, television programs, and advertisements and hear about in our music and other forms of mass communication provide us with information about what males and females should be like in America. These lessons are not as direct as the specific instructions from a mother or father that "girls don't sit that way" or that "this is a boy's toy." A child who grows up watching many hours of television every week is bound to internalize that physicians are males if no physician portrayed on television is ever a female. The lesson will be learned even if no person, real or televised, ever says that "doctors should always be boys." In fact, even if children hear from parents or teachers that "girls can be doctors, too," their beliefs about the world are still likely to be shaped by the cumulative evidence to the contrary that they see in the mass media.

The language children are taught has the same effect. If they learn the words *businessman, chairman,* and *policeman,* they are led to assume that these statuses are for males. Throughout this book I have tried to avoid gender-linked language. If you found it awkward to read words like *chairperson* or constructions like *she and he* (rather than simply *he* in cases of indefinite references), it is probably because you, like I, have been socialized to certain language forms. We prefer the language with which we are familiar, and we have difficulty changing it, even if motivated to do so.

Concept 10: Solidarity—Mechanical and Organic

Organic solidarity is the feeling of belonging that comes from group members' having complementary, or interdependent, roles. In terms of sex and gender, such roles might include division of labor in procreation, in the operation of the household, and in the workplace. While it may be a long time before there is any serious change in the interdependent roles women and men play in the creation of babies, it is clear that feminists, both female and male, are working hard to change the nature of the division of labor elsewhere in society.

By contrast, **mechanical solidarity** is the feeling of belonging that comes from group members' having similar values and ways of life. When men get together in all-male gatherings, such as fishing trips, they are expressing the mechanical solidarity of like-minded people. Women do this also, and are doing so in increasing numbers. One measure of the mechanical solidarity among women is the extent to which women have organized politically. The modern history of feminism is marked by the many groups it has generated. The National Women's Suffrage Association was formed to work for the right of women to vote in America (achieved in 1920). In the 1920s, the National Women's Party drafted an Equal Rights Amendment for women, which is still an item on the political agenda of many women's groups, including the largest of them, the National Organization for Women (NOW). There are also many smaller groups that work on narrow issues of interest to their members, including rape prevention and prosecution, women and the law, working conditions for women, women in specific professions, and domestic violence, to name just a few. In each case, the underlying theme of women's rights creates what can be called "the women's movement," or "feminism."

As with any large-scale social movement, there is no single voice of feminism. Rather, there are different theoretical and political movements within feminism, and debates about what they stand for. For example, *liberal feminists* are usually considered to be people who are working to achieve equality between the sexes. *Socialist feminists,* whose ideas are rooted in Marxism, contend that inequality between the sexes is the inevitable consequence of the fact that capitalists must create inequalities in order to control the cost of labor. *Radical feminists* generally endorse the equality goals of liberal feminists as a good beginning, but also seek separation and independence from the world of male influence. An additional form of radical feminism is the sexual separation sought by some lesbians. In sum, there are many subgroups within feminism, each with its own agenda and degree of internal solidarity, and there is an overall, broadly based movement that knits them together and helps legitimate their activities. A number of surveys have revealed significantly greater support for gender equality in America than was the case twenty years ago (Basow, 1992).

Concept 11: Power and Legitimacy

Power is the ability to control or influence the behavior of others, even against their will, and **legitimacy** is the agreement among people that the exercise of power in a given situation is appropriate. Societies differ in the extent to which men exercise power over women. However, men's domination of societies is essentially universal. In the United States, evidence for this is not tough to unearth. The federal legislature and all fifty state legislatures are overwhelmingly male. In 1977 there were no female senators in Washington. Between 1981 and 1992 there were two (out of 100), and as of 1993 there were six. The percentage tripled, but only to 6 percent. The proportions were not much more even in the House of Representatives. In 1977, 4 percent of all representatives were female. By 1992 it was 7 percent, and in 1993 the percentage jumped to 11 percent women. Thus, it is men who have written—and still write—the laws that distribute much of the opportunity and wealth in society. It was not until 1920, after a strenuous campaign, that American women won the right to vote. The U.S. Supreme Court, which interprets the meaning of our constitution (including, significantly, who will control much surrounding the reproductive process), is currently composed of eight men and one woman. Of the last 106 justices appointed to the Supreme Court, 105 have been male. Sandra Day O'Connor is the first female justice in our history. As of 1991, slightly more than 80 percent of all lawyers and judges in the United States were male (U.S. Bureau of the Census, 1992). And there has never been a female president or vice president of the

United States. So all three branches of the U.S. government are controlled by men.

If there is not much doubt that men have more power in American society than women, what about the legitimacy of that power? When people agree to subject themselves to the power of another, the power is legitimated, as when citizens of a town agree to permit police to ticket them for speeding or judges to send them to jail for assault. In American society male power is often legitimated by unwritten norms that enforce long-standing belief in the appropriateness of male power. Émile Durkheim called this *traditional authority,* in which power is conferred by custom and accepted practice. Though being male is not a formal, written qualification for the American presidency, a female has never even been nominated by one of the major political parties to run for the office. This is due to the power of males within the political structure who control the political process, and the cultural norm that the office should be held by a male. And this belief that males should have authority in a wide range of areas is not limited to males. For example, research was conducted in 1968 in which a sample of college women read and rated a number of professional articles for their persuasiveness, value, and so on. Half the women were told that the articles were written by a man, and half that they were written by a woman. Regardless of the field with which the article dealt, these women rated the articles more highly if they were thought to have been written by men (Goldberg, 1968).

But things seem to be changing somewhat in the legitimation of power. American women increasingly consider themselves as capable of doing work traditionally done by men. For example, one study surveyed a sample of 294 young women between 1970 (when they were high school students) and 1981. During that decade the women's attitudes shifted dramatically away from traditional gender-role ideas. By 1981, the majority rejected the notions that (1) education and work are more important for men, and (2) that women should marry, stay home to raise children, and leave the major decisions to their husband (Tallichet and Willits, 1986). The same pattern of women's increasing belief in equality of gender roles is also seen in the data summarized in the Illustration section for Concept 19 (Thornton, 1989). It is increasingly common to see female justices, executives, and laborers in our films, television programs, and real lives. But as much as attitudes have changed among American women, women do not control the process of legitimation; men still do. And at current rates of change it will take a long time before it is accepted without comment for a physician, a mechanic, or the president of the United States to be female.

Concept 12: Ideology

Earlier in this chapter, in the section on the concepts of *ideal type* and *model,* I mentioned that sociologists speculate about a range of possible gender relations in our future. One of these is the consequence of gender equality for our religious institutions in, for example, ordination of priests and ministers. **Ideologies** express the beliefs of a group of people about how the way the world works, and they are often used to justify that group's actions in pursuing its own interests. The ideology of sexual differences typically is expressed by the belief that women and men are inherently different in their capacities and tastes: males are believed to be naturally aggressive, analytical, and goal-oriented, whereas females are believed to be naturally passive, nurturant and relationship-oriented. In societies in which this ideology is widespread, the roles assigned to each gender are determined by the ideology. Ideologies also exist in specific systems within societies. For example, gender ideologies in some religions teach that God is male, wants religious leaders to be male, or specifies domestic tasks for women. For such believers, their religious ideology would limit women to specific roles in society and within the religious structure. Feminists, who actively promote female equality, have been attacking such ideologies in the attempt to create change.

Underlying any ideology are the data and theories that justify their beliefs. Given the thousands of years that males have been in the positions of power within the church and academic worlds, it should be no surprise that the male ideology is well stocked with powerful sets of self-justifying data and theories. Let me cite just a few. More than 2,000 years ago, at the dawn of our written intellectual history, the Greek philosopher Aristotle explained the inferior position of women in society as the consequence of their natural inferiority. Around the turn of the twentieth century, the influential Austrian physician Sigmund Freud proposed that women should never be allowed to hold positions of authority (see the preceding discussion of the concept of power and legitimacy), such as judge and political leader. He explained in his lecture "Femininity" that in order to mature fully, a child needs to be powerfully motivated to defer gratification and to accept the prohibitions and authority of society (represented by the child's parents). In boys, Freud contended, this motivation is the deep, subconscious fear of castration by the father for the child's sexual desires for the mother. However, Freud concluded that because women are born without a penis, they cannot develop fear of castration. So, lacking motivation to accept societal restrictions when very young, he contends, women can never fully mature. Their psychological and emotional lives are supposed to be dominated by penis envy, in particular, and envy and jealousy of others generally. This is what Freud meant by "Anatomy is destiny." Freud claimed that women's domesticity and lack of power and authority in society are the inevitable consequences of their physical characteristics at birth (Freud, 1949).

However odd you may think Freud's reasoning (it is largely out of favor today), it was taught as truth for the first half of the twentieth century, formed part of the basis of psychoanalysis as it was practiced in Europe and America, and was adopted as part of the male ideology that justified the continued domination of women by men. (For another example of data that have been used to justify the ideology of male superiority, look back at S. J. Gould's (1981) work on intelligence testing in the Illustration section for Concept 3.) In combination with other "facts" generated by males, beliefs about the nature of male and female capacities and talents (such as that, by nature, males are hunter-providers and females domestic caretakers) continue to form the foundation of gender ideologies in America.

Concept 13: Social Status and Social Role

A **status** is a position within a social structure, such as the mother in a family or a student at a college. It may be inborn, that is, assigned by others (an **ascribed status**), or achieved as a consequence of an individual's actions (**achieved status**). The status of female is ascribed; the status of student is achieved. A **role** is the behavior expected of a person in a given status. Most of the sociological literature in this area focuses on gender roles. Sociologists have studied how gender roles are learned via socialization, how roles differ between men and women generally and in specific situations, and how gender roles change. I can only touch on a small part of this discussion here.

Women and men fulfill gender-role expectations in two ways. First, there are generalized expectations for the behavior of males versus females as categories of people. Thus, women historically have been expected to be supportive, passive, nurturant, domestic, and concerned with relationships. By contrast, the general expectations for males emphasize qualities such as assertiveness, independence, competitiveness, and achievement outside the home. These general expectations apply to the behavior of men and women no matter what role they are playing. So a woman who is a school teacher generally would be expected to act in a more nurturing and supportive way toward students than a male in the same role. He, by contrast, would generally be expected to fulfill the male role, including characteristics such as assertiveness and authority. It is important to keep in mind that patterns of behavior discovered and described by sociologists, no matter how consistent, are not without exception. Even in the 1950s, when the expectations for female behavior were much more widely shared and enforced than they are today, there were women who were assertive, independent, competitive, and ambitious in their careers outside the home. And this is certainly the case today. But however much the generalized expectations for male and female role behavior have changed, the patterns just described remain visible.

The second area of interest in gender roles is the range of specific social roles that women and men are led to play in society. For example, women are still expected to play the family roles of mother and wife rather than career roles outside the home. Women are also more likely to become nurses than doctors, secretaries than executives, legal aides than lawyers, elementary school teachers than college professors, social workers than engineers, airline cabin attendants than pilots, hostesses and waiters than chefs, and part-time workers than full-time workers. Often, explanations for this clear pattern associate the roles women and men play with the more general gender-role expectations already listed. That is, women are assumed to be led into domestic and supportive roles *because* it is their general gender role to be nurturant, supportive, and concerned with relationships. By comparison, the tendency of males to occupy the authority roles in society is seen as a natural consequence of male assertiveness, competitiveness, and independence. Much of the sociological study of gender roles has examined how these cultural beliefs developed, the extent to which they can be explained by socialization versus inborn tendencies, and the extent to which such expectations have been changing in the last few decades. Anyone who has not noticed the change in the roles played by females in America has not been looking.

One interesting topic of research has been the problem faced by women who add new career roles to their traditional female roles of wife and mother. The "supermom" syndrome has raised the concern that women who try to fulfill too many demanding roles simultaneously may be damaging their own health, ruining their enjoyment of each role, and diminishing the likelihood of succeeding at them. Apparently, even when husbands go along with their wife's desire to hold a job outside the home, they usually have not accepted their resultant increased responsibility for care of the home and the children (Gunter and Gunter, 1990; Hochschild, 1989).

Concept 14: Group: From Small Group to Society

A **group** is a number of people who share some common interests, interact with one another, accept the rights and obligations of membership, and share a sense of identity and belonging. By contrast, a **category** is a number of people who have some characteristics in common but need not interact. According to these definitions, then, women are clearly a category but not as clearly a group. For example, women do not all interact with one another; it would be physically impossible. But it is also clear that women do get together in groups to share common interests. In the past such interests might have grown out of the roles into which women were channeled in America, such as garden clubs, parents' groups, and church and neighborhood associations. However, at least since the creation of women's groups that fought for the right to vote, women increasingly have developed groups focused on common interests that grow directly out of the experiences of women *as women*. In short, the women's movement has created women's groups that fit all the sociological criteria of a group.

Take, for example, the National Organization for Women. The common *goal* is the advancement of the interests of women. The *interaction* includes meetings and activities in which the programs of the organization are planned and/or executed. The *rights of membership* include access to meetings, events, and interaction with other members. The *obligations* focus on the responsibility to work for the interests of women and to express the beliefs of the organization. And the *shared identity* of the members is what has come to be called "feminist consciousness." Consciousness-raising is a formally established component of the women's movement, with scheduled meetings at which women discuss problems such as gender stereotyping, sex discrimination, and sexual harassment and aims such as assertiveness, independence, and mutual solidarity (Freeman, 1989).

Of course, like women, men have had, and continue to have, groups in which they meet not so much *as men* but as a consequence of the roles that men have played in society. Thus male groups form around business, sports, outdoors activities, and so on. Ironically, the rise of feminism and its widespread organizational efforts led to the development of some specifically male consciousness-raising groups. One example is the meetings of men associated with the ideas expressed in the book *Iron John* by the poet Robert Bly (1990). In it Bly argued that men's development often is stunted by their rigid gender-role socialization, and that men's inner strength, and trust of one another, can be released by acknowledging the "wildness" inherent in masculinity. Gender groups clearly exist in America, and seem to be on the increase.

Concept 15: Formal Organization and Bureaucracy

Formal organizations are organized to achieve specific goals by coordinated, collective effort; **bureaucracies** are types of formal organizations featuring hierarchy of authority and expertise, division of labor, and explicit rules of performance for their members. Though bureaucracies have developed a reputation for inefficiency and harmful impersonality, originally they were created to organize human work efficiently, largely by rewarding the most productive work and knowledge. That is, the design of bureaucracy was rational.

You might expect this rational design for bureaucracy to benefit women in such a way as to make it more possible for women to compete with men. After all, a system of rewards based on merit should eliminate discrimination in the workplace. Bureaucracies are designed to reward managers who can arrange the division of labor so that the work of the organizations is done most efficiently. This is supposed to be accomplished by breaking down the overall production into small subtasks that are easy to learn and perform. If these rational rules really reward merit and create jobs for which it is easy to train any worker, women should benefit to the extent that such rules are focused on performance. There would be no opportunity for biases to enter the system. However, it turns out, in reality, that jobs in bureaucracies reproduce the gender inequalities seen elsewhere in society, with supervisors and heads of bureaucracy overwhelmingly male, and lower-paid clerks, secretaries, and line workers overwhelmingly female. Why?

The main reason for this perpetuation of gender inequality can be found in the irrationality underlying bureaucracy. As was discussed in Concept 15, the seeds of bureaucracy's inefficiencies lie in the fact that humans cannot be manipulated as one manipulates the inanimate parts of a machine. We have loyalties, preferences (often including intense likes and dislikes), and the desire to be recognized for the individuals we are. And these characteristics are not left at the door when we enter our jobs in some bureaucratic organization. So a number of things happen to influence the opportunities of women, even within the bureaucracy. For example, beliefs about gender-related capacities are brought into the bureaucracy intact and used to determine merit. That is, if a manager (usually male, remember) is trying to decide who would be best for a certain job within the bureaucracy, he may use his personal biases to satisfy a bureaucratic goal. "We need a receptionist, and everyone knows that women are more pleasant to strangers." This occurs in the same way that informal systems of reward replace the rational system intended by bureaucratic designs. People cover for the inefficiencies of friends, or punish competent enemies. In the same way, men are promoted and rewarded by men in positions of power within the bureaucracy, and women's advance is blocked (Kanter, 1977).

Concept 16: Institutions

The position of women in American institutions has been discussed throughout this book; most often in the context of the institution of the family. **Institutions** are sets of rules and relationships that establish how group members agree to accomplish universal issues, those tasks that a people must perform if they are to survive. The family is a critical institution in American society because it is charged with accomplishing so many universal functions, including caring for, socializing, and much of the job of educating the young; procreating; providing legitimate sexual access between adults; many of the consumption functions and some of the production functions of the economy; providing answers to many of the existential questions with which religion is also concerned; and providing shelter. However much American families have changed (with our ideal type for the institution of the family changing in response), the family is still closely associated with the influence of women. This is because the traditional role of female has been defined primarily in terms of childrearing and care for the home. So it should be no surprise that women are disproportionately represented in parent-school organizations and community safety or health organizations, and in professions related to the family, such as social welfare and primary and secondary education.

However, American women are very differently represented in institutions charged with accomplishing other universal functions. For example, women are underrepresented at higher levels of power in the political system (such as office holders) and overrepresented in lower-level support positions (such as clerks, aides, and secretaries). In religious institutions, women accounted for only 9.3 percent of all clergy in 1991 (U.S. Bureau of the Census, 1992), and they are prohibited by some religions from church leadership positions. In educational institutions as of 1991, women filled 75 percent of teaching positions below the college and university levels (the years when students are still living at home) but only 41 percent of the teaching positions in higher education (U.S. Bureau of the Census, 1992). And in health care institutions as of 1991, women represented 20 percent of all physicians and 10 percent of all dentists, but they made up 95 percent of all nurses and 98 percent of dental assistants (U.S. Bureau of the Census, 1992). These clear gender differences exist because our institutions are structural expressions of the underlying values and beliefs of our culture. All cultures provide for the rules and relationships by which they will accomplish universal functions in ways that make sense to them. If women are thought to be best suited for work with children, and men for work that wields power, then that is the way the institutions of that society will look—and that's exactly how our institutions look.

Concept 17: Stratification and Social Class

The study of **stratification** focuses on systems used by societies for the distribution of access to, and possession of, the things their members value. Much of the data presented in this chapter documents gender inequality in America. Every finding that working women in America earn less than men (even when they do the same jobs), hold positions of lower prestige, or lack the power to control their own destinies is evidence of gender stratification. In fact, as of 1990 the median income of all American women who worked full-time was $20,586, approximately 70 percent of the $29,172 earned by American men who worked full-time (U.S. Bureau of the Census, 1992). It is also true that men dominate the positions with the highest prestige and the greatest power in American society. Just look back over the data in this chapter and you will find plenty of evidence of gender stratification.

None of this is news. Still, it is alarming to note that despite advances in gender equality in America, there is also evidence that gender inequality is increasing in certain ways. Consider the phenomenon called the "feminization of poverty." Poverty in America increasingly has a female face. There are more and more female-headed families depending on only one income for support, an income that often is inadequate, because, on average, women earn less than men in America. How serious is this?

In 1990 approximately 45 percent of all families with children that were headed by single mothers lived below the poverty line, while only 7 percent of those with both parents present were officially poor (U.S. Bureau of the Census, 1992). Also in 1990, the median income for all family households headed by a married couple was almost $40,000 per year. The median yearly income of family households headed by a male when the wife was absent was $31,552. However, the median yearly income of family households headed by a female when the husband was absent was only $18,069 (U.S. Bureau of the Census, 1992). The problem has worsened as the percentage of households headed by women increased by 57 percent in the last twenty years (from 21 percent in 1970 to 33 percent in 1990), and the divorce rate increased by 35 percent (from 3.5 divorces per thousand population in 1970 to 4.7 per thousand in 1988) (U.S. Bureau of the Census, 1992). So, while some gains toward gender equality are being made in such areas as legal protections against sexual harassment and employment opportunities in the professions, forces are also acting to maintain social class differences between the sexes in America.

Concept 18: Discrimination and Prejudice

Discrimination is negative treatment of people on the basis of their group membership; **prejudice** is negative beliefs and feelings about a group of people that serve to justify and reinforce discrimination against them. Women in America are the targets of both prejudice and discrimination, which means they fit the sociological definition of a minority group.

Although women actually outnumber men in America (in 1991, 51.3 percent of Americans were female), this is not important sociologically because we measure minority status in terms of social power rather than by how many people fit the category. In simplest terms, a minority group is a number of people who lack the power to determine their own destiny, while the majority group has that power plus the power to control the destinies of others. So the 15 percent of South Africa that was white and operated the system of racial segregation of apartheid (now officially ended, but still largely in effect there) was the majority group, since they controlled the economy, the police, the land, and all of the country's most valued resources.

The definition of *minority group* that I used in Concept 18 on discrimination and prejudice was somewhat more detailed. Louis Wirth defined a *minority group* as a "group of people who, because of their physical or cultural characteristics are singled out from others in the society in which they live for differential and unequal treatment, and who therefore regard themselves as objects of collective discrimination" (Wirth, 1945:347). So a minority group is defined by three characteristics: (1) members are identifiable, either physically or culturally; (2) members are treated differently than others, meaning they are the targets of discrimination and prejudice; and (3) members see themselves as a group of people who are treated differently. Do American women fit these criteria as well as other groups that sociologists agree constitute a minority group?

(1) *Women are as identifiable as the members of any group.* That is, in most cases you can tell a male physically from a female just by looking, though this is not always the case. Remember that ideal types in sociology (*minority group* is an example) are abstract definitions focusing on the typical characteristic of some real-world phenomenon. The characteristics of a minority group can be expected to be present generally, though not always.

As for cultural identifiability, women can be said to have learned characteristics in common that identify them as different from men. Earlier I discussed such characteristic differences as language (both verbal and nonverbal), leisure-time activities, and clothing preferences. Once again, as with physical identifiability, characteristics that make women culturally identifiable are generally, though not always, present.

(2) *Women are the targets of discrimination and prejudice.* Earlier in this chapter I discussed the evidence of discrimination in the section on conflict, and there is a good deal more evidence elsewhere, in the data on inequality. So I won't spend more time on it here. Instead, let's consider the issue of prejudice. **Prejudice** exists at the level of attitude. It consists of negative beliefs and feelings about a group of people, and is expressed in the form of stereotypes. Sexist stereotypes are everywhere in America. The image of women presented in advertising is a case in point. Whenever manufacturers try to sell anything, from cars to perfume, they present women as sex objects. More obviously negative is the belief in America that women are emotional rather than analytical, are best suited for domestic tasks, and are in need of the protection of men. What is significant about such stereotypes is that, to the extent they are believed, they are used to justify the treatment of women. Thus, if women are seen as sex objects, they are more likely to be sexually harassed or raped; and if they are believed to be emotional and best suited for domestic tasks, they are less likely to be hired to do traditionally male work outside the home.

(3) *Women see themselves as members of a group that is targeted by discrimination and prejudice,* though there is some debate about the extent of this. Just after World War II, American women did not see themselves as members of a minority group (Hacker, 1951). However, there is reason to believe that female group consciousness is increasing. Things have changed a great deal in the nearly fifty years since World War II. The rise of the women's movement in America has been strongest since the 1960s. The National Organization for Women was founded in 1966, and numerous other groups followed, most focusing on such issues of special interest to their members as abortion rights, employment opportunities, and day care. If it is still not the case that most women define themselves as minority-group members, it may be because minority-group membership carries with it such a negative connotation. It is a

goal of feminism, through efforts called "consciousness raising," to convince women that their second-class citizenship is a result of sexism, not their own supposed inferiority.

Concept 19: Social Change

Social change is defined as change in the beliefs and values that the people in a society hold and in the way they agree to act toward one another. Though this chapter is not specifically about social change, a great deal of it presents data demonstrating that Americans are changing their beliefs, values, and behaviors surrounding sex and gender. The very existence of a women's movement with which most Americans are at least familiar if not always sympathetic is evidence of this fact. In 1981 America's first female Supreme Court justice was appointed, and in 1984 the first woman was nominated by a major political party to run for the office of vice president of the United States. At a broader level, just in the years between 1983 and 1991 women made the following gains in employment (U.S. Bureau of the Census, 1992):

1. A 13 percent increase in the percentage of women in managerial and professional occupations, from 40.9 percent in 1983 to 46.3 percent in 1991
2. A 35 percent increase in the percentage of woman architects, from 12.7 percent in 1983 to 17.1 percent in 1991
3. A 41 percent increase in the percentage of woman engineers, from 5.8 percent in 1983 to 8.2 percent in 1991
4. A 27 percent increase in the percentage of woman physicians, from 15.8 percent in 1983 to 20.1 percent in 1991
5. A 31 percent increase in the percentage of woman lawyers, from 15.3 percent in 1983 to 19 percent in 1991

These figures clearly reveal that things are changing in the gender distributions in American occupations. But there are also some signs that the changes are not as great as they may appear. For one thing, the greatest rates of change are in the occupations with the lowest percentage of women. For example, the 41 percent increase in the percentage of female engineers only brings women up to 8.2 percent of the occupation. (Remember that women account for 51.3 percent of the U.S. population.) Second, the proportion of women in lower-level occupations remained basically unchanged over the same period. For example, the percentage of women remained essentially unchanged in each of the following occupations: secretaries (99 percent in both 1983 and 1991), teachers' aides (93 percent), dental hygienists (99 percent), and bank tellers (90 percent) (U.S. Bureau of the Census, 1992).

Changes in opportunities for women are partly influenced by the efforts of feminists, but they are also influenced by larger changes in society. Between 1960 and 1990 the Consumer Price Index, a measure of the average change in prices over time in a fixed "market basket" of goods and services paid for by consumers more than quadrupled (U.S. Bureau of the Census, 1992: 465). Changing economic conditions, such as inflation and other increases in the cost of living, have forced an increasing number of American women to work in order to maintain the standard of living they, or their parents, had enjoyed before. What is most interesting is that most of the increase in labor force participation was among married women. Single women have always worked in America. In 1960, 58.6 percent of single women were in the labor force, and by 1991 the figure had increased by about 13 percent to 66.5 percent. In 1960, by contrast, only 31.9 percent of all married American women were in the labor force. By 1991, however, this had increased by 83 percent to 58.5 percent.

These are just a few of the changes that have been occurring in the United States that influence, and are influenced by, women. In looking at changes in sex and gender in America it is impossible to separate that which is strictly part of sex and gender from larger, related changes. For example, the average family size has dropped by 21 percent, from 3.33 persons in 1960 to 2.63 persons in 1991. Included in these figures are women (often in agreement with husbands, I assume) who have chosen not to have children. To what extent are decisions to have fewer, or no, children the consequence of gender/feminist beliefs, and to what extent the consequence of other factors, such as the desire to save on expenses or even the availability of birth control? I hope it is apparent by now that it is an oversimplification merely to say that social change is complex.

Concept 20: Deviance

Deviance is behavior that does not conform to society's norms for expected behavior. There is a great deal of data to show that there are clear differences in the rate at which males and females commit, and are victimized by, the formally identified acts of deviance in America. But before I cite some of those data, I want to present a very broad theory of deviance as it applies to sex and gender.

By the definition of deviance just given, it is possible to see that all women in America are labeled as deviant. Here is how this happens: The values of American culture stress qualities such as independence, competitiveness, and achievement, all of which are associated with our norms for male behavior. Since women are not socialized to have these characteristics, they can be seen as deviant. Or, as it is sometimes said in everyday terms, "It's a man's world." It should be no surprise that typically male characteristics are central to our cultural values. Males have been in the positions of power in those institutions with the most power to shape values, such as religion, government, and the media. The consequence is that male behavior is normative and female behavior is deviant.

Edwin Schur has described the process in detail in his *Labeling Women Deviant: Gender, Stigma and Social Control* (1984). Schur notes that when women work or play roles normally held by men, they often are described with hyphenated labels. For example, whereas a man would be called simply a doctor, an engineer, or a mechanic, a woman doing any of those jobs would likely be called a woman-doctor, woman-engineer, or (worse yet) a lady-mechanic. The reason for the hyphenated label is to identify the situation as unusual, or deviant. The same thing occurs whenever a person takes on a role out of the normal range of expectations for them. For example, a male-nurse or a child-actor is clearly being labeled as special by these terms. The child who is an actor is likely not stigmatized by the label, but the male-nurse probably is. Howard Becker (1963) has suggested that when people are labeled as deviant, that particular status comes to dominate all their other statuses. For example, once someone is identified as an exconvict, this becomes a *master status,* and his or her other statuses—as parent, employee, or homeowner—fade in relative importance and are interpreted through the filter of the master status. By this perspective, woman is a master status. Whenever a woman plays a role outside the world prescribed for female roles, she is likely to be labeled with a hyphenated term to announce her deviance.

Even if women can be defined as deviant at the broadest cultural level, the data for crime in America clearly identify the male as deviant. For example, males are arrested for all crimes at four times the rate of females. In 1990, 81.6 percent of arrests for all crimes were of males. Males accounted for 78.1 percent of arrests for all serious crimes, which included murder and nonnegligent manslaughter (89.6 percent of these arrests were of men), robbery (91.7 percent), burglary (91.2 percent), aggravated assault (86.7 percent), larceny (68 percent), motor vehicle theft (90 percent), arson (87 percent), and forcible rape (98.9 percent) (U.S. Bureau of the Census, 1992). Men wind up in state prisons at better than twenty times the rate women do: in 1986, 95.6 percent of all state prison inmates were male (U.S. Bureau of the Census, 1992). Men are also disproportionately the victims of homicide in America: as of 1989 just over 20,000 men were the victims of homicide, four times the number of women who were (U.S. Bureau of the Census, 1992).

Women are, of course, the victims of rape. As noted earlier, *rape* actually is defined in gender terms: "unlawful sexual intercourse with a female person without her consent" (Berger, 1977). And the data show that rape and attempted rape have more than doubled in America between 1970 (18.7 rapes and attempted rapes per 100,000 population) and 1990 (41.2 per 100,000) (U.S. Bureau of the Census, 1992). There is debate about whether these figures reflect real increases in the crime or greater willingness among women to report the crime and greater attention to data collection by criminal justice personnel. But whatever the sources of the increase, it is clear that the problem is receiving increased attention in the nation.

Concept 21: Anomie and Alienation

In discussing this last concept in relation to sex and gender in society I would like to do something a little different from what I have done for the previous twenty. I hope by now that you can see that every one of the concepts in the text have application to sex and gender in American society. In each case it was easy to find data that clearly differentiated females from males in America, and/or sociologists who have studied these issues using the specific concept under discussion. For the concepts of anomie and alienation I would like simply to provide some of the questions about how anomie and alienation might be applied to the topic of sex and gender, and allow you to speculate what the data and social research might say. My guess is that by this point you have become more conversant with the sociological perspective and vocabulary than you suspect, and certainly more than when you started the course. So here are my questions.

Anomie is a condition of social ambiguity in which an individual does not know how to act because the rules for behavior in the society are either unclear, entirely absent, or in some way unsatisfactory. Do Americans today experience any ambiguity or dissatisfaction about gender-role expectations? What evidence of this can you see in your everyday life?*

What are the events that might create anomie about gender-role norms in Americans generally, or might create differing rates of anomie in females versus males? If there is gender-role ambiguity or dissatisfaction, who do you think experiences more uncertainty about how to act in social situations, men or women? How serious do you think gender-role normlessness in America is? Do you think it differs in different places? What kinds of places would you guess it to be more serious, and why?

Alienation is a condition of social ambiguity in which an individual has lost the meaning of his or her participation in social roles. How is it possible to be alienated from one's gender role? Why would it happen, and would it be more likely to happen among men or among women? What would be the signs of gender-role alienation? Almost all of my discussion of alienation was in terms of alienation from labor. Does this have any bearing on the possibility that levels of alienation might

*By the way, I used suicide as an example of a consequence of anomie. However, anomie can influence people's behavior without actually causing them to commit suicide. But if you are interested, suicide rates for males and females in America are easy to find in *Statistical Abstract of the United States* in the table on "Deaths by Selected Cause" in the chapter called "Vital Statistics."

differ between women and men in America? Why? What have you personally seen that would illustrate these differences?

A Final Note on Doing Sociology Without Being a Sociologist

We sociologists have available a large number of concepts to enable us to examine human social behavior. As with any skill, each of us becomes more familiar and able with some of these tools than with others. Specialization is almost inevitable, given the vast literature in our field. There simply is not enough time to read it all, and keeping up to date in more than just a few areas is impossible. An introductory course in sociology is like an extended sampling of our field. If you read sociological research, or take more specialized courses, such as the sociology of sex and gender, you will certainly see that the entire repertoire of sociological concepts is never brought to bear by one person. The best work combines them creatively, and with care, to generate insights, to support hypotheses and theories with evidence, and to communicate these to a wider audience beyond our journals and classrooms. I hope you have become interested enough by what you have learned about sociology to use sociological concepts.

In the introduction to this book I described a kind of surgery I performed when I was nine years old: I pulled apart my grandfather's pocketwatch with a knife and a tiny screwdriver. I was curious, and it pleased me, if not my parents. Perhaps you are curious and it will please you to take up a few of the sociological concepts you liked and to try dissecting some social behavior.

References

Adler, Patricia A., P. Adler, and J. M. Johnson, eds. 1992. *Journal of Contemporary Ethnography* 21(1).

Alger, Horatio. 1974. *Cast upon the Breakers.* New York: Doubleday.

Arendt, Hannah. [1963] 1977. *Eichmann in Jerusalem: A Report of the Banality of Evil.* New York: Penguin.

Arluke, Arnold. 1991. "Going into the Closet with Science: Information Control Among Animal Experimenters." *Journal of Contemporary Ethnography* 20: 306–30.

Bandura, Albert, and Richard H. Walters. 1959. *Adolescent Aggression.* New York: Ronald Press.

Banfield, Edward C. 1968. *The Unheavenly City Revisited.* Boston: Little, Brown.

Barron, Milton L. 1953. "Minority Group Characteristics of the Aged in American Society." *Journal of Gerontology* 8: 477–82.

Basow, Susan A. 1992. *Gender: Stereotypes and Roles.* Pacific Grove, Calif.: Brooks/ Cole.

Becker, Howard. 1963. *Outsiders: Studies in the Sociology of Deviance.* New York: Free Press.

Bem, Daryl. 1970. *Beliefs, Attitudes, and Human Affairs.* Pacific Grove, Calif.: Brooks/Cole.

Berger, Peter L. 1971. "Sociology and Freedom." *American Sociologist* 6 (February): 1–5.

Berger, Peter L., and Thomas Luckmann. 1967. *The Social Construction of Reality.* Garden City, N.Y.: Doubleday.

Berger, V. 1977. "Man's Trial and Woman's Tribulation: Rape Cases in the Courtroom." *77 Columbia Law Review* 1:7–10.

Bernard, Jessie. 1972. *The Future of Marriage.* New York: Bantam Books.

Black, Cyril E. 1976. *Comparative Modernization.* New York: Free Press.

Blau, Peter M., and W. Richard Scott. 1962. *Formal Organizations.* San Francisco: Chandler.

Blauner, Robert. 1964. *Alienation and Freedom.* Chicago: University of Chicago Press.

Blumer, Herbert. 1969. *Symbolic Interactionism: Perspective and Method.* Englewood Cliffs, N.J.: Prentice-Hall.

Bly, R. 1990. *Iron John: A Book About Men.* Reading, Mass.: Addison-Wesley.

Boelen, M. W. A. 1992. "Street Corner Society: Cornerville Revisited." *Journal of Contemporary Ethnography* 21 (April): 11–51.

Bogardus, Emory. 1959. *Social Distance.* Yellow Springs, Ohio: Antioch Press.

Bonacich, Edna. 1972. "A Theory of Ethnic Antagonism: The Split-Labor Market." *American Sociological Review* 37:547–59.

Bowles, Samuel, and Herbert Gintis. 1982. "The Crisis of Liberal Democratic Capitalism: The Case of the United States." *Politics and Society* 11:51–93.

Burgess, Ernest W. 1960. *Aging in Western Societies.* Chicago: University of Chicago Press.

Burt, M. R. 1980. "Cultural Myths and Supports for Rape." *Journal of Personality and Social Psychology* 38:217–30.

Butler, Robert. 1975. *Why Survive? Being Old in America.* New York: Harper & Row.

Chandler, Robert. 1972. *Public Opinion.* New York: R. R. Bowker.

Cole, Stephen. 1980. *The Sociological Method.* 3d ed. Chicago: Rand McNally.

Comte, Auguste. [1848] 1957. *A General View of Positivism.* New York: Speller.

Cooley, Charles Horton. [1902] 1964. *Human Nature and the Social Order.* New York: Schocken Books.

———. 1909. *Social Organization.* New York: Charles Scribner's.

Coser, Lewis. 1956. *The Functions of Social Conflict.* New York: Free Press.

Cumming, Elaine. 1963. "Further Thoughts on the Theory of Disengagement." *International Social Science Journal* 15:377–93.

Cumming, Elaine, and W. E. Henry. 1961. *Growing Old: The Process of Disengagement.* New York: Basic Books.

Dahrendorf, Ralf. 1959. *Class and Class Conflict in Industrial Society.* Stanford, Calif.: Stanford University Press.

Davis, Kingsley, and Wilbert Moore. 1945. "Some Principles of Stratification." *American Sociological Review* 10:242–49.

Davis, Nancy J., and Robert V. Robinson. 1988. "Class Identification of Men and Women in the 1970s and 1980s." *American Sociological Review* 53: 103–12.

Deseran, F. A., and W. W. Falk. 1982. "Women as Generalized Other and Self Theory: A Strategy for Empirical Research." *Sex Roles* 8:283–97.

Dollard, John, Leonard W. Doob, Neal E. Miller, O. H. Mowrer, and Robert R. Sears. 1939. *Frustration and Aggression.* New Haven, Conn.: Yale University Press.

Domhoff, William G. 1967. *Who Rules America?* Englewood Cliffs, N.J.: Prentice-Hall.

Durkheim, Émile. [1893] 1958. *The Rules of Sociological Method.* New York: Free Press.

————. [1893] 1960. *The Division of Labor in Society,* trans. G. Simpson. New York: Free Press.

————. [1897] 1951. *Suicide: A Study in Sociology,* trans. J. A. Spaulding and G. Simpson. New York: Free Press.

Eitzen, Stanley. 1980. *Social Problems.* Boston: Allyn & Bacon.

Ellison, Christopher G. 1991. "An Eye for an Eye? A Note on the Southern Subculture of Violence Thesis." *Social Forces* 69:1223–39.

Epstein, H. T. 1978. "Growth Spurts During Brain Development: Implications for Educational Policy and Practice." In *Education and the Brain,* ed. J. S. Chall and A. F. Mirsky, 343–70. 77th Yearbook, National Society for the Study of Education. Chicago: University of Chicago Press.

Etzioni, Amitai. 1975. *A Comparative Analysis of Complex Organizations.* New York: Free Press.

Festinger, Leon. 1954. "A Theory of Social Comparison Processes." *Human Relations* 7:117–40.

Freeman, J. 1989. "Feminist Organization and Activities from Suffrage to Women's Liberation." In *Women: A Feminist Perspective* (4th ed.), ed. J. Freeman, 541–55. Mountain View, Calif.: Mayfield.

Freud, Sigmund. [1909] 1957. "The Origin and Development of Psychoanalysis." In *A General Selection from the Works of Sigmund Freud,* ed. John Rickman. Garden City, N.Y.: Doubleday.

————. 1923. *The Ego and the Id.* Vol. 29 of *The Standard Edition of the Complete Psychological Works of Sigmund Freud.* London: Hogarth Press.

————. 1930. *Civilization and Its Discontents,* trans. Joan Riviere. Garden City, N.Y.: Doubleday.

————. 1949. *New Introductory Lectures on Psychoanalysis.* London: Hogarth Press.

Fromm, Erich. 1941. *Escape from Freedom.* New York: Avon Books.

Gans, Herbert J. 1972. "The Positive Functions of Poverty." *American Journal of Sociology* 78 (September): 275–89.

Garfinkel, Harold. 1967. *Studies in Ethnomethodology.* Englewood Cliffs, N.J.: Prentice-Hall.

Gerbner, George, Larry Gross, Michael Morgan, and Nancy Signorielli. 1986. *Television's Mean World: Violence Profile No. 14–15.* Philadelphia: Annenberg School of Communications, University of Pennsylvania.

Giele, J. Z., and A. C. Smock (eds.). 1977. *Women: Roles and Status in Eight Countries.* New York: Wiley.

Gilligan, Carol. 1982. *In a Different Voice: Psychological Theory and Women's Development.* Cambridge, Mass.: Harvard University Press.

Glenn, Norval, and Michael Supancic. 1984. "The Social and Demographic Correlates of Divorce and Separation in the United States: An Update and Reconsideration." *Journal of Marriage and the Family* 46:563–75.

Glock, Charles Y., ed. 1973. *Religion in Sociological Perspective.* Belmont, Calif.: Wadsworth.

Goffman, Erving. 1959. *The Presentation of Self in Everyday Life.* Garden City, N.Y.: Doubleday.

———. 1961a. *Asylums: Essays on the Social Situation of Mental Patients and Other Inmates.* Chicago: Aldine.

———. 1961b. *Encounters.* Indianapolis: Bobbs-Merrill.

———. 1963. *Stigma: Notes on the Management of a Spoiled Identity.* Englewood Cliffs, N.J.: Prentice-Hall.

———. 1971. *Relations in Public.* New York: Basic Books.

Goldberg, P. 1968. "Are Women Prejudiced Against Other Women?" *Transaction* 5:28–30.

Golding, William. 1954. *Lord of the Flies.* New York: G. P. Putnam's.

Goode, W. J. 1960. "A Theory of Role Strain." *American Sociological Review* 25 (August): 483–96.

Gould, Stephen J. 1981. *The Mismeasure of Man.* New York: Norton.

Gouldner, Alvin W. 1970. *The Coming Crisis of Western Sociology.* New York: Basic Books.

Gracey, Harry L. 1968. "Learning the Student Role: Kindergarten as Academic Boot Camp," in *Readings in Introductory Sociology,* ed. Dennis H. Wrong and Harry L. Gracey. New York: Macmillan.

Granfield, Robert. 1991. "Making It by Faking It: Working-Class Students in an Elite Academic Environment." *Journal of Contemporary Ethnography* 20: 331–51.

Graves, Robert. 1965. *Mammon and the Black Goddess.* Garden City, N.Y.: Doubleday.

Gunter, N. C., and B. G. Gunter. 1990. "Domestic Division of Labor Among Working Couples: Does Androgyny Make a Difference?" *Sex Roles* 14:355–70.

Hacker, H. M. 1951. "Women as a Minority Group." *Social Forces* 30:60–9.

Haller, John S., Jr. 1971. *Outcasts from Evolution: Scientific Attitudes of Racial Inferiority, 1859–1900.* Urbana: University of Illinois Press.

Harris, Louis, et al. 1975. *The Myth and Reality of Aging in America.* Washington, D.C.: National Council on the Aging.

Henslin, James M. 1975. *Introducing Sociology.* New York: Free Press.

Hochschild, A. 1989. *The Second Shift: Working Parents and the Revolution at Home.* New York: Viking.

Hodge, Robert, Paul Siegel, and Peter Rossi. 1964. "Occupational Prestige in the United States: 1925–1963." *American Journal of Sociology* 60:286–302.

Holmes, David S. 1972. "Aggression, Displacement and Guilt." *Journal of Personality and Social Psychology* 21 (March): 296–301.

Hyman, Herbert H., and Eleanor Singer. 1968. *Readings in Reference Group Theory and Research.* New York: Free Press.

Johnson, Don. 1984. *The Importance of Visible Scars.* Green Harbor, Mass.: Wampeter Press.

Jolly, E. J., and C. G. O'Kelly. 1980. "Sex Role Stereotyping in the Language of the Deaf." *Sex Roles* 6:85–92.

Kahn, A. 1984. "The Power War: Male Response to Power Loss Under Equality." *Psychology of Women Quarterly* 8:234–47.

Kanter, R. M. 1977. *Men and Women of the Corporation.* New York: Basic Books.

Kluckhohn, Clyde. 1967. *To the Foot of the Rainbow.* Glorieta, N.M.: Rio Grande Press.

Kluckhohn, Florence, and Fred L. Strodtbeck. 1961. *Variations in Value Orientations.* New York: Harper & Row.

Knowles, Louis L., and Kenneth Prewitt, eds. 1969. *Institutional Racism in America.* Englewood Cliffs, N.J.: Prentice-Hall.

Kohlberg, L. A. 1966. "A Cognitive-Developmental Analysis of Children's Sex-Role Concepts and Attitudes." In *The Development of Sex Differences,* ed. E. E. Maccoby, 82–173. Palo Alto, Calif.: Stanford University Press.

Kuhn, Thomas. 1970. *The Structure of Scientific Revolutions.* Chicago: University of Chicago Press.

Lakoff, R. 1975. *Language and Woman's Place.* New York: Harper & Row.

————. 1990. *Taking Power: The Politics of Language in Our Lives.* New York: Basic Books.

Lemert, Edwin. 1951. *Social Pathology.* New York: McGraw-Hill.

Lemon, B. W., K. L. Bengtson, and J. A. Peterson. 1972. "An Exploration of the Activity Theory of Aging: Activity Types and Life Satisfaction Among In-Movers to a Retirement Community." *Journal of Gerontology* 27:511–23.

Lenski, Gerhard. 1966. *Power and Privilege.* New York: McGraw-Hill.

Lenski, Gerhard, and Jean Lenski. 1978. *Human Societies: An Introduction to Macrosociology.* New York: McGraw-Hill.

Levin, Jack, and William C. Levin. 1982. *The Functions of Discrimination and Prejudice.* New York: Harper & Row.

Levin, William C. 1988. "Age Stereotyping: College Student Evaluations." *Research on Aging* 10:134–48.

Lévi-Strauss, Claude. 1963. *Totemism.* Boston: Beacon Press.

Lewis, Oscar. 1968. "The Culture of Poverty." In *On Understanding Poverty,* ed. Daniel P. Moynihan, 187–200. New York: Basic Books.

Linton, Ralph. 1936. *The Study of Man.* New York: Appleton-Century-Crofts.

Lipsky, Michael. 1980. *Street Level Bureaucracy.* New York: Russell Sage Foundation.

Lombroso, Cesare. 1911. *Crime: Its Causes and Remedies.* Boston: Little, Brown.

Longino, C. F., and C. Kart. 1982. "Explicating Activity Theory: A Formal Replication." *Journal of Gerontology* 36:713–22.

Mannheim, Karl. 1936. *Ideology and Utopia,* trans. Louis Wirth and Edward Shils. New York: Harcourt Brace Jovanovich.

Mao Tse-tung. 1963. *The Political Thought of Mao Tse-tung,* ed. R. R. Schram. New York: Praeger.

Marx, Karl. [1844] 1964. *The Economic and Philosophical Manuscripts of 1844.* New York: International Publishers.

————. [1867] 1967. *Capital: A Critique of Political Economy,* ed. Friedrich Engels. New York: New World.

————. 1964. *Karl Marx: Early Writings,* ed. and trans. T. B. Bottomore. New York: McGraw-Hill.

Marx, Karl, and Friedrich Engels. [1848] 1964. *The Communist Manifesto.* New York: Appleton-Century-Crofts.

Mead, G. H. 1934. *Mind, Self and Society,* ed. C. W. Morris. Chicago: University of Chicago Press.

Merton, Robert K. [1949] 1968. *Social Theory and Social Structure.* 2d ed. New York: Free Press.

————. 1976. "Discrimination and the American Creed." In *Sociological Ambivalence and Other Essays,* 189–216. New York: Free Press.

Milgram, Stanley. 1974. *Obedience to Authority: An Experimental View.* New York: Harper & Row.

Miller, Neal E., and Richard Bugelski. 1948. "Minor Studies of Aggression: The Influence of Frustrations Imposed by the In-Group on Attitudes Expressed Toward Out-Groups." *Journal of Psychology* 25:437–42.

Miller, Walter. 1958. "Lower Class Culture as a Generating Milieu of Gang Delinquency." *Journal of Sociological Issues,* 14:5–19.

Mills, C. Wright. 1956. *The Power Elite.* New York: Oxford University Press.

————. 1959. *The Sociological Imagination.* London: Oxford University Press.

Moore, Wilbert. 1963. *Social Change.* Englewood Cliffs, N.J.: Prentice-Hall.

Morgan, Lewis H. 1877. *Ancient Society.* New York: Holt, Rinehart & Winston.

Murdock, George P. 1949. *Social Structure.* New York: Macmillan.

Myrdal, Gunnar. 1944. *An American Dilemma.* New York: Harper & Row.

Nakao, Keiko, and J. Treas. 1990. "Computing 1989 Occupational Prestige Scores." *General Social Survey Methodological Report Number 70.* Chicago, Ill.: National Opinion Research Center.

Nisbet, Robert. 1966. *The Sociological Tradition.* New York: Basic Books.

O'Kelly, Charlotte G., and Linda E. Bloomquist. 1979. "Women and Blacks on TV." In *Mass Media and Society,* ed. Alan Wells. Palo Alto, Calif.: Mayfield.

Ondaatje, Michael. 1979. *There's a Trick with a Knife That I'm Learning to Do: Poems 1963–1978.* New York: W. W. Norton.

Orshansky, R. 1978. "Poverty Among America's Aged." Testimony presented before the U.S. Senate Select Committee on Aging, 95th Cong., 2d sess. August 9, 1978. Publication no. 95–154.

Orum, Anthony M. 1978. *Introduction to Political Sociology: The Social Anatomy of the Body Politic.* Englewood Cliffs, N.J.: Prentice-Hall.

Parelius, Ann P., and Robert J. Parelius. 1978. *The Sociology of Education.* Englewood Cliffs, N.J.: Prentice-Hall.

Parsons, Talcott. 1937. *The Structure of Social Action.* New York: McGraw-Hill.

————. 1954. *Essays in Sociological Theory.* New York: Free Press.

————. 1966. *Societies: Evolutionary and Comparative Perspectives.* Englewood Cliffs, N.J.: Prentice-Hall.

————. 1971. *The System of Modern Societies.* Englewood Cliffs, N.J.: Prentice-Hall.

Parsons, Talcott, and R. F. Bales. 1955. *Family Socialization and Interaction Process.* New York: Free Press.

Parsons, Talcott, and E. A. Shils. 1951. *Toward a General Theory of Action.* Cambridge, Mass.: Harvard University Press.

Parsons, Talcott, Edward Shils, Kaspar D. Naegele, and Jesse R. Pitts, eds. 1961. *Theories of Society.* New York: Free Press.

Peter, L. F., and R. Hull. 1969. *The Peter Principle.* New York: William Morrow.

Prindle, David F. 1988. "Labor Ideology in the Screen Actors Guild." *Social Science Quarterly* 69:675–86.

Reid, S. T. 1979. *Crime and Criminology.* 2d ed. New York: Holt, Rinehart & Winston.

Riesman, David. 1961. *The Lonely Crowd.* New Haven, Conn.: Yale University Press.

Robinson, Edward Arlington. [1897] 1978. "Richard Cory." In *An Introduction to Poetry,* 4th ed., ed. X. J. Kennedy, 104. Boston: Little, Brown.

Rubin, Beth. 1986. "Class Struggle American Style: Unions, Strikes and Wages." *American Sociological Review* 51:618–31.

Ryan, William. 1971. *Blaming the Victim.* New York: Vintage Books.

Schulz, James H. 1982. "Inflation's Challenge to Aged Income Security." *Gerontologist* 22:115–16.

Schur, Edwin. 1984. *Labeling Women Deviant: Gender, Stigma and Social Control.* Philadelphia: Temple University Press.

Seashore, Stanley E. 1954. *Group Cohesiveness in the Industrial Work Group.* Ann Arbor: Survey Research Center, Institute for Social Research, University of Michigan.

Shakespeare, William. 1975. *The Complete Works of William Shakespeare.* New York: Avenel Books.

Sheldon, William. 1949. *Varieties of Delinquent Youth.* New York: Harper & Row.

Simmel, Georg. [1908] 1955. *Conflict and the Web of Group Affiliation,* trans. Kurt Wolff. New York: Free Press.

———. [1908] 1964. *The Sociology of Georg Simmel,* ed. and trans. Kurt Wolff. New York: Free Press.

Smelser, Neil J. 1975. *The Sociology of Economic Life.* 2d ed. Englewood Cliffs. N.J.: Prentice-Hall.

Sorokin, Pitirim A. 1937. *Social and Cultural Dynamics.* New York: American Books.

South, Scott J., and Glenna Spitze. 1986. "Determinants of Divorce over the Marital Life Course." *American Sociological Review* 51: 583–90.

Spengler, Oswald. 1932. *The Decline of the West.* New York: Knopf.

Srole, Leo. 1956. "Social Integration and Certain Corollaries: An Exploratory Study." *American Sociological Review* 21: 709–16.

Stouffer, Samuel A. 1949. *The American Soldier: Studies in Social Psychology in World War II.* Princeton, N.J.: Princeton University Press.

Straus, Murray A. 1991. "Discipline and Deviance: Physical Punishment of Children and Violence and Other Crime in Adulthood." *Social Problems* 38: 133–54.

Strauss, Helen May. 1968. "Reference Group and Social Comparison Processes Among the Totally Blind." In *Readings in Reference Group Theory and Research,* ed. Herbert H. Hyman and Eleanor Singer, 222–37. New York: Free Press.

Sumner, William G. 1906. *Folkways: A Study of the Sociological Importance of Usages, Manners, Customs, Mores, and Morals.* Boston: Ginn.

Sutherland, E. 1924. *Criminology.* New York: Lippincott.

Tallichet, S. E., and F. K. Willits. 1986. "Gender-Role Attitude Change of Young Women: Influential Factors from a Panel Study." *Social Psychology Quarterly* 49: 219–27.

Thomas, W. I., and D. S. Thomas. 1928. *The Child in America.* New York: Knopf.

Thorne, Barrie. 1986. "Girls and Boys Together . . . But Mostly Apart." In W. W. Hartup and Z. Rubin (Eds.) *Relationships and Development* (pp. 108–122). Hillsdale, N.J.: Lawrence Erlbaum Associates.

Thornton, Arland. 1989. "Changing Attitudes Toward Family Issues in the United States." *Journal of Marriage and the Family* 51: 873–93.

Toennies, Ferdinand. [1887] 1957. *Community and Society,* trans. Charles P. Loomis. East Lansing, Mich.: Michigan State University Press.

Toffler, Alvin. 1970. *Future Shock.* New York: Random House.

Turnbull, Colin. 1962. *The Forest People.* New York: Simon & Schuster.

U.S. Bureau of the Census. 1992. *Statistical Abstract of the United States.* Washington, D.C.: U.S. Government Printing Office.

Walters, Ray. 1983. "The Coming of the Computer." *New York Times,* Book Review Section, July 24, 12–13.

Weber, Max. [1905] 1958. *The Protestant Ethic and the Spirit of Capitalism,* trans. Talcott Parsons. New York: Charles Scribner's.

————. [1925] 1946. From *Max Weber: Essays in Sociology,* ed. and trans. Hans H. Gerth and C. Wright Mills. New York: Oxford University Press.

————. [1925] 1964. *The Theory of Social and Economic Organization,* trans. Talcott Parsons. New York: Free Press.

Westie, Frank R. 1964. "Race and Ethnic Relations." In *Handbook of Modern Sociology,* ed. R. E. L. Faris. Skokie, Ill.: Rand McNally.

Whyte, William F. [1943] 1981. *Street-Corner Society: The Social Structure of an Italian Slum.* 3d ed. Chicago: University of Chicago Press.

Whyte, William H. 1956. *The Organization Man.* Garden City, N.Y.: Doubleday.

Wilkie, J. R. 1991. "The Decline in Men's Labor Force Participation and Income and the Changing Structure of Family Economic Support." *Journal of Marriage and the Family* 53: 111–22.

Williams, Robin. 1970. *American Society: A Sociological Interpretation.* 3d ed. New York: Knopf.

Wilson, E. O. 1975. *Sociobiology: The New Synthesis.* Cambridge, Mass.: Belknap Press.

Wilson, William. 1978. *The Declining Significance of Race.* Chicago: University of Chicago Press.

Wirth, Louis. 1945. "The Problem of Minority Groups." In *The Science of Man in the World Crisis,* ed. Ralph Linton, 347–72. New York: Columbia University Press.

Witkin, Herman A., et al. 1976. "Criminality in XYY and XXY Men." *Science* 193 (August): 547–55.

Glossary

Achieved status A social status that a person occupies as a result of what he or she has done.

Aggregate A number of people who are in the same place but need have nothing else in common.

Alienation A condition of social ambiguity in which an individual has lost or no longer believes in the meaning of his or her participation in social roles.

Anomie A condition of social ambiguity in which an individual does not know how to act because the rules for behavior (norms) are either unclear, entirely absent, or in some way unsatisfactory.

Anomie theory of deviance The belief that deviance is a response to disparities between social norms for goals and the means for achieving them.

Anticipatory socialization The process of taking on the expectations for behavior in a new role before actual occupancy of the role.

Ascribed status A social status into which a person is born.

Association (correlation) The tendency of the values of two or more variables to change together in a consistent pattern.

Authority Legitimated power.

Balance of forces The form of social stability that results when neither of two contending forces can improve its position with respect to the other.

Biological theory of deviance The theory that some people are born with a physiological predisposition to deviance.

Blaming the victim The tendency to attribute a social problem to the characteristics of the people who are its principal victims.

Bureaucracy A type of formal organization featuring extensive hierarchy of authority and expertise, division of labor, and explicit rules of performance for its members.

Category A number of people who have some characteristic in common but need not be in contact at all.

Causal hypothesis A guess not only that variables change together but that a change in the value of one variable actually causes a change in the value of another variable.

Central tendency The extent to which data points share some common characteristics; statistical measures include the mean, the median, and the mode.

Coding The process in research in which information is placed into clearly defined categories of meaning.

Cohesion (solidarity) The ability of a group to hold together in spite of obstacles.

Comparative reference group A group used by an individual for evaluating his or her performance.

Concept An abstraction drawn from observed events.

Conflict Any condition of disagreement between individuals, groups, or categories of people that can be expressed in attitudes or behaviors.

Conflict theory The sociological view of conflict as a source of both social order and social change.

Conflict theory of deviance The theory that the definition of deviance is a tool used by conflicting social forces (either individuals or groups) as they contend for the resources they value.

Conflict theory of social change The theory that social change is the result of conflicts between individuals and groups as they contend for the valued resources of the society.

Conflict view of stratification The theory that social inequality is the consequence of a struggle between contending forces for the things that are valued in a society.

Consensus Strong agreement among the members of a group about how to think or act.

Content analysis The systematic, quantitative analysis of textual (written) or pictorial data.

Correlation *See* Association.

Correlational descriptive hypothesis A guess about the simultaneous distributions of two or more variables.

Counterculture A category of subculture, consisting of groups within the larger culture whose beliefs, values, and styles of life are not only different in some respects from those of the larger culture but are also in conflict with them.

Cross-sectional study A study comparing measurements made on two or more simultaneously existing groups.

Cultural relativism The belief that no culture is superior to any other in providing the particular knowledge, values, and forms of symbolic expression adaptive for its own people.

Culture A way of life that is learned and shared by human beings and is taught by one generation to the next.

Cyclical theory of social change The theory that social change can proceed in more than one direction, repeating in a back-and-forth or a cycled pattern.

Dependent variable The variable in a causal relationship whose value is assumed to be a consequence, or effect, of the value of the independent variable.

Descriptive hypothesis *See* Correlational descriptive hypothesis; Simple descriptive hypothesis

Descriptive statistics Techniques for the nongraphic description of data at the interval level of measurement, such as the mean, the median, range, and mean deviation.

Deviance Behavior that does not conform to society's norms for expected behavior.

Discrimination Negative treatment of people on the basis of their group membership rather than their individual qualities.

Distribution A pattern of consistent variations in the qualities being observed by scientists.

Dominance The form of social stability that results when one contending force wins and imposes its will on the other.

Dramaturgy An approach to social analysis in which individual social behavior is seen as analogous to an actor's presentation of self on stage.

Dysfunction (negative function) A consequence that diminishes the stability of a system.

Ego The conscious self, which regulates the id.

Elaboration The process of examining a relationship between two variables under each level of a third variable in order to determine if the original relationship is spurious.

Empirical knowledge The accumulated information that a culture shares about how the world is constructed and how it operates.

Empiricism The process of testing hypotheses against observable events in the real world.

Ethnocentrism The belief that the way of life within one's own culture is superior to that of any other culture.

Ethnomethodology A method of studying social behavior that focuses on the way interacting individuals create the meanings of the specific interaction.

Evolutionary theory of social change A linear theory proposing that societies develop, as do biological systems, through a process of natural selection.

Existential knowledge The body of ideas that focus on questions of existence and its meaning.

Experimental research A method of research designed to study causal hypotheses. The researcher manipulates one or more independent variables to measure their effect on the value of a dependent variable.

Folkway A norm that provides guidelines for everyday social interactions; violation of folkways is punished informally.

Formal organization A group that is organized to achieve some specific goal or goals by coordinated, collective effort.

Frequency polygon A technique for the graphic description of data at the interval level of measurement.

Frequency table A technique for the nongraphic description of data at the nominal or ordinal level of measurement.

Function (positive function) A consequence that aids the stability or adjustment of a system.

Functional importance The degree to which an element of a system contributes to its stability.

Functionalism An approach to social analysis in which society is seen as a complex system of interrelated parts, each of which has consequences for the operation of the system.

Game In G. H. Mead's theory of socialization, the stage of development in which children take on the roles of a group of people simultaneously.

Gender The set of learned behaviors and beliefs that distinguish males from females.

Generalized other G. H. Mead's term for the group or community standards for social behavior that a child adopts in the final stage of socialization.

Group A number of people who share some common interests, interact with one another, accept the rights and obligations of membership (including rules for behavior) within the group, and share a sense of identity and belonging with others in the group.

Histogram (or bar graph) A technique for the graphic description of data at the nominal or ordinal level of measurement.

Hypothesis A guess about the nature of the measurable world.

"I" The aspect of the self that is spontaneous, creative, impulsive, and individualistic.

Id The reservoir of instinctual desires.

Ideal type An abstract definition of some phenomenon in the real world, focusing on its typical characteristics.

Ideology A set of ideas that explains how the world operates and is used to justify a group's actions in pursuing its own interests.

Independent variable The variable in a relationship that is assumed to cause changes in the dependent variable. It must occur first in such a relationship and so is also called the *precedent variable.*

Instinct An automatic, inborn instruction (or set of instructions) for behavior; passed from generation to generation genetically.

Institution A set of rules that establishes how group members agree to accomplish universal issues of survival, such as procreation, socialization, and care of the young or distribution of power.

Institutional discrimination Preferential treatment of one group over another that is built into the rules and structures of a society.

Internalization The adoption of social norms by individuals.

Interval measurement The level of measurement in which information is divided into equal intervals that reflect the exact degree to which a given characteristic is present.

Labeling theory of deviance The theory that powerful people apply the label of deviant to others by defining acceptable and unacceptable behavior in the society.

Latent function or dysfunction A consequence that is generally unrecognized, unintended, or both.

Learned deviance theory The theory that deviance is caused by learning subcultural norms that differ from those of the larger culture.

Legitimacy The agreement among people that the exercise of power in a given situation is appropriate.

Life-chance The opportunity a person gets to apply her or his talents and energies.

Linear theory of social change The theory that social change proceeds in a uniform, recognizable direction.

Longitudinal study A study comparing measurements made on one group at two or more different times.

Looking-glass self C. H. Cooley's term for the process whereby an individual develops self-identity by internalizing the reactions of others.

Majority Those in a position to distribute rewards in society.

Manifest function or dysfunction A consequence that is generally recognized or intended.

"Me" The aspect of the self that accomplishes conformity to societal expectations.

Meaningful behavior The typically human pattern of response to the environment, in which learned guidelines are used to evaluate the meaning of a stimulus before deciding on any reaction to it.

Measures of central tendency *See* Central tendency.

Measures of variability *See* variability.

Mechanical solidarity Durkheim's term for the type of group cohesion that results from members' having similar values and ways of life.

Meritocracy The system of success based on merit alone.

Mind G. H. Mead's term for the human ability to symbolically evaluate reality and possible courses of action.

Minority A group that has been singled out from others in society for differential and unequal treatment, on the basis of physical or cultural differences, and whose members therefore regard themselves as objects of collective discrimination.

Model A mental picture of the real world designed to promote speculation about what effect specific conditions might have on the social world as we understand it.

Moderating variable A variable whose potential influence in a causal relationship must be controlled.

Modernization The process by which traditional societies become more complex and industrial, including all the social changes that accompany that process.

More A norm thought by a group to be vital to its welfare and survival. The violation of mores is therefore severely punished.

Motivation The provision of positive sanctions for engaging in socially approved behavior.

Negative function *See* Dysfunction.

Negative sanction Punishment.

Net aggregate of consequences The balance of functions and dysfunctions for a specific action, idea, or object.

Nominal measurement The level of measurement in which information is named, that is, categorized into mutually exclusive groupings between which there are no distinctions of degree.

Norm A widely accepted rule for social behavior that specifies how to act in a given situation.

Normative reference group A group used by an individual for evaluating his or her beliefs.

Operational definitions Specifications for how a concept will be quantitatively measured.

Ordinal measurement The level of measurement in which information is ranked, that is, ordered into categories in terms of the extent to which a given characteristic is present, but with no precise knowledge of the distinctions of degree.

Organic solidarity Durkheim's term for the type of group cohesion that results from members' having the complementary, or interdependent, roles.

Panel study *See* Longitudinal study.

Paradigm A collection of normally unstated assumptions that shape the ideal types and models we create to describe and explain human social behavior.

Participant observation A research technique in which an investigator tries to develop and test hypotheses by direct participation in and observation of a group's activities.

Particular other G. H. Mead's term for the specific individual whose behavior is imitated by children and then later used as the model for learning the meaning of social behavior.

Peer group A group of people who share some important characteristic(s), such as age or employment.

Play In G. H. Mead's theory of socialization, the stage of development in which children take on the roles of significant other people, one at a time.

Population In research terms, all the people who conform to a set of characteristics.

Positive function *See* Function.

Positive sanction Reward.

Positivism The assumption that what is being studied has a stable reality that can be measured from the outside by an objective observer.

Power The ability to control or influence the behavior of others, even against their will.

Precedent variable *See* Independent variable.

Prejudice Negative beliefs and feelings about a group of people that serve to justify and reinforce discrimination against them.

Prescription A rule that focuses on approved behavior.

Primary deviance An instance of norm breaking, whether or not it is labeled deviant.

Primary group An enduring small group with diffuse, noninstrumental, emotionally based relationships.

Primary sex characteristics The male versus female genitalia used in the reproductive process.

Problem oriented Having a tendency to focus on problems.

Proscription A rule that focuses on disapproved behavior.

Psychological theory of deviance The theory that some individuals, through their experiences, become incapable of adequately learning or following social rules.

Qualitative research A research method in which the qualities of objects, behavior, or relationships are evaluated in textual terms (words) rather than quantitative terms (numbers).

Quantitative research A research method that attaches numbers to the qualities of objects, behavior, or relationships.

Random sampling A method of sampling designed to ensure that every element in a population has an equal chance of being chosen for the sample.

Reference group A group to which people compare themselves when evaluating their own beliefs or performance. A person does not have to belong to a particular group to use it as a reference.

Relative deprivation A feeling of dissatisfaction arising from comparison of one's condition with that of a reference group or with some more favorable condition in history or imagination.

Remedial resocialization A process whereby society attempts to compel an individual's beliefs, values, and social behavior.

Resocialize To adapt to changes in society (or in their situation within society) by internalizing new expectations for social behavior.

Role conflict Conflicting expectations for a person's behavior that arise when he or she occupies two or more role sets simultaneously.

Role confusion Confusion about what role to play when two or more status structures apply in a social situation.

Role distancing The practice of separating oneself from the expectations of a role by expressing one's individuality within it.

Role set All the roles associated with a particular status.

Role strain Conflicting expectations for the behavior of an individual within one role set.

Role taking Assumption of the expectations of only one person at a time (in G. H. Mead's terms, particular others).

Sample A number of cases (often people) chosen from a larger population.

Sanction *See* Negative sanction, Positive sanction.

Scientific perspective An approach to studying the observable world that stresses systematic, objective measurement aimed at the discovery and explanation of stable order in that world.

Secondary deviance An instance of norm breaking that causes the violator to be labeled as deviant by the society.

Secondary group A group of any size or permanence whose relationships are formal, instrumental, and segmental.

Self G. H. Mead's term for the idea a person has of who he or she is, which is constructed from the actions of others toward the individual.

Secondary sex characteristics The physical traits other than those of the reproductive organs; many of these are used to identify the sex of an individual in normal, public interaction.

Sex The set of biological characteristics that distinguish females from males.

Simple descriptive hypothesis A guess about the distribution of a single variable.

Small group A group with few enough members to allow for personal contact among all of them.

Social change Change in the beliefs and values that the people in a society hold and in the way they agree to act toward one another.

Social class A category of people within a system of stratification who share a similar style of life and socioeconomic status.

Social control The provision of negative sanctions for violating social norms.

Social fact The force that constrains (or controls) human behavior and results from membership in groups rather than from individual characteristics.

Social forces theory of deviance A theory that deviance is not the result of flaws in individuals but, rather, of the way social forces operate.

Socialization The process by which an individual internalizes the rules of the social order in which he or she is reared and by which the character of the society is maintained from one generation to the next.

Social moral A statement of what ought to be that has the force of social support.

Social role The behavior expected of an individual because of the social status she or he occupies in a given social structure.

Social status A position or place within a social structure to which certain rights and obligations apply.

Social structure A pattern of social relationships that forms the stable framework within which social interaction takes place.

Social theory Systematic and testable sets of statements about the social world that explain a variety of seemingly diverse behaviors.

Society The largest possible group (because it is not a subgroup of any other group). Its broad goal is the satisfaction of the basic survival needs of its members; its beliefs and rules for behavior are expressed in a shared culture; and its membership is generally territorially bounded and recruited by the automatic inclusion of the children of members.

Sociological perspective A view of human behavior that focuses on the patterns of relationships among individuals rather than solely on the individuals themselves.

Solidarity *See* Cohesion, Mechanical solidarity, Organic solidarity.

Status *See* Achieved status, Ascribed status, Social status

Status inconsistency A condition in which a person has high status on one or more measures of social class standing but low status on others.

Stereotype A generalization about the characteristics of a group of people that depicts them as inferior.

Spurious relationship A relationship between two variables that is believed to be causal but is shown not to be. *See* Elaboration.

Stratification A system of ranking individuals in terms of their access to, and possession of, the things valued by their society.

Structural functionalist theory of social change The theory that society tends naturally toward equilibrium and that social change is evidence of the social system's adjustment to new elements introduced into the society.

Structural functionalist view of stratification The theory that social inequality is a consequence of the need to heavily reward those positions that are most important for the smooth operation of the society as a system.

Subculture A group within the larger culture whose values, beliefs, and styles of life differ in some respects from those of the larger culture.

Superego The form of conscience that develops from socialization.

Survey research A method of research used to study descriptive hypotheses. Respondents are generally asked to answer questions designed to measure the variables being studied.

Symbol Anything (but especially a word or gesture) that stands for or represents the abstracted qualities of something else.

Symbolic expression Creative activity that results in the production of art, literature, dance, architecture, music, and various other art forms in a culture.

Symbolic interaction The view that social reality is constructed at the level of symbols in each human interaction.

System A set of interacting, interdependent elements that form a relatively stable overall order.

System stability The capacity of a system to maintain the operating relationship of all its elements to one another and, when appropriate, to adapt to changes in its environment.

Theological or philosophical moral A statement of what ought to be that has the force of truth in a theological or philosophical system of judgment.

Universal issue A task that must be accomplished if the members of a group are to survive.

Value A shared idea about the most abstract goals that a group believes are worth achieving.

Variability The extent to which data points differ from one another; measures include range, mean deviation, and standard deviation.

Variable Anything that is free to vary and, therefore, to take on differing values.

Verstehen An approach to social behavior that stresses subjective, empathic understanding of the meaning of an interaction to the participants.

Voluntary association A type of formal organization that develops out of the shared interests of its members.

Suggested Readings

Concept 1: The Sociological Perspective

Berger, Peter L. *Invitation to Sociology: A Humanistic Perspective.* Garden City, N.Y.: Doubleday, 1963.

Mills, C. Wright. *The Sociological Imagination.* London: Oxford University Press, 1959. Both of these fine books are brief, well-written overviews of the sociological perspective. If someone asks you what sociology is, refer the person to either of these books for a shortcut answer. Use them that way yourself.

Concept 2: Ideal Type, Model, and Paradigm

Gouldner, Alvin W. *The Coming Crisis of Western Sociology.* New York: Basic Books, 1970. This is a fairly difficult but extremely worthwhile analysis of the social theories that dominated American sociology through the 1950s. Gouldner focuses on the dominant paradigms of social theorists such as Talcott Parsons, suggests how they developed, and states the paradigms that he believes ought to replace them.

Kuhn, Thomas. *The Structure of Scientific Revolutions.* Chicago: University of Chicago Press, 1970. Kuhn's analysis of the role of paradigms in the conduct of science has become a classic. He destroys the common belief that science is an unbroken line of discoveries, each of which builds on the knowledge that preceded it.

Concept 3: The Scientific Perspective and Quantitative Research

Cole, Stephen. *The Sociological Method.* 3d ed. Chicago: Rand McNally, 1980. A fine brief introduction to the use of quantitative methods in sociology and their application to the development of broader theories. An especially good treatment of the procedures for distinguishing between correlated and causally related variables.

Gould, Stephen J. *The Mismeasure of Man.* New York: Norton, 1981. A wonderful book by one of the most engaging and convincing writers on science I have ever read. As Gould points out how science has been badly done, he illustrates good science by his own example. He even manages to be generous to his "victims."

Greer, Scott A. *The Logic of Social Inquiry.* Chicago: Aldine, 1962. A systematic discussion of the methods of social research and their relationship to the study of society. More useful as a reference book for students of introductory sociology than a book you might try to read right through.

Concept 4: Symbolic Interaction and Qualitative Research

Becker, Howard, et al. *Boys in White.* Chicago: University of Chicago Press, 1961. This study of medical students is a classic in the literature of symbolic interactionist research and of participant observation. The writing is clear, and the text is as absorbing as a good novel.

Shaffir, William, Robert Stebbins, and Allan Turowetz. *Fieldwork Experience: Qualitative Approaches to Social Research.* New York: St. Martin's Press, 1980. Insights into the experiences of participant observers (such as police officers and artists) of a wide variety of fascinating social behaviors.

Spradley, James. *Participant Observation.* New York: Holt, Rinehart & Winston, 1980. More of a text on participant observation methods than those above, but very sensibly organized in logical steps appropriate for those not familiar with this method of research.

Concept 5: Function and Dysfunction

Coser, Lewis. *The Functions of Social Conflict.* New York: Free Press, 1956. A brief, systematic summary of the work of the German sociologist Georg Simmel on the positive functions of conflict for group solidarity.

Merton, Robert K. *Social Theory and Social Structure.* 2d ed. New York: Free Press, [1949] 1968. This is the original statement from Merton on the nature of dysfunctions, manifest and latent functions and dysfunctions, and the net aggregate of consequences. It is not necessary to read the whole book. The section on manifest and latent functions is clearly identified in the table of contents.

Turner, J. H., and A. Maryanski. *Functionalism.* Redwood City, Calif.: Benjamin/Cummings, 1979. A good, brief summary of the history and basics of functionalism.

Concept 6: Conflict: Individual, Group, Class, and Societal

Coser, Lewis. *The Functions of Social Conflict.* New York: Free Press, 1956. This summary of the work of Georg Simmel on the positive functions of conflict for group solidarity focuses mostly on conflict at the level of large-scale groups and societies.

Marx, Karl, and Friedrich Engels. *The Communist Manifesto.* New York: Appleton-Century-Crofts, [1848] 1964. This is probably the most famous and influential statement of conflict theory ever written. It is also a political statement predicting and advocating massive social change in the Western world. It is worth reading as social theory and to understand what the word *communism* refers to beyond its everyday, political sense.

Concept 7: Culture

Benedict, Ruth. *Patterns of Culture.* Boston: Houghton Mifflin, 1961. A classic study of the way culture shapes personality, comparing the process in three distinct cultures.

Harris, Marvin. *Cultural Materialism.* New York: Random House, 1979. The author of several works on the concept of culture argues in detail that culture can best be understood as a complex series of adaptations to the resources of an environment.

Liebow, Elliot. *Tally's Corner: A Study of Negro Streetcorner Men.* Boston: Little, Brown, 1967. Another of the brief classics of sociological literature, this one examining the way subcultures develop as adaptations to the special circumstances in which minority-group members find themselves. What is most important, Liebow shows that poor blacks share the values of the dominant culture (such as the work ethic) even though they are often accused of lacking such beliefs.

Williams, Robin M. *American Society: A Sociological Interpretation.* 3d ed. New York: Knopf, 1970. In this extensive analysis of American culture, Williams presents the list of American values that has become a standard reference for sociologists.

Concept 8: Norms: Folkways and Mores

Goffman, Erving. *The Presentation of Self in Everyday Life.* Garden City, N.Y.: Doubleday, 1959.

———. *Interaction Ritual: Essays on Face-to-Face Behavior.* Garden City, N.Y.: Doubleday, 1967.

———. *Relations in Public.* New York: Basic Books, 1971. These are a few of the many books by Goffman in which he explores the way everyday interactions are regulated by the folkways of American culture.

Sumner, William Graham. *Folkways: A Study of the Sociological Importance of Usages, Manners, Customs, Mores, and Morals.* Boston: Ginn, 1906. The classic work on the distinction between folkways and mores, explaining how each operates in the cultural order.

Concept 9: Socialization and the Development of Adult Character

Freud, Sigmund. *Civilization and Its Discontents,* trans. Joan Riviere. Garden City, N.Y.: Doubleday, 1930. Freud's classic statement illustrating the process of socialization, in which conflict between the individual's internal drives and the demands of society for conformity are played out.

Mead, G. H. *Mind, Self and Society,* ed. C. W. Morris. Chicago: University of Chicago Press, 1934. Mead's statement of the process by which the individual comes to develop an adult character through interaction with the social environment.

Rosow, Irving. *Socialization to Old Age.* Berkeley: University of California Press, 1974. A fine examination of one type of adult socialization in which individuals are taught the behaviors that will be expected of them as older members of the society.

Concept 10: Solidarity—Mechanical and Organic

Coser, Lewis. *The Functions of Social Conflict.* New York: Free Press, 1956. A brief, systematic summary of the work of the German sociologist Georg Simmel on the positive functions of conflict for group solidarity. This is the book from which the Illustration for this concept was summarized.

Durkheim, Émile. *The Division of Labor in Society,* trans. G. Simpson. New York: Free Press, [1893] 1960. In this work, Durkheim made the original distinction between mechanical and organic sources of social cohesion.

Concept 11: Power and Legitimacy

Milgram, Stanley. *Obedience to Authority: An Experimental View.* New York: Harper & Row, 1974. A summary of a series of experimental studies conducted at Yale University in which Milgram illustrates powerfully that obedience to authority may well reside in the structure of social situations rather than in the pliant personalities of obedient individuals.

Mills, C. Wright. *The Power Elite.* New York: Oxford University Press, 1956. A fascinating study of power in the United States, in which Mills proposes that a small group of military, industrial, and political elites controls much of the power in the country.

Riesman, David. *The Lonely Crowd.* New Haven, Conn.: Yale University Press, 1961. Riesman's theory (contradicting Mills's theory) that power in the United

States is distributed among many groups, each of which is capable of exercising a veto.

Weber, Max. *The Theory of Social and Economic Organization,* trans. Talcott Parsons. New York: Free Press, [1925] 1964. In this work Weber outlines the three types of authority: traditional, charismatic, and rational-legal.

Concept 12: Ideology

Alger, Horatio. *Jed the Poorhouse Boy.* New York: Amereon, 1976. One of over a hundred books by Alger expressing powerful American ideology of the late nineteenth century. Any library is likely to have several, and any one of the books can be read for its ideological content.

Mannheim, Karl. *Ideology and Utopia,* trans. Louis Wirth and Edward Shils. New York: Harcourt Brace Jovanovich, 1936. Mannheim's work is the classic on the sociology of knowledge, the study of how an individual's social experiences influence the way he or she is likely to explain the operation of society.

Marx, Karl, and Friedrich Engels. *The Communist Manifesto.* New York: Appleton-Century-Crofts, [1848] 1964. Perhaps the most influential statement of an ideology of our time, at least in terms of the conflict between the current global powers.

Concept 13: Social Status and Social Role

Linton, Ralph. *The Study of Man.* New York: Appleton-Century-Crofts, 1936. This work contains Linton's now-famous discussion of social status as a component of social structure.

Merton, Robert K. *Social Theory and Social Structure.* 2d ed. New York: Free Press, [1949] 1968. On pages 422–38, Merton extends Linton's work on status and role to a consideration of role sets, the conflicts within them, and their place within larger social structures.

Concept 14: Group: From Small Group to Society

Merton, Robert K., and Alice S. Rossi. "Contributions to the Theory of Reference Group Behavior." In *Social Theory and Social Structure,* 2d ed., by Robert K. Merton, 279–334. New York: Free Press, [1949] 1968. Merton and Rossi explore the concept of the reference group and related issues, such as relative deprivation and reference-group membership.

Nisbet, Robert. *The Social Bond.* New York: Knopf, 1970. An examination of the forces that contribute to the overall cohesion of society.

Simmel, Georg. *The Sociology of Georg Simmel,* ed. and trans. Kurt Wolff. New York: Free Press, [1908] 1964. This volume summarizes much of Simmel's work, including his pioneering examination of the workings of small groups.

Concept 15: Formal Organization and Bureaucracy

Merton, Robert K. "Bureaucratic Structure and Personality." In *Social Theory and Social Structure,* 2d ed., by Robert K. Merton, 249–60. New York: Free Press, [1949] 1968. Merton focuses here on dysfunctions in bureaucratic operation, especially those deriving from the clash between the formal structure of a bureaucracy and the individual characters of its members.

Weber, Max. "Bureaucracy." In *From Max Weber: Essays in Sociology,* ed. and trans. Hans H. Gerth and C. Wright Mills, 196–244. New York: Oxford University Press, [1925] 1946. Weber's classic statement of the model of bureaucracy.

Whyte, William H. *The Organization Man.* Garden City, N.Y.: Doubleday, 1956. Whyte's book has become the classic statement of the way formal organizations crush originality and spontaneity by their demands for conformity among members.

Concept 16: Institutions

Durkheim, Émile. *The Elementary Forms of Religious Life.* New York: Collier, [1915] 1961. Durkheim's classic statement of the social origins of religion as the expression of a people's collective consciousness.

Orum, Anthony M. *Introduction to Political Sociology: The Social Anatomy of the Body Politic.* Englewood Cliffs, N.J.: Prentice-Hall, 1978. A fine introduction to the operation of political institutions; includes theories and research findings.

Parelius, Ann P., and Robert J. Parelius. *The Sociology of Education.* Englewood Cliffs, N.J.: Prentice-Hall, 1978. An excellent summary of the history, social theories, and data on American educational institutions, compared with those in other cultures.

Queen, Stuart, Robert Habenstein, and John Adams. *The Family in Various Cultures.* Philadelphia: J. B. Lippincott, 1974. An examination of the institution of the family, illustrating the many ways in which universal issues are dealt with by family structures in various cultures.

Smelser, Neil. *The Sociology of Economic Life.* 2d ed. Englewood Cliffs, N.J.: Prentice-Hall, 1975. Smelser examines economic institutions and their relationship to the operation of other institutions of the larger society.

Concept 17: Stratification and Social Class

Bendix, Reinhard, and Seymour M. Lipset, eds. *Class, Status, and Power.* New York: Free Press, 1966. An excellent collection of articles on social inequality, including the Davis and Moore article on the functions of inequality that is discussed in the Definition section of this concept.

Bottomore, T. B. *Classes in Modern Society.* New York: Pantheon Books, 1965.

A useful introduction to systems of stratification in modern and industrial societies, including the United States, the Soviet Union, and Great Britain.

Gans, Herbert. *More Equality.* New York: Pantheon Books, 1973. An excellent discussion of social class in the United States, its functions, and the prospects for change in the system of inequality.

Marx, Karl, and Friedrich Engels. *The Communist Manifesto.* New York: Appleton-Century-Crofts, [1848] 1955. Perhaps the most influential argument on the subject of social class ever published. The manifesto proposes a theory of conflict between workers and owners of the means of production that has been used as a blueprint for political revolution in many societies, including the Soviet Union.

Concept 18: Discrimination and Prejudice

Allport, Gordon W. *The Nature of Prejudice.* Reading, Mass.: Addison-Wesley, 1954. Allport's book has become the classic reference for understanding the development of the prejudiced personality (authoritarianism). It can be difficult to read in some spots, but it is well worth the effort for those with a special interest in the area.

Butler, Robert. *Why Survive? Being Old in America.* New York: Harper & Row, 1975. A Pulitzer Prize-winning examination of the problems of aging in the United States.

Friedan, Betty. *The Feminine Mystique.* New York: W. W. Norton, 1974. A powerful study of the inequalities of sex in the United States, which was an early manifesto for the current women's movement.

Levin, Jack, and William C. Levin. *The Functions of Discrimination and Prejudice.* New York: Harper & Row, 1982. A brief text arguing that discrimination and prejudice persist because they have important system-maintaining consequences for a variety of groups in the United States.

Myrdal, Gunnar. *An American Dilemma.* New York: Harper & Row, [1944] 1962. This is the shortened version of the Nobel Prize-winning Swedish social economist's great work on race inequality in the United States.

Concept 19: Social Change

Moore, Wilbert. *Social Change.* Englewood Cliffs, N.J.: Prentice-Hall, 1963. A brief summary of theories of social change that serves as a good reference.

Ogburn, William F. *Social Change.* New York: Viking Press, 1950. A classic on the subject of social change, in which Ogburn introduces the concept of culture lag.

Toffler, Alvin. *Future Shock.* New York: Random House, 1970. The best-selling examination of the tremendous rate of social change in our times and its effects on our institutions and daily lives.

Concept 20: Deviance

Becker, Howard. *Outsiders: Studies in the Sociology of Deviance.* New York: Free Press, 1963. A very well-written discussion of deviance from the labeling perspective.

Clinard, Marshall B., and Robert F. Meier. *Sociology of Deviant Behavior.* 5th ed. New York: Holt, Rinehart & Winston, 1979. An excellent introduction to the sociology of deviance, including discussions of theories and the data supporting them.

Rubington, Earl, and Martin Weinberg, eds. *Deviance: An Interactionist Perspective.* New York: Macmillan, 1978. A very good collection of essays on the sociology of deviance, with helpful clarifications by the editors.

Concept 21: Anomie and Alienation

Durkheim, Émile. *Suicide: A Study in Sociology,* trans. J. A. Spaulding and G. Simpson. New York: Free Press, [1897] 1951. The classic study of suicide in which Durkheim, using statistical evidence, first demonstrated that suicide occurred at different rates among different populations and then proposed a social theory to account for such differences. One type of suicide he identified was attributed to extreme feelings of anomie, or normlessness.

Marx, Karl. *The Economic and Philisophic Manuscripts of 1844.* New York: International Publishers, [1844] 1964. In this work Marx made his original statement of the character of alienation and its roots in the structure of industrial capitalism.

Merton, Robert K. "Social Structure and Anomie." In *Social Theory and Social Structure,* 2d ed., by Robert K. Merton, 185–214. New York: Free Press, [1949] 1968. Merton examines deviant behavior as an adaptation to anomie, which he views in terms of a discrepancy between goals and socially approved methods of attaining them.

Chapter 22: Sex and Gender in Society

Ehrenreich, B., and D. English. *For Her Own Good: 150 Years of the Experts' Advice to Women.* Garden City, NY: Anchor Press, 1978. This is a fascinating historical study of how medicine and health care was once the domain of women, but was wrested from them and came to be dominated by the (largely) male institutions of medicine we see today in the West.

Friedan, Betty. *The Feminine Mystique.* New York: Dell, 1963. This is an early manifesto of the women's movement that retains the power and appeal that led it to become a rallying call for feminism in America.

Gilligan, Carol. *In a Different Voice: Psychological Theory and Women's Development.* Cambridge, Mass.: Harvard University Press, 1982. Gilligan was a student of the developmental psychologist Lawrence Kohlberg when she found that females did not do as well as males on his tests of moral development. This book explains why and suggests her alternative theory of the moral development of women.

Goffman, Erving. *Gender Advertisements.* Cambridge, Mass.: Harvard University Press, 1979. Goffman presents evidence of the way males and females are presented in everyday life in advertising images, one of the most important ways we form and reinforce the symbolic meanings of the culture.

Index

Methods Index